Topics in Applied Physics   Volume 14

W0235174

Topics in Applied Physics   Founded by Helmut K. V. Lotsch

Laser Monitoring of the Atmosphere

Edited by E. D. Hinkley

With Contributions by
R. T. H. Collis   E. D. Hinkley   H. Inaba
P. L. Kelley   R. T. Ku   S. H. Melfi   R. T. Menzies
P. B. Russell   V. E. Zuev

With 84 Figures

Springer-Verlag Berlin Heidelberg GmbH 1976

Dr. E. DAVID HINKLEY

Lincoln Laboratory, Massachusetts Institute of Technology, Lexington, MA 02173, USA
and
Jet Propulsion Laboratory, California Institute of Technology, Pasadena, CA 91103, USA
(Present address: Laser Analytics, Inc., Lexington, MA 02173, USA)

ISBN 978-3-662-31255-1   ISBN 978-3-540-38239-3 (eBook)
DOI 10.1007/978-3-540-38239-3

Library of Congress Cataloging in Publication Data. Main entry under title: Laser monitoring of the atmosphere.
(Topics in applied physics; 14). Includes bibliographies and index. 1. Atmosphere—Laser observations. I. Hinkley,
Everett, D., 1936–. QC975.9.A1L37   551.5′11′028   76-13000

© by Springer-Verlag Berlin Heidelberg 1976
Originally published by Springer-Verlag Berlin Heidelberg New York in 1976
Softcover reprint of the hardcover 1st edition 1976

Preface

There is a growing need for continuous monitoring of atmospheric constituents. This has been emphasized recently by the controversy over potentially adverse effects of high-flying aircraft, space shuttles, and even aerosol spray propellants on the life-sustaining ozone layer in the stratosphere. Moreover, the widespread use of more polluting fuels due to dwindling energy resources may produce serious consequences (such as reducing the amount of useful solar energy reaching the earth's surface, changing the earth's heat balance, and degrading air quality in the lower atmosphere) which have yet to be fully understood. Mathematical models of the atmosphere are needed in order to be able to predict the environmental impact of increased source pollution and take the necessary corrective action in a timely and cost-effective manner. The development of these models and the associated continuous surveillance require a wide range of monitoring capabilities.

Surveillance-type monitoring can usually be performed best, and at least expense, with optical techniques. Laser technology has made great strides recently, and the role of lasers in atmospheric monitoring applications has steadily increased; indeed, all of the major laser schemes proposed for monitoring (based on scattering, fluorescence, absorption, and emission) have now been demonstrated experimentally. It is appropriate, therefore, to devote a volume in this Topics series to the subject of laser monitoring of the atmosphere.

This book describes, in a comprehensive and tutorial manner, the fundamental techniques of laser detection of gases and particles. Each chapter contains basic information such as mathematical expressions for the processes and typical values of relevant parameters, in addition to examples of actual measurements made in the field. Consequently, this book should be a useful reference for working scientists and engineers, and a supplementary text book for graduate and undergraduate courses in environmental studies. The broad scope of laser monitoring can be appreciated if we consider that atoms existing in a layer *above* the stratosphere have been detected and their concentrations measured by ground-based laser systems, and gases have been monitored *within* the stratosphere using a balloon-borne tunable laser system. In the lower atmosphere, pollutant gases, particles, wind speed, and atmospheric temperature have been monitored remotely using laser techniques; and the development of mathematical models for pollutant transport, dispersion, and conversion has been aided by long-path laser measurements.

Each chapter is devoted to a particular type of laser monitoring; but there is strong continuity between chapters, with many cross-references and a uniform set of symbols, defined in the Introduction (Chapt. 1). Chapter 2 considers the structure of the atmospheric and how laser techniques can fulfill some of the air quality management needs of the future. Chapter 3 discusses atmospheric transmission and the selection of appropriate laser wavelengths. Chapters 4–7 concentrate on specific laser techniques. Safety requirements, which must be considered for active laser monitoring systems, are also discussed.

We now recognize that life on earth is affected, both directly and indirectly, by atmospheric constituents not only near ground level but throughout the rest of the troposphere and stratosphere, and possibly above. For proper surveillance of this vast region at reasonable cost, it is becoming increasingly evident that the capabilities offered by laser monitoring techniques may be the only answer. The authors hope that this book will be a useful guide for the evaluation and development of these future laser monitoring systems.

I am indebted to several colleagues for their helpful ideas and comments; particularly, P. L. KELLEY and W. E. BICKNELL of M. I. T. Lincoln Laboratory, V. E. DERR of the National Oceanographic and Atmospheric Administration, and P. B. RUSSELL of Stanford Research Institute. Sincere appreciation is also expressed to the contributors and their families, and to my own wife and children for their patience and encouragement.

Concord, Mass.
May 1976   E. D. HINKLEY

Contents

3. Laser-Light Transmission Through the Atmosphere. By V. E. ZUEV
(With 7 Figures)

6. Techniques for Detection of Molecular Pollutants by Absorption of Laser Radiation. By E. D. HINKLEY, R. T. KU, and P. L. KELLEY (With 22 Figures)

Contributors

COLLIS, RONALD T. H.
  Atmospheric Sciences Laboratory, Stanford Research Institute,
  Menlo Park, CA 94025, USA

HINKLEY, E. DAVID
  Lincoln Laboratory, Massachusetts Institute of Technology,
  Lexington, MA 02173, USA, and
  Planetary Atmospheres Section, Jet Propulsion Laboratory,
  California Institute of Technology, Pasadena, CA 91103, USA
  (Present address: Laser Analytics, Inc., Lexington, MA 02173, USA)

INABA, HUMIO
  Research Institute of Electrical Communication, Tohoku University,
  1-1, Katahira 2-Chome, Sendai, 980, Japan

KELLEY, PAUL L.
  Lincoln Laboratory, Massachusetts Institute of Technology,
  Lexington, MA 02173, USA

KU, ROBERT T.
  Lincoln Laboratory, Massachusetts Institute of Technology,
  P.O. Box 73, Lexington, MA 02173, USA

MELFI, SAMUEL H.
  Remote Sensing Division, Environmental Monitoring and Support
  Laboratory, U.S. Environmental Protection Agency, P.O. Box 15027,
  Las Vegas, NV 89114, USA

MENZIES, ROBERT T.
  Jet Propulsion Laboratory, California Institute of Technology,
  Pasadena, CA 91103, USA

RUSSELL, PHILIP B.
  Atmospheric Sciences Laboratory, Stanford Research Institute,
  Menlo Park, CA 94025, USA

ZUEV, V. E.
  Institute of Atmospheric Optics, Siberian Branch of the USSR
  Academy of Sciences, Tomsk, 634055, USSR

1. Introduction

E. D. Hinkley

The importance of lasers to modern scientific research, and their resulting technological applications, has been highlighted in earlier volumes of this *Topics in Applied Physics* series. From initial experiments based on the high intensity of laser radiation emerged studies of the nonlinear properties of materials and useful spectroscopic information on the nature of solids, liquids, and gases. With the development of *tunable* lasers, important advances have been made in the field of high-resolution spectroscopy, as highlighted in Volume 2.

The interaction between electromagnetic radiation and atoms and molecules serves as the basis for using lasers to detect and continuously monitor atmospheric constituents and properties. With the steady increase in industrial activity, power generation, transportation, and other potential sources of air pollution, new techniques for atmospheric monitoring are clearly needed to augment those already in use [1.1, 2]. Most of the present instrumentation is based on sample-extraction methods; there are some notable exceptions, however, involving correlation [1.3], dispersive [1.4], and multi-spectral [1.5] techniques, which involve incoherent (non-laser) electromagnetic radiation. With lasers, however, one generally has greater flexibility of operation and can monitor a wider variety of pollutants due to higher resolution.

All of the basic laser techniques which had been proposed for monitoring atmospheric gases and particles have now been shown to work experimentally. The purpose of this volume is to present a unified, tutorial discussion of these techniques, their applications, and their limitations. Included are examples of results obtained so far (to 1976), so that the reader can evaluate the present capabilities of laser monitoring for specific applications and make extrapolations to the future when improved equipment is available. The reader is also referred to a monograph edited by DERR [1.6] which contains a large number of papers on remote sensing up to 1972, and a recent study by WRIGHT et al. [1.7].

This volume covers the application of laser techniques to the detection and continuous monitoring of particulate matter, aerosols, atoms, and molecules in the atmosphere. The term "atmosphere" is taken to encompass the troposphere (or lower atmosphere) which is between ground level and the tropopause at 10–15 km altitude, and the stratosphere (lower portion of the upper atmosphere) which is bounded below by the tropopause, and extends upward to 28–30 km. Monitoring of pollutants near the ground level is necessary in order to determine the quality of the air we breathe; and both ground-level and higher-level monitoring (at least to the inversion layer height) must be performed in order to develop mathematical models for predicting air quality for varying circumstances. Stratospheric gases and particles affect us in a less direct, but equally-important way; and laser techniques should prove to be especially useful for continuous surveillance of the stratosphere [1.8]. Other important atmospheric parameters, such as temperature and wind velocity, can also be measured remotely by lasers.

Considering the many types of monitoring instruments now available commercially, why should laser instrumentation be developed for atmospheric monitoring? This question is addressed in Chapter 2, which begins with a discussion of the present structure of the atmosphere and continues with an overview of the present capabilities of laser monitoring instrumentation. Remote sensing is not generally considered to be a substitute for point sampling, but an adjunct to it. Nevertheless, in some cases, remote sensing represents the only economical or technically feasible technique. An important application of laser monitoring is expected to be in the area of surveillance to check compliance with source emissions regulations. Chapter 2 considers this application and others which appear most promising for the future.

Chapter 3 describes in detail the transmission of laser radiation through the atmosphere. Not all wavelengths which would be optimal for detecting certain pollutants can be used due to strong absorption by normal atmospheric gases (this restriction is not as severe at high altitudes); and since the laser beam must sometimes travel distances of several km or more, careful consideration must be given to absorption, scattering, nonlinear effects, turbulence, and scintillation. For some laser techniques the monitoring process depends upon one or more of these interactions, and Chapter 3 can provide useful information for optimizing them.

The most advanced laser monitoring technique has been lidar (laser radar), which involves the detection of particles and aerosols by measurement of the laser radiation scattered by them. Lidar systems are currently in use in many countries, and applications include the mapping of particles for operational meteorology, atmospheric research, as well as air pollution studies. Chapter 4 concentrates on these lidar applications involving backscattered laser radiation, and also covers the newly-emerging technique using tunable lasers in a differential-absorption lidar system to remotely measure gaseous pollutants.

With more sophisticated instrumentation it is possible to detect characteristic shifts in the wavelength of the backscattered radiation due to specific scattering molecules. This phenomenon, known as Raman scattering, can be used to monitor a variety of gases using a single, fixed-frequency laser, as discussed in Chapter 5. The Raman-scattering cross sections are not usually large, however, so this technique will probably be limited to major atmospheric constituents and source monitoring. Chapter 5 also covers detection of atoms and molecules by induced fluorescence using a tunable laser, where the cross sections are typically ·higher than for Raman scattering, but where application is generally limited to higher altitudes (lower pressures) where quenching of the fluorescent signal is reduced.

The most sensitive laser technique is based on the principle of resonance absorption, which occurs when the laser radiation is at the same wavelength as a major absorbing transition of the molecular species to be detected. This technique has already provided very high specificity in point sampling applications. It has also been the basis for instrumentation for *in situ* source monitoring (without taking a sample) and for long-path (to several km) ambient-air monitoring. The wide variety of applications of the absorption technique, involving detection of laser power returned from remote retroreflectors, buildings, and even natural foliage, is described in Chapter 6.

One unique characteristic of narrow-linewidth laser radiation is that it can be very sensitively detected by heterodyne techniques. Heterodyne detection permits the monitoring of backscattered radiation from non-cooperative targets and the passive, single-ended, remote monitoring of gases using their own emission lines. Chapter 7 presents a thorough treatment of all phases of heterodyne detection, with several examples from experiments already performed. It concludes with a discussion of expected future developments.

Thoughout this book an attempt has been made to achieve uniformity both in the text and in the symbols used. Following conventional usage, the terms "wavenumber" and "frequency" are used interchangeably, but the individual symbols are quite explicit in their meaning. A complete list of the symbols used, their definitions, and typical units, follows:

Glossary of Symbols (Typical Units)

a	Particle radius [μm], laser beam radius [mm]
$â$	Anisotropic part of polarizability tensor
a_{ij}	Oscillator strength between states i and j
A	Area [cm^2, m^2]

A_{ij} Einstein coefficient (transition probability) between states i and j [s^{-1}]

$A(v)$ True spectral absorption

$A'(v)$ Measured spectral absorption

$\mathscr{A}$ Absorption function

b_j Zero-point amplitude of j-th vibrational mode [erg$^{1/2}$/s]

B Bandwidth [Hz]

$\boldsymbol{B}$ Molecular rotational constant [cm^{-1}]

$\boldsymbol{B}'$ Combined scattering parameter in differential absorption equation

$B(v, T)$ Brightness [W/cm^2-Hz-sr]

c Speed of light in vacuum (2.998×10^{10} cm/s)

c_j Relative line strength

c_p Pressure-broadening coefficient [cm^{-1}/atm]

$C_E(\tau)$ Autocorrelation function for constant-amplitude field with random phase fluctuations [W/cm^2]

$C_i(\tau)$ Autocorrelation function for current [A^2]

C_n^2 Atmospheric structure constant [m$^{-2/3}$]

C_v, C_p Specific heat at constant volume, pressure [cal/gm-K]

C_D Photodiode capacitance [F]

d_c Coherence diameter for heterodyne detection [mm]

D Diameter [cm]

D_c Diameter of collecting aperture [cm]

D_p Particle diameter ($= 2a$) [μm]

D_{LO} Diameter of local oscillator beam [mm]

D_{aa} Sum diameter for optical collisions $a \leftrightarrow a$ [μm]

D_{ab} Sum diameter for optical collisions $a \leftrightarrow b$ [μm]

$D_s(d)$ Variance of phase difference over diameter of transmitting aperture

D_λ^* Detectivity [cm-Hz$^{1/2}$-W^{-1}]

e Electronic charge (1.602×10^{-19} C)

$\mathscr{E}$ Electrical field strength [V/cm]

$E(t)$ Complex optical field [W$^{1/2}$ cm^{-1}]

$E_S(t)$ Signal optical field [W$^{1/2}$ cm^{-1}]

$E_{LO}(t)$ Local oscillator optical field [W$^{1/2}$ cm^{-1}]

E_i Energy level of quantum state i [eV]

E_t Threshold illuminance, eye detectability [W/cm^2]

E_N Noise energy per quantum mode [erg]

f Frequency [Hz]

Δf Width (FWHM) of optical power spectrum [Hz]

f_p Laser pulse repetition frequency [Hz]

f_t Turbulence fluctuation frequency [Hz]

f_{LO} Local oscillator frequency ($= cv_{LO}$) [Hz]

$f(a)$ Particle size distribution function [cm^{-1}]

F Coefficient in photocurrent equation [cm^2/W-s]

g Gravitational force on unit mass [dyne]

g_i Degeneracy of i-th vibrational mode

g_M Transmission factor for scanning Michelson interferometer

$g^{(1)}(\tau)$ Normalized first-order correlation function

$g^{(2)}(\tau)$ Normalized second-order correlation function

G Gain

G_p Geometrical cross-section per unit volume for particles [cm^2/cm^3]

G_q Density of quantum modes [cm^{-3}]

G_D Incremental shunt conductance of photodiode [mho]

h Planck's constant (6.625×10^{-34} J-s)

$\hat{i}$	Isotropic part of polarizability tensor
$i(t)$	Photocurrent [A]
$i(\omega)$	IF photocurrent [A]
i_{LO}	Local oscillator current [A]
$I(v, t)$	Laser beam intensity [W/cm^2]
$\bar{I}$	Time average of optical field intensity [W/cm^2]
$I_s(\lambda, \theta)$	Intensity of scattered radiation of wavelength λ, at angle θ from direction of propagation [W/cm^2]
$I_S(v)$	Radiation intensity at top of atmosphere [W/cm^2]
$(\delta I_F)^2$	Relative variance of intensity fluctuations in the Fresnel zone of diffraction of a collimated beam
$(\delta I_R)^2$	Relative variance of intensity fluctuations due to random refraction
$I'(v)$	Intensity per unit spectral interval [W/cm^{-1}]
J	Rainfall rate [mm/h]
J	Rotational quantum number
k	Boltzmann's constant $(1.380 \times 10^{-23}$ J/K)
$k(v)$	Differential absorption coefficient [cm^{-1} ppm^{-1}]
k	Wave vector $(k = 2\pi/\lambda)$ [cm^{-1}]
K'	Optical system efficiency
$K(v, \psi)$	Total atmospheric absorption at wavenumber v and angle ψ with respect to zenith
ℓ	Inter-electrode detector lead spacing [mm]
ℓ_c	Coherence length of laser radiation [m]
L	Depth of pollutant layer or cloud [m]
L	Cell length [cm]
L_p	Spatial pulse length $(= c\tau_p/2)$ [m]
m	Mass of molecule or particle [gm]
m	Complex refractive index $(= n - i\kappa)$
m	Temperature exponent of Lorentzian linewidth
m_e	Mass of electron $(9.109 \times 10^{-31}$ gm)
M	Molecular weight [gm]
$\hat{n}$	Relative refractive index $(= n_1/n_2)$
n_e, n_B	Effective transducer noise due to electronics, Brownian molecular motion [atm/Hz$^{1/2}$]
n_0, n_r	Number of photons emitted, received per pulse
n_b, n_d	Number of photoelectrons due to background, dark current
$\bar{n}_k$	Average number of photons in quantum state k
N	Number density of molecules or particles [cm^{-3}]
$N'_p(a)da$	Concentration of particles with radii between a and $a + da$ [cm^{-3}]
N_i	Concentration of molecules in state i [cm^{-3}]
N_0	Number density of air molecules per atmosphere pressure at 15° C $(2.55 \times 10^{19}$ cm^{-3} atm^{-1})
N_A	Avogadro's number $(6.022 \times 10^{23}$ mol^{-1})
NEP	Noise-equivalent power [W/Hz$^{1/2}$]
p	Pressure [atm, Torr]
$p(I)$	Intensity probability distribution function
P_0, P_r	Transmitted, received laser power [W]
$P(\theta)$	Phase function for angular dependence of scattered radiation
P_s	Scattered power [W]
$P_i(\omega)$	Power spectrum of photocurrent [A^2/Hz]
P_{LO}	Local-oscillator power [W]
P_S	Signal power [W]

δP	Differential absorbed power [W]
P_Ω	Scattered power per unit solid angle [W/sr]
q	Halfwidth parameter in Γ-distribution of particle sizes
q_j	Fluorescence quenching coefficient [atm^{-1}]
q_n	Coefficient in intensity fluctuation equation
Q	Total partition function
Q_j	Normal coordinate of j-th vibrational mode
Q_A	Wave parameter of aperture $(= ka^2/z)$
Q_B	Total backscatter efficiency [sr^{-1}]
Q_F	Fluorescence quenching factor
Q_S	Total scattering efficiency
Q_W	Water content [g/m^3]
r	Radial distance [m]
R	Range [m]
$\mathscr{R}$	Detector responsivity [V/W]
R_{ij}	Transition dipole matrix element between states i and j [erg$^{1/2}$ cm$^{3/2}$]
R_o, R_s	Mixer IF output, series resistance [Ohm]
R_p	Radius of phase-front curvature [m]
R_{IF}	IF amplifier input impedance [Ohm]
S	Integrated spectral line intensity [cm]
S_{ij}	Transition intensity between states i and j [cm]
$S(R)$	"S-value" for lidar performance
$(S/N)_{PC}$	Signal-to-noise ratio for pulse-gated photon-counting system
$(S/N)_{BI}$	Signal-to-noise ratio for boxcar integration detection system
t	Time [s]
t_s, t_r	Sampling, transit time [s]
Δt	Lifetime for re-emission [s]
T	Absolute temperature [K]
$T(v)$	Transmittance
$\mathscr{T}$	Transmission function
T'	Combined absorption parameter in differential-absorption equation
T_s	System noise temperature [K]
T_A, T_M, T_{IF}	Antenna, mixer, IF-input noise temperature [K]
$(\Delta T)_m$	Minimum detectable temperature change for an ideal radiometer [K]
$(\Delta T')_m$	Minimum detectable temperature change for an actual radiometer [K]
$U(v)$	Apparatus spectral transmission function [cm]
v	Vibrational quantum number
v_r	Radial velocity of scatterers [m/s]
$v_\perp$	Wind velocity perpendicular to laser beam [m/s]
V	Volume [m^3]
V_L	Visual range of lights (at night) [km]
V_M	Meteorological range [km]
w_j	Nuclear spin weight
$W^{(1)}(t)$	Rate of photocarrier generation [s^{-1}]
$W^{(2)}(t)$	Joint probability of photocarrier generation between t and $t+\tau$ [s^{-2}]
x	Dimensionless particle size parameter $(= 2\pi a/\lambda)$
y	Altitude-pressure variable $(= -\ln p)$
$Y(R)$	Geometrical factor to account for overlap of transmitted and received beam paths
z	Pathlength variable [km]
Z	Specific distance [km]
α	Attenuation, extinction coefficient $(= 4\pi\kappa/\lambda)$ [cm^{-1}, km^{-1}]

α_s, α_a	Extinction coefficients due to scattering, absorption $[\text{cm}^{-1}]$
α_M, α_R	Extinction coefficients due to Mie, Rayleigh scattering $[\text{cm}^{-1}]$
α_p, α_g	Attenuation coefficient due to particles, gases $[\text{cm}^{-1}, \text{km}^{-1}]$
β	Volume backscattering coefficient $[\text{m}^{-1}\,\text{sr}^{-1}]$
β	Fraction of laser energy transferred into translational energy in an opto-acoustic cell $[=\frac{3}{2}(C_p/C_v - 1)]$
β_R	Rayleigh volume backscattering coeffcient $[\text{m}^{-1}\,\text{sr}^{-1}]$
γ	Halfwidth (HWHM) of spectral line at half maximum absorption coefficient $[\text{cm}^{-1}]$
γ_D, γ_L	Doppler, Lorentzian halfwidths $[\text{cm}^{-1}]$
γ_0	Halfwidth characteristic of apparatus spectral resolution $[\text{cm}^{-1}]$
Γ	Full spectral width (FWHM) of line at half maximum absorption coefficient $[\text{cm}^{-1}]$
δ	Thermal diffusivity $(=\kappa/\varrho_m C_v)$ $[\text{cm}^2/\text{s}]$
δ	Scattering depolarization ratio
δ_p	Aerosol scattering phase function asymmetry coefficient
$\delta^2_{\ln I}$	Variance of log intensity
δ^2_I	Variance of intensity
δ^2_d	Variance of shifts of laser beam position
Δ	Beam overlap parameter $[\text{cm}]$
ε	Emissivity
$\zeta(z)$	Scattering ratio
$\zeta_z(v)$	Fraction of radiation reaching top of atmosphere from lower altitude z
η	Detector quantum efficiency
η_F	Fluorescence efficiency
θ	Angular separation, commonly scattering angle $[\text{rad}]$
κ	Index of absorption (imaginary part of refractive index, m)
κ	Thermal conductivity $[\text{W/cm-K}]$
$\kappa(y)$	Weighting function $[=\text{d}\zeta(y)/\text{d}y]$
λ	Wavelength $[\mu\text{m}, \text{nm}]$
Λ	Junge model curve-fitting parameter
Λ	Resonance Raman enhancement factor
μ	Electric dipole moment vector $[\text{C-m}]$
μ	Chemical potential energy $[\text{erg}]$
μ_h	Hole mobility $[\text{cm}^2/\text{V-s}]$
μ_S, μ_N	Photomultiplier noise factors for signal, dark current pulses
v	Wavenumber (or "frequency") of electromagnetic radiation $(=f/c)$ $[\text{cm}^{-1}]$
v_0	Wavenumber at line center $[\text{cm}^{-1}]$
v_{ij}	Transition wavenumber between states i and j $(|E_i - E_j|/hc)$ $[\text{cm}^{-1}]$
v_j	Wavenumber corresponding to j-th vibrational mode $[\text{cm}^{-1}]$
δv	Laser frequency modulation amplitude $[\text{cm}^{-1}]$
ξ	Parameter in Junge scattering equation $(=\Lambda - 2)$
ξ	Optical efficiency
ξ	Laser power distribution parameter
ξ_S, ξ_N	Counting efficiency for photoelectron signal, dark current pulses
ϱ	Reflectivity
ϱ	Transverse coordinate vector $[\text{mm}]$
$\varrho(f)$	Spectral energy density $[\text{erg/cm}^3\text{-Hz}]$
ϱ_m	Mass density $[\text{gm/cm}^3]$
$\sigma(v)$	Absorption cross-section per molecule or atom $[\text{cm}^2]$
σ_s, σ_a	Cross section for scattering, absorption $[\text{cm}^2]$
σ_p, σ_g	Cross section due to particles, gases $[\text{cm}^2]$

σ_M, σ_R Cross section due to Mie, Rayleigh scattering $[cm^2]$

$(\sigma_a)_T$ Total absorption cross section $[cm^2]$

$\sigma_B(a, m, \lambda)$ Backscattering cross-section for particles of radius a refractive index m, at wavelength λ $[cm^2]$

$d\sigma/d\Omega$ Differential cross section $[=\sigma(\theta, \phi)]$ $[cm^2/sr]$

τ Optical depth

τ Lifetime, time constant $[s]$

τ_p Pulse length $[s]$

τ_t Thermal relaxation time $[s]$

ϕ Polarization angle $[rad]$

ϕ Generalized aerosol parameter

$\phi(t)$ Time-dependent phase angle $[rad]$

ψ Angle of sun with respect to zenith $[rad]$

ψ_i, ψ_j Eigenfunctions for i-th, j-th molecular states

ω Circular frequency $[rad/s]$

ω_c Photodiode rolloff frequency $[rad/s]$

Ω Solid angle $[sr]$

Ω_r Receiver field of view $[sr]$

References

1.1 T. H. MAUGH, II: Science **177**, 685 (1972); Science **177**, 1090 (1972)

1.2 R. BYERLY: IEEE Trans. NS-**22**, 856 (1975)

1.3 A. J. MOFFAT, J. R. ROBBINS, A. R. BARRINGER: Atmosph. Environment **5**, 511 (1971)

1.4 M. GRIGGS, C. B. LUDWIG, E. R. BARTLE, C. N. ABEYTA: 63rd Annual Meeting of the Air Pollution Control Association, St. Louis, Missouri (June 1970) Paper 70–125

1.5 M. GRIGGS: J. Air Poll. Control Assoc. **25**, 622 (1975)

1.6 *Remote Sensing of the Trosposphere*, ed. by V. E. DERR (U.S. Government Printing Office, Washington, D.C. 1972), Cat. No. C55.602:T75, Stock No. 0323-0011

1.7 M. L. WRIGHT, E. K. PROCTOR, L. S. GASIOREK, E. M. LISTON: "*A preliminary study of air-pollution measurement by active remote-sensingtechniques*"; Final Report to the National Aeronautics and Space Administration, Langley Research Center, No. CR-132724 (June 1975)

1.8 For a detailed and comprehensive analysis of the earth's stratosphere, see *The Natural Stratosphere of 1974*, CIAP Monograph 1, A. J. GROBECKER, editor-in-chief, Final Report for the U.S. Department of Transportation, DOT-TST-75-51 (September 1975)

2. Remote Sensing for Air Quality Management

S. H. MELFI

With 3 Figures

In recent years, it has become clear that man's activities are producing significant and measurable effects on the environment. Two isolated events dramatically brought this to the attention of the world: The air pollution incidents in Donora, Pennsylvania, during 1948 (20 dead and several hundred ill), and London during December 1952 (4000 dead in a few days). It became clear that the environment is not a bottomless sink, but is fragile and could not continue to be abused. As a result of these tragedies and other significant environmental insults, public opinion demanded that the environment be considered a valued resource and as such, be protected. The U.S. Congress passed a number of laws which dictate that environmental pollution be reduced to socially accepted levels.

To accomplish this mandate for atmospheric pollutants requires a detailed assessment of a number of interrelated factors. First, an evaluation of the effects of specific pollutants is performed which points out the critical receptor of that pollutant and the maximum allowable pollutant level to minimize effects on health or welfare. Next, pollutant sources and their respective strengths are identified. This information is incorporated in pollutant transport and dispersion models to predict ambient concentrations. High ambient concentrations may require controls on the sources, and the efficacy of these controls must be measured. These processes are iterated until ambient levels are below the established maximum level. Most of these factors require pollutant concentration monitoring.

Historically, air pollution monitoring has been performed using wet chemical techniques and grab sampling for later laboratory analysis. During the last decade, physical and automated methods have been developed which satisfy many monitoring requirements. These instruments are generally precise and accurate, but are limited to single-point measurements. Remote monitoring instrumentation, especially active techniques incorporating laser sources, holds promise for providing three-dimensional pollution concentration measurements. These systems have the potential for being used in:

a) Monitoring sources—Rapid, non-interfering measurement of pollutant effluents.

b) Measuring transport of pollutants—Monitoring at ground level and aloft over wide geographical areas.

c) Monitoring ambient pollutant concentrations—Spatial resolution similar to model predictions and other applications which are discussed in Section 2.4.

The following sections of this chapter will provide a review of some basic information concerning laser probing of the atmosphere.

A general discussion of the atmosphere is presented in Section 2.1 providing a review of atmospheric constituents, their significance, and their concentration ranges. This is followed in Section 2.2 with a brief overview of mathematical modeling for predicting pollutant transport and diffusion. The next Section 2.3 provides a general overview of laser probing techniques and systems to monitor the atmosphere; and finally, the last Section 2.4 describes advantages of remote monitoring and points out specific applications for laser sensing systems.

2.1 The Atmosphere

This section will briefly describe the atmosphere in relationship to laser probing. A brief description of the medium (the air) in which the measurements are made is in order, along with a description of the atmospheric constituents which are candidates for laser monitoring.

The atmosphere consists primarily of nitrogen and oxygen. These molecules make up 99% of the dry air, with argon comprising the majority of the remainder (nearly 1%). Water vapor, which is highly variable, is the next major gaseous constituent, followed by carbon dioxide. Many other gases are present in trace concentrations and will be discussed later in this section. In addition to these, the atmosphere consists of solid particles, aerosols and liquid and/or solid water.

Most of the constituents modify the transmission of electromagnetic radiation through the atmosphere. It is just this modification (attenuation, absorption, and scattering) which is the measurable parameter in the application of laser probing; and it is, therefore, important to have a clear understanding of the air to appreciate such applications. The discussion will treat separately the two major regions of the lower atmosphere—the troposphere and stratosphere.

2.1.1 Structure of the Atmosphere

The atmosphere is generally divided into a number of layers, each of which is characterized by its temperature structure. The mean temperature profile of the atmosphere, shown in Fig. 2.1, demonstrates this

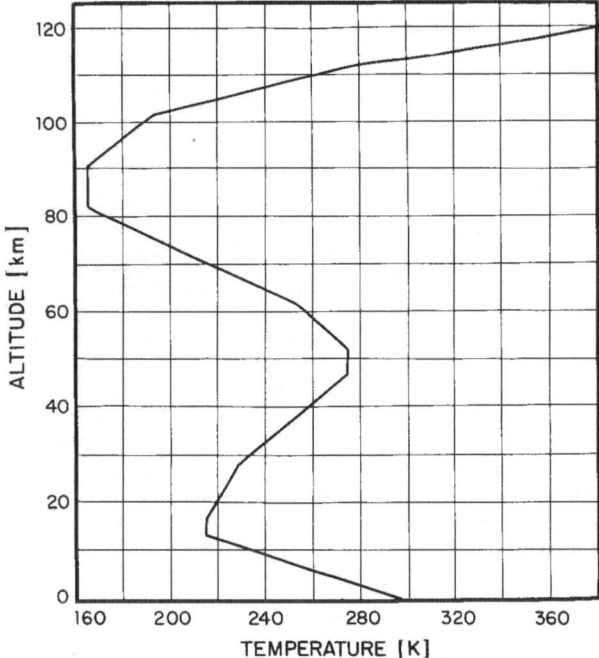

Fig. 2.1. Mean temperature as a function of altitude at 45° North latitude during July. Adapted from [2.1]

division. In the lowest layer, the temperature decreases from the Earth's surface to an altitude of approximately 10–12 km. Above this altitude, temperature increases with increasing altitude. The altitude at which this change occurs is referred to as the *tropopause* and defines the boundary between the *troposphere* (the lowest layer) and the *stratosphere* above. Within the troposphere, the air is generally near neutral stability (controlled by the decreasing temperature with altitude), and turbulent mixing is good; whereas, in the stratosphere, because of the increasing temperature with altitude, the air is very stable and mixing is discouraged. Because of the difference in stability and, thereby, difference in mixing, and since washout due to rainfall generally occurs in the troposphere, the residence time for minor constituents in the stratosphere is significantly longer. Residence time in the stratosphere may range from a few weeks to years, whereas in the troposphere, residence time may be as low as a few hours to days. The other significant difference is the high level of ultra-violet radiation available to promote photochemical reactions in the upper portions of the stratosphere.

Table 2.1. Composition of "clean" dry air near sea level. Adapted from [2.2] and [2.3]

Component [% by volume]		Content [ppm]	Component [% by volume]		Content [ppm]
Nitrogen	78.09	780 900	Hydrogen	0.00005	0.5
Oxygen	20.94	209 400	Methane	0.00015	1.5
Argon	0.93	9 300	Nitrogen	0.0000001	0.001
Carbon	0.0318	318	dioxide		
dioxide			Ozone	0.000002	0.02
Neon	0.0018	18	Sulfur	0.00000002	0.0002
Helium	0.00052	5.2	dioxide		
Krypton	0.0001	1	Carbon	0.00001	0.1
Xenon	0.000008	0.08	mon-		
Nitrous	0.000025	0.25	oxide		
oxide			Ammonia	0.000001	0.01

Note: The concentrations of some of these gases may differ with time and place, and the data for some are open to question. Single values for concentrations, instead of ranges of concentrations, are given above to indicate order of magnitude, not specific and universally accepted concentrations.

2.1.2 Troposphere

The normal dry-air constituents of the troposphere and their respective concentrations are listed in Table 2.1. As was mentioned earlier, nitrogen and oxygen comprise 99% of the air in the lower atmosphere, followed in order by argon (1%) and carbon dioxide (0.03%). Within the troposphere, water vapor is highly variable, ranging in concentration from 1 to 3%.

Some of the minor constituents listed in Table 2.1 as well as other pollutants, vary both temporally and spatially in the troposphere. The significance of these constituents to air quality mangement, along with their sources and concentration ranges, will be discussed within one of the following categories:
1) Aerosols and particles
2) Oxides of carbon
3) Sulfur compounds
4) Nitrogen compounds
5) Hydrocarbons
6) Ozone.

Aerosols and Particles

The most noticeable effect aerosols and particles have on the troposphere is their contribution to the reduction in visibility. In addition, they provide reaction sites for pollutant gases, modify precipitation by

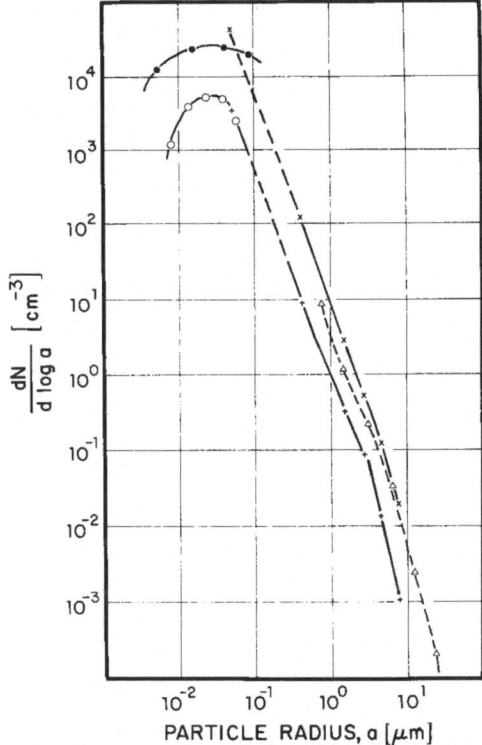

Fig. 2.2. Complete size distributions of aerosol particles at various stations in Germany. Data for the lower curve were obtained from ion mobility measurements, whereas those for the upper curve were based on impactor measurements. Individual sets of marked data points represent averages of many measurements which were not necessarily simultaneous. N is the total concentration of aerosol particles of radius smaller than a. After JUNGE [2.4]

functioning as condensation nuclei, and act as carries to transport pollutants into the lungs, eyes, and other sensitive organs of man and onto the leaves of plants and crops. The natural sources of aerosols and particles include sea spray, wind-blown soil, pollen, forest fires, volcanoes, and the reaction products of naturally occurring chemicals (e.g. the Great Smokey Mountains). Man contributes to this natural burden through the combustion of fossil fuels, the reactions of man-made pollutants, and other activities such as construction and agriculture.

Particulate matter in the troposphere varies in composition from solid particles such as fly ash, lead, sulfates, and nitrates, to liquid aerosols, including sulfuric and nitric acid and other dissolved salts. In between these two extremes are solid particles with liquid cladding.

Table 2.2. Summary of NASN (National Air Surveillance Network) suspended-particle samples for urban stations, by population class, during the period 1967-63. From [2.5]

Pop. class	No. of samples	No. of stations[a]	Min. $[\mu g/m^3]$	Max. $[\mu g/m^3]$	Arith. mean $[\mu g/m^3]$	Geo. mean $[\mu g/m^3]$
3 million and over	316	2	57	714	182	167
1–3 million	519	3	34	597	161	146
0.7–1.0 million	1191	7	14	658	129	113
0.4–0.7 million	3053	19	18	977	128	112
0.1–0.4 million	9531	92	10	1706	113	100
50000–100000	5806	81	6	982	111	93
25000–50000	1606	23	5	679	85	71
10000–25000	484	6	11	539	80	63
<10000	150	5	22	396	100	84

[a] 64 Stations participate every year: the remaining stations participated 1 or more years during the 7-year period.

The shapes of the aerosols range from the very irregular fly ash to the spherical liquid droplets. Also, the size of the aerosols vary, but under steady-state conditions normally range from 0.01 to 20 µm in radius. These limits are generally controlled by natural processes: the lower limit because smaller particles tend to coagulate, and the upper limit due to gravitational settling. A typical distribution of aerosol sizes in the atmosphere, measured by JUNGE [2.4], is shown in Fig. 2.2. Concentrations of aerosols and particles vary considerably and depend on meteorological conditions and the locations of local sources. Two *in-situ* techniques are used to monitor concentration: An electro-optical method which counts particles in certain size ranges, and provides a measure of aerosol number density; and a filter technique which monitors the total mass of particles collected (in $\mu g/m^3$). Concentrations vary from background levels of 5 $\mu g/m^3$ to high levels of over 1000 $\mu g/m^3$ in urban and industrialized areas. Table 2.2 presents a summary of particulate mass concentration measurements in urban areas of varying sizes.

Oxides of Carbon

Carbon dioxide is not generally considered a pollutant, but because its concentration is increasing globally, it may be a significant factor in the Earth's heat balance. Carbon monoxide, on the other hand, is a toxic pollutant and is especially dangerous to the health of humans.

Carbon dioxide and CO are produced both naturally and by man's activities. Natural sources of CO_2 and CO include, volcanoes, forest fires, and vegetation decay. In addition, CO is produced in nature by photochemical oxidation of organic matter. Man's activities contribute significantly to the concentration of CO and, to a lesser extent, to the levels of CO_2. In both cases, burning of fossil fuels produces the greatest impact. It has been estimated that the automobile is responsible for nearly 80% of the CO in the troposphere [2.3].

The background level of CO_2 (neglecting the direct effects of man's activities) is approximately 320 ppm, and it is generally agreed to be increasing at a rate of approximately 0.7 ppm/year. In urban and industrialized areas concentrations as high as 500 ppm have been observed [2.2]. The background level for CO is considerably lower, and is estimated to be in the range 0.05 to 0.1 ppm. Urban extremes for CO are closely correlated with vehicular traffic patterns, and may range from a few ppm in residential areas to a few tens of ppm in urban centers. Maximum instantaneous measurements of CO have been as high as 100 ppm [2.2].

Sulfur Compounds

The sulfur compounds in the atmosphere considered to be pollutants include SO_2, H_2S and sulfate aerosols. These compounds are toxic to man in high concentrations and also produce vegetation damage due both to direct contact and as a result of acid rainfall. In nature, volcanoes produce the majority of SO_2, while decaying vegetation is responsible for most of the H_2S. Man's contribution to the levels of sulfur compounds stems mainly from fossil fuel combustion. It has been estimated that nearly 50% of the man-made SO_2 is produced by fossil-fuel-burning electrical generating plants [2.6].

The residence time in the atmosphere for SO_2 is typically less than one week—its concentration is, therefore, strongly dependent on source locations. Even in industrialized regions the concentration is generally less than 0.1 ppm. However, much higher average concentrations have been observed. During the London episode in December 1952, the average concentration over a two-day period reached 1.34 ppm [2.2]. Hydrogen sulfide's background concentration is much lower, and has been estimated to be approximately 0.2 ppb [2.6]. In isolated locations H_2S has been measured at concentrations greater than 1 ppm [2.4]. Sulfate aerosol concentrations vary widely from a background of a few $\mu g/m^3$ to a high of over 600 $\mu g/m^3$ during the 1952 London episode [2.5].

Nitrogen Compounds

a) *Nitrogen Oxides.* The oxides of nitrogen considered most toxic include both NO and NO_2, often referred to cumulatively as nitrogen oxides (NO_x). They are toxic to man in the ppm range and cause plant damage even at lower concentrations. NO_x are the basis for the formation of nitrate aerosols and contribute to the formation of photo-chemical smog through their reaction with hydrocarbons. In nature, NO_x are formed as NO through biological activity. Man's contribution to the levels of NO_x is primarily due to the production of NO during high-temperature combustion. Both natural and man-made NO are further oxidized in the atmosphere to NO_2.

The concentration levels of NO_x are highly variable in the atmosphere, depending on source strengths and locations, and the conditions of the atmosphere. For example, the concentration of NO_x in Los Angeles ranges between 0.05 ppm (non-smoggy day) to recorded maxima of nearly 4.0 ppm [2.2].

b) *Ammonia.* The significance of ammonia is its role in the formation of ammonium sulfate aerosols. Its major source is the natural decay of animal and vegetable matter. It has been estimated that man produces less than 1% of the total ammonia in the atmosphere, mainly in combustion by-products and chemical manufacturing. Ammonia concentrations range from a low of a few ppb to highs of a few ppm near local sources [1.2].

c) *Organic Nitrogen Compounds.* Organic nitrogen compounds such as PAN and other peroxyacetyl nitrates are considered to be the principal irritants in photo-chemical smog. Their concentrations range downward from 35 ppb [2.2].

Hydrocarbons

The term hydrocarbons refers to a wide range of hydrogen-carbon-containing compounds. Hydrocarbons are generally classed in terms of their reactivity—with the olefins being most reactive, benzene being slightly reactive, and methane considered to be nearly non-reactive. The significance of hydrocarbons is the role they play in the formation of photo-chemical smog. They are often referred to as oxidant precursors because of the reaction $HC + NO_x + h\nu \rightarrow O_3 +$ byproducts. The natural sources of hydrocarbons include terpines from forests and vegetation, and methane from decomposition of organic matter. On the average, man's activities produce nearly 15% of the total hydrocarbons present in the atmosphere through the combustion of fossil fuels and the processing and utilization of petroleum products. Because of the low reactivity of methane, our primary interest is with non-methane hydrocarbons. The non-methane component may be as high as 50% of the total concentration. Background levels of methane have been measured to be approximately 1.5 ppm [2.3]. However, in urban areas, total hydrocarbons range from a few ppm to maximum instantaneous concentrations in Los Angeles of 40 ppm [2.2].

Ozone⁻

The significance of ozone in the troposphere is its toxicity to man in concentrations as low as 100 ppb. The natural sources of ozone include transport from the stratosphere, lightning, and volcanic eruptions. Man contributes to this natural level by releasing into the atmosphere hydrocarbons and NO_x which react in the presence of sunlight to produce ozone by the reaction discussed in Subsection 2.1.2 under *Hydrocarbons*.

Background levels of ozone at sea level range between 10 and 30 ppb; however, during air pollution episodes, concentrations as high as 500 ppb have been observed [2.6].

2.1.3 Stratosphere

The basic constituents of the atmosphere, N_2, O_2, argon and CO_2, are mixed uniformly from the earth's surface through the stratosphere to altitudes of approximately 100 km. Water vapor, other trace gases, and aerosols or particles are variable in concentration both spatially and temporally within the stratosphere. The variable constituents of the stratosphere will be discussed in the following sub-sections.

Aerosols and Particles

Aerosols and particles exist throughout the stratosphere. The concentration peaks near an altitude of 20 km, referred to as the "20 km" or "Junge" aerosol layer. Examples of early measurements of this layer are shown in Fig. 2.3. The primary significance of aerosols in the stratosphere is their possible effects on radiation transfer through the atmosphere and subsequent effect on the total radiation balance of the earth. Chemically the particles have been found to be primarily sulfates, probably due to oxidation, at stratospheric altitudes, of sulfur-containing gases such as SO_2 and H_2S. The source of SO_2 and H_2S is a combination of man-made and naturally occurring gases transported to the stratosphere from the troposphere. Another major source of particulate matter in the stratosphere is sulfur-containing gases and particles emitted by volcanic eruptions and injected into the atmosphere up to stratospheric altitudes. It has been suggested that volcanoes are the principle source of particulate matter in the stratosphere: An example is

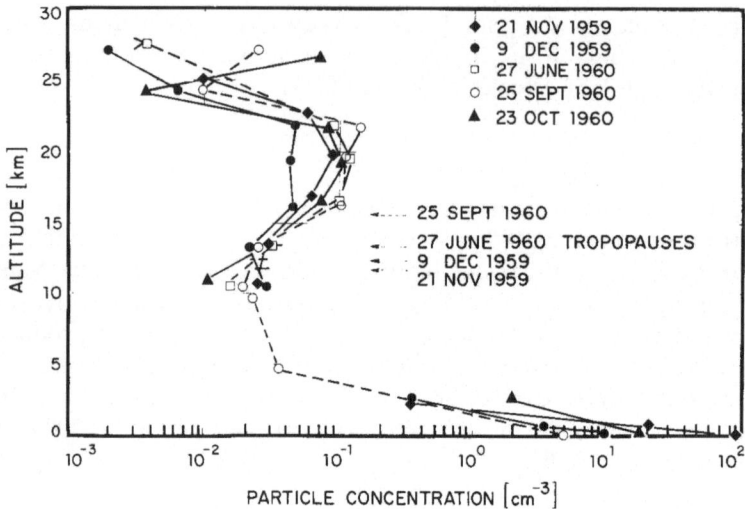

Fig. 2.3. Vertical distribution of stratospheric particles collected by inertial impactors. From [Ref. 2.6, p. 199]

the eruption of Mt. Agung in March 1963. During the winter of 1963–64, the aerosol concentrations were measured to be as much as a factor of 20 above the data shown in Fig. 2.3. By 1970 the stratospheric aerosol concentration had receded to the pre-Agung levels. More recently, the eruption of Volcan de Fuego in Guatamala again produced an elevated stratospheric aerosol concentration [2.7]. Fluctuations of this magnitude which are correlated with volcanic activity strongly indicate that natural processes determine the major quantity of particulate matter in the upper atmosphere.

Minor Molecular Constituents

a) *Sulfur Components*. As mentioned in the preceding subsection, the significance of sulfur compounds such as SO_2 and H_2S is the role they play in the formation of the sulfate aerosol.

b) NO_x. The significance of NO and NO_2 in the stratosphere is their possible contribution to the depletion of ozone. The concentration of NO_x varies from approximately 10^8 to 10^{10} molecules/cm³ in the altitude range from 10 to 30 km.

c) The significance of nitric acid in the stratosphere is its role in the removal of NO_2 and its formation of nitric acid aerosols. The distribution of nitric acid is similar to ozone, peaking in concentration at approximately 20 km. The concentration is between 10^9 and 10^{10} molecules/cm^3 between 10 and 30 km.

d) *Methane*. It has been suggested that methane in the stratosphere may be involved in ozone photo-chemistry [2.6]. The reaction results in methane being oxidized to form water vapor and CO_2. The most significant effect, then, is methane's contribution to the increase in stratospheric water vapor. Methane varies in concentration between 10^{11} and 10^{13} molecules/cm^3 in the altitude range 10 to 30 km [2.8].

e) *Ozone*. The importance of ozone in the stratosphere is that it provides the earth with a shield against high energy ultraviolet radiation from the sun. The mixing ratio of ozone peaks at approximately 8 ppm in the altitude range between 25 and 30 km. The peak value and distribution of ozone, however, is highly variable both spatially and temporally.

f) *Chlorofluorocarbons*. It has been postulated that chlorofluorocarbons in the stratosphere may undergo photo-dissociation due to the presence of high-energy ultraviolet radiation, and their by-products may significantly deplete the ozone concentration. Chlorofluorocarbons, particularly the freons, are used as refrigerants and aerosol can propellants. Because of their inertness in the troposphere, the global concentration of chlorofluorcarbons has been steadily increasing, approximately doubling in the three years between 1972 and 1975; and it is feared that an increasing amount is being transported into the stratosphere [2.9]. Recent measurements indicate that the stratospheric chlorofluorocarbon concentration is of the order of a few hundred parts per trillion at the present time [2.8].

2.2 General Pollution Transport Modeling

As mentioned in the introductory section of this chapter, the connection between sources of air pollution and the location of a critical receptor of the pollutant involves transport, transformation, and dispersion through the atmosphere. A complete and thorough treatment of the transport and diffusion of a number of pollutants in a given region is desirable in order to uniquely determine the cumulative concentrations at all points within the region as a function of time, but generally impossible to obtain since it requires data approaching the spatial scale of molecules and temporal scale of molecular collision times. Fortunately, for most applications an understanding of atmospheric transport and diffusion between sources and receptor to this extent is not required. In an attempt to understand and predict transport and diffusion through the atmosphere, a number of mathematical models have been developed. These models are used to:

a) Predict ambient concentrations due to existing pollutant sources.

b) Establish maximum source strengths while maintaining safe ambient levels.

c) Assess the impact of proposed new or modified sources on ambient concentrations.

2.2.1 Description of Mathematical Models

The simplest mathematical model for predicting the transport and dispersion of pollutants through the atmosphere is referred to as the box model. In the development of this model it is assumed that the area under study can be approximated as rectangular. The atmosphere above this area is contained in a hypothetical box whose height above ground level corresponds to the height of the mixed layer. The sides of the box correspond either to significant geological features, such as mountain ranges, or are positioned so that two sides are perpendicular to the prevailing winds. In more advanced models the area box is divided into a large number of smaller boxes whose square bases are located at ground level and whose height either matches the mixed layer height or some fraction of that height, thereby providing a number of stacked boxes over each square base. Thus, the larger area box is divided into a large number of small boxes both horizontally and vertically.

In the simplest application of the multi-box model to predict pollutant transport, the format below is followed:

1) An accurate pollutant source inventory is made providing source strengths and locations.

2) Pollutant source strengths in each small box are summed to yield an approximate average concentration for the small box.

3) Wind data and estimated atmospheric stability are used to provide the trajectory and dispersion of pollutants from each small box.

4) The resultant average concentration of pollutants in any given small box, as a function of time, is estimated to be the summation of its own average source strength and the contribution of pollutants to and from other small boxes, transported by the wind.

This model, and other more sophisticated ones, will continue to be improved, thus providing a better understanding of pollutant transport. The treatment of chemical reactions and other transformation processes during transport, which act as either sources or sinks of pollutants, is an example of recent improvements in some mathematical models.

One important characteristic of this class of models, as far as laser monitoring is concerned, is that the *average* pollutant concentration is predicted for spatial dimensions equal to the size of the small boxes. In most applications of these models, the dimensions are typically 1 km horizontally and 0.2–1 km vertically.

2.2.2 Monitoring Needed for Model Development and Validation

In order to provide data for the validation and development of models, dedicated methods for monitoring pollutants, which provide spatial and temporal resolution similar to model predictions, are required. Present *in-situ* techniques monitor at a single point in most cases, and thus do not provide data which are representative of the *average* concentration over the spatial extent of the model predictions. Integrated horizontal measurements at ground level and integrated vertical measurements through the mixed layer matched to the spatial scales of the models will provide data to either validate developed models or act as feedback to improve model predictions. Measurements of this type can be performed either by mobile *in-situ* instruments which are moved horizontally and/or vertically over the prediction grid in a time comparable to the temporal scale of the model or, more preferably, by remote-monitoring instrumentation which measures integrated concentration (long-path laser absorption), or a profile concentration which can subsequently be integrated (various LIDAR applications).

Until air pollution monitoring provides data on the proper spatial and temporal scales, the utility of atmospheric transport models cannot be firmly established. Remote monitoring techniques hold promise for cost-effectively providing these data.

2.3 Remote Monitoring Capabilities of Laser Systems

This section will present a brief overview of the capabilities of remote-monitoring laser instrumentation. Many of the techniques discussed will be covered in detail in subsequent chapters.

Remote monitoring can be defined as sensing qualitatively and/or quantitatively a chemical or physical parameter in the environment where the monitoring instrument and the parameter under investigation are spatially separated. In the atmosphere, remote sensing techniques can be divided into two broad categories: active and passive, depending on the source of radiation. Passive techniques make use of available radiation in the atmosphere (e.g., solar and earth-reflected or emitted radiation). The interaction of this radiation with the species under investigation (absorption and scattering) or the thermal emission of the species is observed to infer concentration. Active techniques, on the other hand, are characterized by the introduction of specific radiation into the atmosphere, typically incorporating lasers as the source of this radiation. The interaction (scattering, absorption, fluorescence) with the atmosphere is observed to infer species concentration. The remaining discussion of this section will be limited to active techniques which use lasers as the source of radiation[1]. Active techniques, for the purpose of this discussion, will be further subdivided into single-ended and double-ended systems.

2.3.1 Single-Ended Systems

Single-ended systems, as the name implies, co-locate the laser source and telescope receiver. Generally single-ended systems depend on scattering as the primary means for interrogating the atmosphere. This may be either direct scattering by the species of interest, or scattering by other atmospheric constituents, with the species of interest modifying the transmission of the laser beam in a measurable manner. Direct scattering techniques include LIDAR, Raman LIDAR and resonance LIDAR, whereas, indirect scattering techniques include: differential absorption scattering LIDAR and an additional application of Raman LIDAR. Each of these techniques will be discussed separately.

LIDAR

LIDAR is an acronym for Light Detection And Ranging. All single-ended LIDAR systems consist of a laser and telescope whose optical axes are aligned parallel such that the telescope field-of-view includes the laser beam as it propagates through the atmosphere. The differences in the LIDAR systems mentioned above are in the selection of the laser wavelength, the receiver wavelength, and the data analysis and interpretation.

In the most basic system, referred to simply as LIDAR, the detector wavelength is matched to the laser wavelength. As the laser radiation propagates through the atmosphere, it interacts with aerosols and molecules. The principle interaction of interest is elastic scattering by the aerosols. Some of the scattered radiation is collected by the telescope

---

[1] Lasers can also be used in the passive mode as local oscillators to detect radiation by high-resolution, sensitive heterodyne techniques, as discussed in Chapter 7.

and detected by a sensitive photomultiplier. The detected signal is recorded as a function of time to provide a range-resolved measure of atmospheric scattering. Analysis of the recorded data is then performed to provide an indication of aerosol distribution in both the troposphere and stratosphere.

Raman LIDAR

In Raman LIDAR the receiver is made sensitive, through the use of a spectrometer or interference filter, to Raman-shifted wavelengths. Most molecules scatter electromagnetic radiation not only at the excitation wavelength but also at specific shifted wavelengths by the phenomenon referred to as Raman scattering. The magnitude of the shift is unique to the scattering molecule, and the intensity of the Raman band is proportional to the scattering molecule's concentration. Raman LIDAR has been applied to the measurement of a number of atmospheric molecules, including H_2O, SO_2 and CO_2. The most significant limitation of the technique arises from the low cross section for Raman scattering. Thus, the application requires the use of high-power lasers, large telescopes, long integration times, and is generally limited to measurements where high molecular concentrations are present.

Resonance LIDAR

Resonance LIDAR is characterized by a careful selection of both laser and receiver wavelength so that they are matched to an absorption wavelength of the atmospheric species of interest. Stimulating a molecule at an absorption frequency produces resonant scattering which may be significantly more intense than non-resonant scattering. The utility of resonance LIDAR is limited in the lower atmosphere because radiationless quenching reduces the scattering intensity in a nonlinear fashion dependent on the concentration of the quenching molecules. However, this technique has been successfully applied to the measurement of high-altitude concentrations of atomic sodium and potassium.

Differential Absorption Scattering LIDAR

This system is the first of the indirect scattering techniques to be discussed. It depends on scattering by aerosols in the atmosphere but measures selected molecular species due to their absorption. The technique makes use of at least two laser beams at different wavelengths which are sequentially or simultaneously transmitted along the same path in the atmosphere. One laser beam is absorbed by the molecular species of interest, whereas the other at a nearby wavelength is not strongly absorbed. Since the beams are spectrally separated by a small wavelength increment, the cross-section for aerosol scattering may be essentially identical for each. To the extent that this is valid, the difference in scattered intensity of the beams as they propagate is due to the difference in absorption by the molecule under investigation. Analysis of the detected signals from both beams as a function of time provides a range-resolved measurement of the absorbing molecule's concentration.

Raman LIDAR for Visibility

Raman LIDAR, as described above, can also be applied to the measurement of atmospheric attenuation. In this application Raman signals from either N_2 or O_2 in the atmosphere are monitored. Since the distribution of these major constituents is well known, a comparison of the Raman signal, which includes attenuation, with the known molecular distribution provides a range-resolved measure of double-path attenuation. In many cases, especially under low-visibility, attenuation can be related to visibility.

2.3.2 Double-Ended Systems

Double-ended systems are characterized by either having the laser transmitter and receiver telescope located separately or having the laser and telescope co-located with a physical reflector located at a distance. The two principle double-ended systems are Bistatic LIDAR and Long Path Absorption. These techniqeus will be discussed separately.

Bistatic LIDAR

Bistatic LIDAR is similar to the LIDAR system discussed above expect that the laser and telescope are separated by a distance. In the utilization of this technique, both the laser and telescope are aimed toward the same point in the atmosphere. The beam divergence of the laser and the field-of-view of the telescope together define the scattering volume at the point of intersection. As the laser pulse propagates through the scattering volume, some of the energy scattered elastically by atmospheric aerosols is collected by the telescope and detected by a photomultiplier. The output of the photomultiplier is recorded. By carefully varying the pointing angle of both the laser and telescope, measurements of scattering at a fixed altitude can be observed for a variety of scattering angles. The bistatic LIDAR technique can provide valuable scattering angle data which can be interpreted to infer the refractive index and/or size distribution of atmospheric aerosols.

Long Path Absorption

This technique utilizes the absorption of a laser beam as it propagates through the atmosphere as the measurable parameter to infer molecular concentration. The laser source and telescope receiver can be separated and pointed toward each other; but, for ease of operation, the laser and telescope are generally co-located with their optical axes aligned and pointing to a retro-reflector or topographical target. As in the Differential Absorption Scattering LIDAR discussed above, at least two laser beams at slightly different wavelengths are used to interrogate the intervening atmosphere between the system and the reflector. The two beams are adjusted in wavelength to be "on" and "off" an absorption band of the molecule under investigation. Comparison of the two signals collected by the telescope provides a measure of the integrated concentration of the molecular species along the transmission path. In some applications of the technique more than two beams of differing wavelength are required because of multiple molecular absorption. The number of beams required is at least one more than the number of significant molecular absorbers. A promising outgrowth of this technique is the use of topographical targets instead of cooperative retro-reflectors. The targets may be either trees, hills, or buildings when using a horizontal ground-based system, or the earth's surface for an airbone system.

   With this discussion of the large variety of active remote-sensing techniques and systems which are available and/or being developed for monitoring the atmosphere, the question remains: How will these and other remote-monitoring systems be used in air quality management? The answer to this question is the subject of the next section.

2.4 Remote Monitoring in Air Quality Management

As mentioned in the introductory section of this chapter, remote monitoring holds promise for providing three-dimensional pollution concentration data of the atmosphere. The large number of pollutants of concern and their significance, distribution, and concentration levels were reviewed in Section 2.1. Section 2.2 described mathematical transport models which attempt to predict the movement of pollutants from their source to the critical receptor, and a discussion of the various laser techniques for remotely sensing the atmosphere was presented in the last Section 2.3. The question to be addressed here is: What are the unique advantages of remote atmospheric monitoring, when compared to present methods, which will insure that these techniques will be incorporated into air quality management systems of the future? To answer this question adequately, the unique advantages will be considered, and specific applications based on these advantages will be listed.

2.4.1 Advantages of Remote Monitoring

Although it is reasonable to assume that remote monitoring instrumentation may generally provide as accurate or precise measurements as *in-situ* instrumentation, they do, as a class of techniques, provide certain unique advantages: namely, they permit specific applications which would be difficult, if not impossible, using standard instrumentation. Some of these advantages are listed below and will be discussed individually:

1) Non-interfering for source effluent monitoring.
2) Integrated-path measurements.
3) Measurements at ground level and aloft.
4) Perspective in monitoring.
5) Measurements over large geographical areas.

Non-Interfering

Remote monitoring of pollutant emissions at their source is non-interfering in two ways. Since it is a probeless technique, no extraction of source gases is required, thus eliminating the possibility of modifying the sample during the measurement. In addition, it has the potential for providing specific measurements of effluent concentration without interfering with the operation of the industrial facility under investigation. Measurements made off the property of the facility provide an opportunity to perform unannounced inspections of emission concentrations.

Integrated-Path Measurements

The utility of integrated measurements for pollution transport model verification and development has been previously discussed (Sec. 2.2). In this case, remote-monitoring laser techniques provide measurements on a spatial scale comparable with model predictions. Another need for integrated measurements is related to the accurate estimation of total pollutant dose to humans. Remote monitoring can provide this information with integrated measurements over an area which approximates the motion area of the population under study.

Measurements at Ground Level and Aloft

To obtain a better understanding of air quality, it is becoming increasingly recognized that measurements at ground level must be supplemented by measurement aloft. Laser monitoring may prove to be the cost-effective method to monitor pollutant concentrations aloft. Three-dimensional measurements are also needed to assess the impact of photo-chemical reactions and to trace pollutants from their source to the receptors. Measurements as a function of altitude are required for model development and verification, since most models do not specifically predict ground-level pollution concentration.

Perspective in Monitoring

Conventional *in-situ* instruments normally provide an accurate measurement at only one location. Pollutant concentrations in the atmosphere surrounding this location may vary significantly, thus invalidating the measurement as being representative of the larger area. Local sources, meteorology, topography and man-made structures all contribute to possible variability. A single remote monitoring instrument located in the area has the potential for providing accurate representative measurements. In addition, remote sensing can be used in initial surveys to optimally locate *in-situ* monitoring networks.

Measurement Over Large Geographical Areas

There are many large air sheds which require monitoring. This monitoring would be performed initially to establish a baseline so that, in the future, changes in air quality could be determined. Laser instrumentation mounted on mobile platforms (aircraft, trucks) may prove to be the most cost-effective technique to measure air quality over these large geographical areas.

2.4.2 Applications of Remote Monitoring

As mentioned and discussed in the previous section, there are a number of unique advantages in employing laser remote-sensing instrumentation for monitoring air quality. By making use of these advantages, remote monitoring techniques can significantly complement the present *in-situ* methods. A number of specific applications are listed below:

    1) Monitoring opacity of a source plume.

    2) Monitoring specific effluents at the source.

    3) Measuring plume transport and diffusion.

    4) Monitoring to determine representativeness of point measurements (pollutant variability studies).

    5) Surveying to design optimum *in-situ* monitoring networks.

    6) Monitoring for verification and development of pollution transport models.

7) Measuring mixed height over large air sheds.

8) Monitoring long-range transport of pollutants from urban and industrialized areas.

9) Monitoring expansive wilderness areas to determine air quality trends.

10) Assessing proposed sites for new sources.

11) Monitoring during air pollution episodes.

These are some of the applications of laser remote-monitoring instrumentation currently under consideration. It is certain that others will be recognized as the technology is applied to the requirements of air quality management.

2.5 Conclusion

There is now, and will continue to be, a need for understanding atmospheric processes in connection with the transport and conversion of pollutants from both natural and man-made sources. Not only is the lower atmosphere which we breathe important, but so is the upper atmosphere which contains gases and particles which affect the spectrum and intensity of solar radiation impinging on the Earth's surface. A case has been made here for the use of laser remote-sensing techniques for achieving some of the monitoring goals associated with model development and atmospheric surveillance. Specific monitoring techniques based on the use of laser radiation will be considered in detail in the following chapters.

References

2.1 U.S. Standard Atmosphere Supplements (1966), U.S. Government Printing Office, Washington, D.C.

2.2 B.D. TEBBENS: In *Air Pollution*, Vol. 1, ed. by A.C. STERN (Academic Press, New York 1968) Chap. 2

2.3 *Cleaning our Environment—The Chemical Basis for Action*. American Chemical Society, Washington, D.C. (1969)

2.4 C.E. JUNGE: *Air Chemistry and Radioactivity* (Academic Press, New York 1963); for later review, see [2.5]

2.5 M. CORN: In *Air Pollution*, Vol. 1, ed. by A.C. STERN (Academic Press, New York 1968) Chap. 3

2.6 *Remote Measurement of Pollution*, National Aeronautics and Space Administration, Special Publication 285, Washington, D.C. (1971)

2.7 M.P. McCORMICK, W.H. FULLER: Appl. Opt. **14**, 4 (1975)

2.8 *The Natural Stratosphere of* 1974, ed. by F.R. REITER. CIAP Monograph 1, U.S. Department of Transportation, Washington, D.C. (1975)

2.9 R.J. CICERONE, R.S. STOLARSKI, S. WALTERS: Science **185**, 1165 (1974)

3. Laser-Light Transmission Through the Atmosphere

V. E. ZUEV

With 7 Figures

Transparency of the atmosphere for laser radiation is one of the most important parameters in the laser monitoring equations, whose solutions enable extraction of quantitative information on the profiles of atmospheric gases and particles. The energy loss of a laser beam propagating through the atmosphere is mainly the result, as a rule, of the following simultaneously-acting phenomena: 1) molecular absorption; 2) molecular scattering; 3) particulate scattering. In this connection we shall now consider these phenomena, paying special attention to their quantitative descriptions; and will subsequently select the most effective laser wavelengths from the standpoint of energy losses under various meteorological conditions. The data given in this chapter refer to a wide range of wavelengths, including ultraviolet, visible, and infrared spectra of the electromagnetic wave scale. We are, however, mainly concerned with the region from approximately 0.2 to 20 μm, since beyond these limits optical radiation is nearly completely absorbed by small thicknesses of the atmosphere. At short wavelengths, such absorption is due to oxygen and ozone, while at longer wavelengths, to water vapor.

3.1 Molecular Absorption

3.1.1 Basic Definitions

Attenuation Coefficient. The attenuation coefficient $\alpha(v)$ of a medium for radiation of wavenumber v is a proportionality factor in Bouguer's law[1] characterizing transmission properties of the medium. Its units are conventionally cm^{-1} (linear attenuation coefficient), or cm^2/cm^3 (volume extinction coefficient); the former terminology preferred for transmission studies, and the latter for the remote sounding of specific volumes of the atmosphere. Written in differential form, Bouguer's law becomes, for a plane wave propagating along the z direction,

$$dI(v) = -I(v)\alpha(v, z)dz , \qquad (3.1)$$

---

[1] Also known as Beer's law or the Beer-Lambert equation.

where $dI(v)$ is the change in intensity after the radiation has passed through a medium layer of depth dz. Integration of (3.1) over a layer depth L, between distances Z_1 and Z_2 from the laser transmitter yields, for a one-way path,

$$I(v) = I_0(v) \exp\left[-\int_{Z_1}^{Z_2} \alpha(v, z)dz\right],\tag{3.2}$$

which, in the case of a homogeneous medium, converts to the following simple expression of the exponential attenuation law

$$I(v) = I_0(v) \exp\left[-\alpha(v)L\right].\tag{3.3}$$

In (3.2) and (3.3), $I_0(v)$ is the initial radiation intensity $[\text{W/cm}^2]$, and the exponent $\alpha(v)L$ is called the *optical thickness* $\tau(v)$ of the medium layer—a dimensionless quantity.

Spectral Absorption and Spectral Transmission. The following equations define spectral absorption $A(v)$ and spectral transmission $T(v)$

$$\begin{aligned} A(v) &= [I_0(v) - I(v)]/I_0(v)\,; \\ T(v) &= I(v)/I_0(v)\,, \end{aligned}\tag{3.4}$$

which characterize the fraction of monochromatic radiation absorbed by a given medium layer, or passing through it, when the attenuation is caused by absorption alone. (Scattering will be considered in Subsects. 3.2 and 3.3 below.)

Absorption Function and Transmission Function. The absorption function $\mathscr{A}$ denotes that portion of non-monochromatic radiation absorbed by a medium layer in a given spectral interval $\Delta v = v_2 - v_1$. The transmission function $\mathscr{T}$ is similarly defined. Thus,

$$\mathscr{A} = \int_{v_1}^{v_2} I_0(v)A(v)dv / \int_{v_1}^{v_2} I_0(v)dv\,,\tag{3.5}$$

$$\mathscr{T} = \int_{v_1}^{v_2} I_0(v)T(v)dv / \int_{v_1}^{v_2} I_0(v)dv\,.\tag{3.6}$$

As seen from (3.5) and (3.6), the absorption and transmission functions in the spectral interval $[v_1, v_2]$ characterize *average* values of spectral absorption and transmission in this interval.

In the particular case when $I_0(v)$ is constant within the spectral interval $[v_1, v_2]$, we obtain, from (3.5) and (3.6),

$$\mathscr{A} = \int_{v_1}^{v_2} A(v)dv / (v_2 - v_1)\,,\tag{3.7}$$

$$\mathscr{T} = \int_{v_1}^{v_2} T(v)dv / (v_2 - v_1)\,.\tag{3.8}$$

If $N(z)$ is the number density $[\text{cm}^{-3}]$ of absorbing molecules at location z, and $\sigma(v, z)$ is the absorption cross section $[\text{cm}^2]$ per molecule at wavenumber v, then the attenuation coefficient becomes $\alpha(v, z) = N(z)\sigma(v, z)$. If we now rewrite (3.5–8) by substituting (3.2–4), with $\alpha = N\sigma$, we obtain

$$\mathscr{A} = \int_{v_1}^{v_2} I_0(v)\{1 - \exp[-\int_{Z_1}^{Z_2} N(z)\sigma(v, z)dz]\} dv / \int_{v_1}^{v_2} I_0(v)dv, \tag{3.9}$$

$$\mathscr{T} = \int_{v_1}^{v_2} I_0(v) \exp[-\int_{Z_1}^{Z_2} N(z)\sigma(v, z)dz] dv / \int_{v_1}^{v_2} I_0(v)dv, \tag{3.10}$$

$$\mathscr{A} = \int_{v_1}^{v_2} \{1 - \exp[-\int_{Z_1}^{Z_2} N(z)\sigma(v, z)dz]\} dv / (v_2 - v_1), \tag{3.11}$$

$$\mathscr{T} = \int_{v_1}^{v_2} \exp[-\int_{Z_1}^{Z_2} N(z)\sigma(v, z)dz] dv / (v_2 - v_1). \tag{3.12}$$

Equations (3.9) and (3.10) represent the absorption and transmission functions for the general case of radiation propagation in the spectral interval $[v_1, v_2]$ in a non-homogeneous absorbing medium for an arbitrary source radiation intensity spectrum in the interval. Equations (3.11) and (3.12) represent the case for a neutral spectral dependence of the source intensity over the interval $[v_1, v_2]$.

For a homogeneous absorbing medium, (3.9–12) take a simpler form

$$\mathscr{A} = \int_{v_1}^{v_2} I_0(v)\{1 - \exp[-N\sigma(v)L]\} dv / \int_{v_1}^{v_2} I_0(v)dv, \tag{3.13}$$

$$\mathscr{T} = \int_{v_1}^{v_2} I_0(v) \exp[-N\sigma(v)L] dv / \int_{v_1}^{v_2} I_0(v)dv, \tag{3.14}$$

$$\mathscr{A} = \int_{v_1}^{v_2} \{1 - \exp[-N\sigma(v)L]\} dv / (v_2 - v_1), \tag{3.15}$$

$$\mathscr{T} = \int_{v_1}^{v_2} \exp[-N\sigma(v)L] dv / (v_2 - v_1), \tag{3.16}$$

where $L = Z_2 - Z_1$.

As is seen from the above expressions, for quantitative determination of monochromatic radiation energy losses owing to molecular absorption in a homogeneous atmosphere, we must know the absorption cross section for the propagated radiation frequency, the density or concentration of absorbing molecules, and the geometric depth of the medium layer. Solution of an analogous problem for the case of an inhomogeneous atmosphere requires additional knowledge of the density distribution of absorbing molecules along the path.

Calculation of absorption and transmission functions of non-monochromatic radiation in the simplest case of a homogeneous atmosphere and a neutral source spectrum (in the spectral interval of interest) requires knowledge of the spectral dependence of the cross section in this interval as well as the density values of the absorbing molecules and the geometric depth of the absorbing layer. In the more complicated case involving determination of the absorption and transmission functions when the atmosphere is homogeneous, but the source spectrum is not neutral, data on the density distributions of absorbing molecules along the path are necessary. Finally, for the most general case of determining the absorption (transmission) function for a source with arbitrary spectrum propagated in an inhomogeneous atmosphere, knowledge is required of the spectral dependence of the molecular absorption cross sections and source radiation intensity, the density profile of absorbing molecules, the geometric depth of the absorbing layer, and the width of the spectral interval. At the same time, it is necessary to keep in mind that the cross sections depend not only on the radiation wavelength but also on the medium coordinates, as will be shown later.

When estimating energy losses of laser radiation propagated in the atmosphere, resulting from molecular absorption, any of the above formulas can be used, depending on the character of the laser radiation spectrum. If the variation of $\sigma(v)$ with frequency is negligible within the spectral interval investigated, the radiation can be considered monochromatic; and for estimating its losses, (3.2) or (3.3) should be used. Equations (3.11), (3.12), (3.15), and (3.16) are applied in those cases where laser power is independent of wavelength.

3.1.2 Absorption by a Separate Line

When considering the absorption of laser radiation by atmospheric gases, we have to deal in most cases with the absorption of separate lines of one or another gas. Each absorption line is characterized by the following three parameters: 1) position; 2) width; 3) intensity. In addition, each line has a definite shape.

In lower regions of the atmosphere, broadening of spectral lines is mainly due to collisions between molecules. The line contour determined by these effects is called dispersion. Its simplest form was obtained by Lorentz [3.2] in 1906

$$\sigma(v) = (S/\pi) \frac{\gamma_L}{(v - v_0)^2 + \gamma_L^2}, \tag{3.17}$$

where S is the line intensity (to be defined below), v_0 is the line center frequency (wavenumber), v the radiation frequency, and γ_L the line-halfwidth, which is one-half the linewidth between points v_1 and v_2 satisfying the condition

$$\sigma(v_1) = \sigma(v_2) = \sigma(v_0)/2 . \tag{3.18}$$

Note that γ_L is the halfwidth at half-maximum absorption coefficient (HWHM), whereas Γ $(=2\gamma)$ is defined as the full width at half-maximum (FWHM).

For the case of a two-component gaseous mixture, the following expression for line halfwidth is obtained from kinetic theory

$$\gamma_L = \{8\pi kT[N_a D_{aa}^2 (2/m_a)^{1/2} + N_b D_{ab}^2 (1/m_a + 1/m_b)^{1/2}]\}^{-1/2}, \qquad (3.19)$$

where k is Boltzmann's constant, T is the absolute temperature, and the indices a and b refer to absorbing and non-absorbing molecules; N_a and N_b are their respective number densities; D_{aa} and D_{ab} are the sum diameters of optical collisions between molecules $a \leftrightarrow a$ and $a \leftrightarrow b$; and m_a and m_b are the molecular masses.

The formula for a two-component mixture is often expressed in the following simple form

$$\gamma_L = \gamma_L^0 (p/p_0)(T_0/T)^{1/2}, \qquad (3.20)$$

where p is the pressure. γ_L^0 is the line halfwidth under standard conditions (usually, $p_0 = 1$ atm, $T_0 = 273$ K).

In spite of the fact that more precise theories of spectral line broadening due to collisions between molecules have been developed, (3.17), (3.19), and (3.20) are still widely used since they satisfactorily provide a real picture of spectral lines within a few halfwidths of line center. For this reason, we will not discuss other theories in this chapter.

Summing up the available calculations of absorption line halfwidths for atmospheric gases, it should be noted that their values in the ground atmospheric layer are of the order of 0.01–0.1 cm^{-1}. If pressure and temperature variations with height are known, it is not difficult to calculate the value of γ_L at any altitude using (3.20) if its value under standard conditions is known.

Spectral line broadening is also produced by the Doppler effect. Assuming thermodynamic equilibrium for the translational degrees of freedom of a molecule, and a Maxwellian velocity distribution of the molecules, then for a spectral line contour for which broadening is due to the Doppler effect only, the following expression is obtained

$$\sigma(v) = (S/\gamma_D)(\ln 2/\pi)^{1/2} \exp[-(v - v_0)^2 \ln 2/\gamma_D^2], \qquad (3.21)$$

where S is the line intensity, as before, and

$$\gamma_D = (v_0/c)(2kT \ln 2/m)^{1/2} \qquad (3.22)$$

is the Doppler line halfwidth, m is the mass of the molecule, and c is the velocity of light.

As is seen from (3.22), the Doppler width does not depend on pressure, but is a function of line center position. Since average temperature variations throughout the troposphere and stratosphere do not exceed a factor of 1.5, the change in γ_D due to atmospheric temperature variations is usually less than $\pm 15\%$. It should be mentioned that variations in the pressure-broadened linewidth γ_L due to atmospheric temperature changes are also less than $\pm 15\%$; however, variations of γ_L and γ_D with tempera-ture have different signs [cf. (3.19) and (3.22)].

Calculations show that the value of γ_D for atmospheric gases in the wavelength range of interest to us, including the ultraviolet (from 0.2 μm), visible, and infrared (to 50 μm) regions, is approximately within the limits from 6×10^{-2} cm^{-1} (for O_2 lines near 0.2 μm) to 3×10^{-4} cm^{-1} (H_2O vapor lines near 50 μm).

The representative values of γ_L and γ_D quoted above show that in the ground atmospheric layer molecular collisions are the dominating cause of spectral line broadening. Since γ_L depends linearly on pressure, whereas γ_D is independent of pressure, for every spectral line γ_L becomes equal to γ_D at a certain altitude. With a subsequent increase in altitude, the Doppler effect will play an increasingly dominating role in spectral line broadening. For the overwhelming majority of absorption lines of atmospheric gases in the visible and infrared, however, line broadening due to collisions prevails over broadening due to the Doppler effect, at least up to altitudes approaching 15 km. In the ultraviolet, consideration of joint action of both effects is necessary in some cases, not only in the upper layers of the troposphere but in the ground layer of the atmosphere as well.

The *intensity* of a spectral line is expressed as an integral of the absorption cross section over all frequencies

$$S = \int_{-\infty}^{\infty} \sigma(v)dv, \tag{3.23}$$

where, as before, $v = f/c$, where f [Hz] is the frequency of the radiation and v[cm^{-1}] the corresponding wavenumber. Since the units for $\sigma(v)$ are cm^2, S is in units of cm.

Quantum mechanics gives the following expression for the intensity of separate transitions of a molecule from the state i into the state j (e.g., see [Ref. 3.56, p. 158])

$$S_{ij} = (8\pi^3 v_{ij}/3hc)(N_i/g_i N)|R_{ij}|^2[1 - \exp(-hcv_{ij}/kT)], \tag{3.24}$$

where v_{ij} is the transition frequency (wavenumber), N_i/N is the fractional number density in state i with statistical weight g_i, $|R_{ij}|^2$ is the square of the matrix element of the dipole moment, and h is Planck's constant. The fractional number density portion of (3.24) may be rewritten in terms of the energy level E_i of state i and the total partition function Q as

$$N_i/g_i N = Q^{-1} \exp(-E_i/kT), \qquad (3.25)$$

which, when substituted into (3.24) shows the temperature dependence of the transition intensity.

The matrix element of the electric dipole moment μ is expressed by the wave functions of molecular states Ψ_i and Ψ_j

$$R_{ij} = \int \Psi_i^* \mu \Psi_j \, dV, \qquad (3.26a)$$

where integration is carried out over all configurational space. The wave functions, obtained from solution of the Schrödinger equation, obey the orthogonality condition

$$\int \Psi_i^* \Psi_j \, dV = 0 \qquad (i \neq j). \qquad (3.26b)$$

Thus, if $\mu = $ constant, then $R_{ij} = 0$; consequently, transitions of a molecule from one energy state to another by dipole radiation cannot take place if the dipole moment is invariant.

It should be noted that emission or absorption of light can also occur with a change of *magnetic* dipole or electric *quadrupole* of a molecule; however, intensities of lines involving these transitions are rather small. For example, transitions associated with variation of the electric dipole moment of a molecule yield line intensities roughly five orders of magnitude greater than those of magnetic dipole transitions, and eight orders greater than electric quadrupole transitions.

The matrix elements R_{ij} are associated with the known Einstein coefficients A_{ij} characterizing probabilities of induced radiation and absorption between molecular levels i and j

$$A_{ij} = (64\pi^4 v_{ij}^3/3hg_i)|R_{ij}|^2. \qquad (3.27)$$

For electric dipole radiation the value A_{ij} is of the order of 10^8, 10, and $1 \, s^{-1}$ for electronic, vibrational, and rotational transitions, respectively. The most difficult problem in line intensity calculations is the determination of $|R_{ij}|^2$.

Completing our consideration of absorption by individual lines, we stress that every absorption line of a molecule in the atmosphere has, generally speaking, its own value of center position, width, and intensity. Without precise knowledge of these parameters, a theoretical determination of absorption coefficient is out of the question, even with correct knowledge of the dispersing contour. Furthermore, the line parameters depend in a complicated manner on the variable macroconditions of the medium—notably, temperature and pressure (total and partial) of the component gases. All this creates exceptional difficulties in making a quantitative estimate of energy loss of an optical wave due to an individual absorption line.

3.1.3 General Characteristics of Absorption Spectra of Atmospheric Gases

The main absorbing gases in the atmosphere in the ultraviolet, visible, and infrared are water vapor, carbon dioxide, ozone, and oxygen. The atmosphere contains minor constituents that also absorb radiation in the wavelength range considered. These are primarily carbon monoxide, methane, and nitrogen oxides. In addition, various gases of industrial origin are present in localized regions, and these can absorb optical radiation. In this chapter we describe only the absorption spectra of gases existing in the atmosphere on a planetary scale, since these will generally be most important in determining optimal wavelengths for laser monitoring.

Water vapor and ozone are asymmetric top molecules; hence, their electronic, vibrational, and rotational spectra are very rich in absorption lines. The molecules of CO_2, N_2O, NO, CO, O_2, and N_2, being linear, generally have fewer lines per spectral interval; however, the spectra are rather complicated, as is the absorption spectrum of methane (CH_4)—a symmetric top molecule.

Water Vapor. The electronic absorption spectrum of water vapor is located in the far ultraviolet (wavelengths less than 186 nm). The vibrational-rotational spectrum contains three main bands v_1, v_2, and v_3 with centers at 3657.05, 1594.78, and 3755.92 cm^{-1}, respectively, and overtones, combination, and hot bands in the infrared and visible.

Fine structure of the vibrational-rotational spectrum of H_2O is extremely complex and intricate. Each of the bands consists of hundreds, even thousands, of lines recorded in high resolution experiments, and there is no doubt that the number of recorded lines will increase with increasing spectral resolution made possible by better spectrometers and by tunable lasers.

The lines of the following water vapor isotopes were detected in the solar spectrum: H_2O^{16}, H_2O^{18}, H_2O^{17}, and HDO, existing in the atmosphere in proportions: 99.73%, 0.20%, 0.04%, and 0.03%, respectively. Absorption lines of the H_2O^{18} molecule are shifted relative to the H_2O^{16} lines by 1–11 cm^{-1}; whereas for the H_2O^{17} molecule this shift is twice as large. The intensity ratios of the corresponding lines of these isotopes are proportional to their concentrations. The absorption bands of HDO differ noticeably from the H_2O^{16} bands.

The most intensive and broad vibration-rotation band of H_2O is the main ν_2 band with center near 6.3 μm. This band completely absorbs solar radiation in the vertical column of the atmosphere within the spectral range 5.5–7.5 μm [3.3]. Since the ν_3 (2.66 μm), ν_1 (2.73 μm), and ν_2 (3.14 μm overtone) bands overlap, there is effectively total absorption of solar radiation in the vertical column of the atmosphere between 2.6 and 3.3 μm. The other vibration-rotation bands of water vapor form absorption bands with centers close to 1.87, 1.38, 1.10, 0.94, 0.81, and 0.72 μm. There are also some weak bands in the visible region of the spectrum.

The large dipole moments of the water molecule and its isotopes give rise to an intensive rotational spectrum that occupies a very broad region beginning near 8 μm and extending to wavelengths of several centimeters.

Carbon Dioxide. The CO_2 electronic absorption spectrum is in the far ultraviolet. The vibration-rotation bands are in the range between approximately 0.78 and 20 μm. Carbon dioxide has no purely rotational spectrum.

Of the three main vibrations ν_1, ν_2, and ν_3 of the CO_2 molecule, ν_1 is optically inactive—because of symmetry its dipole moment remains constant during vibrations [cf. (3.25) and the ensuing discussion]. Centers of the main ν_2 and ν_3 vibration-rotation bands are located near $667.40\ cm^{-1}$ (~ 15 μm) and $2349.16\ cm^{-1}$ (~ 4.3 μm), respectively. In addition to the main bands, CO_2 has overtones, combination frequencies, and hot bands, resulting in complex absorption bands with centers near 10.4, 9.4, 5.2, 4.3, 2.7, 2.0, 1.6, and 1.4 μm, and a series of weak bands in the range 0.78–1.24 μm.

The main ν_2 band, together with 14 hot bands, occupies a rather broad spectral interval between 12 and 20 μm. Near the central part of this band ($\sim 13.5–16.5$ μm) the vertical column of the atmosphere completely absorbs solar radiation.

The main ν_3 band, overlapping two others, forms a complex structure known as the 4.3-μm band in the literature. Absorption in its central region is so great that in the 4.2–4.4 μm range solar radiation propagating in the vertical column of the atmosphere fails to penetrate below an altitude of 20 km.

The CO_2 molecule has the following isotopes occurring in the atmosphere: $C^{12}O_2^{16}$, $C^{13}O_2^{16}$, $C^{12}O^{16}O^{17}$, and $C^{12}O^{16}O^{18}$, with respective percentages: 98.42%, 1.11%, 0.06%, and 0.41%. Centers of the main vibration-rotation bands of these isotopes are shifted relative to one another by a few cm^{-1} to some tens of cm^{-1}.

Ozone. Electronic transitions in the O_3 molecule produce Hartley and Huggins bands located in the ultraviolet (wavelengths shorter than 340 nm) and Chappuis bands in the 450–740 nm region. The maximum value of absorption coefficient in the Chappuis bands causes only 7% absorption of solar radiation when the atmospheric mass equals two.

All three main vibration frequencies of O_3 are active in absorption. They form three basic vibration-rotation bands with centers: $v_1 = 1110$ cm^{-1} (9.0 μm), $v_2 = 710 cm^{-1}$ (14.1 μm), and $v_3 = 1043 cm^{-1}$ (9.6 μm). The v_1 band is very weak and completely overlaps the v_3 band. Overtones and combination frequencies of the vibrations produce bands in the regions 5.75, 4.75, 3.59, 3.27, and 2.7 μm, wherein the band at 4.75 μm is the most intensive one.

The strongest absorption band of ozone, with its center near 9.6 μm, is in the center of a long-wave (8–13 μm) atmospheric transparency window. Its central part, with a width of approximately 1 μm, absorbs approximately half the solar radiation, within this spectral interval, in a vertical column of the atmosphere.

The following three isotopic modifications of the ozone molecule: O_3^{16}, $O^{16}O^{18}O^{16}$, and $O^{16}O^{16}O^{18}$ appear in the atmosphere with percentages of 99.4%, 0.2%, and 0.4%, respectively. The centers of the main absorption bands of these modifications are shifted relative to one another by frequencies of from several cm^{-1} to several tens of cm^{-1}.

A purely rotational absorption spectrum of O_3 is located in the microwave spectral region.

Nitrous Oxide. The strong electronic bands of the linear asymmetric molecule N_2O are in the far ultraviolet region. All three main oscillation frequencies: $v_1 = 1285.6 cm^{-1}$ (7.8 μm), $v_2 = 588.8 cm^{-1}$ (17.0 μm), and $v_3 = 2223.5 cm^{-1}$ (4.5 μm), are active in the infrared. The N_2O molecule has many overtone bands, combination frequencies, and hot bands; however, the majority of these are of low intensity and do not appear in the solar spectrum of atmospheric absorption. It also has twelve stable isotopes produced by combinations of the atoms N^{14}, N^{15}, O^{16}, O^{17}, and O^{18}. Only the spectrum of the basic isotopic modification $N_2^{14}O^{16}$ has been investigated thoroughly.

The purely rotational absorption spectrum of N_2O is in the far infrared spectral region.

Methane. The electronic absorption spectrum of CH_4 is located in the far ultraviolet. A high degree of molecular symmetry causes strong degeneracy of the vibrational energy levels. Among nine basic vibration frequencies, one is twice degenerated and two are triply degenerated.

Thus, the molecule has only four basic oscillations differing in frequencies, while the v_1 and v_2 frequencies are optically inactive. The centers of the basic v_3 and v_4 vibration-rotation bands are close to 3020.3 cm^{-1} (3.3 μm) and 1306.2 cm^{-1} (7.7 μm), respectively. Methane has many overtone bands and combination frequencies, nine of which have been detected in the solar spectrum of atmospheric absorption. There is no purely rotational spectrum of CH_4.

Carbon Monoxide. The spectrum of CO has been thoroughly investigated. The basic vibration-rotation band is close to 2143.2 cm^{-1} (4.67 μm). Centers of the second, third, and fourth overtones are near 4260.1, 6350.4, and 8414.5 cm^{-1}, respectively. Hot bands also appear in the spectrum, corresponding to transitions between the levels 1–3, 2–4, 3–5, 4–6, and 5–7, the centers of which are located in an interval of frequencies from 4207.2 cm^{-1} to 3996.9 cm^{-1} (2.38–2.50 μm).

The CO molecule has two isotopic modifications: $C^{12}O^{16}$ and $C^{13}O^{16}$. The purely rotational spectrum of CO is in the far infrared.

Oxygen. Molecular oxygen has strong electronic bands in the ultraviolet region, and weak ones in the red and near infrared. The electronic transitions in the O_2^{16} molecule between the levels 0–1, 0–2, 0–3, 0–4, 1–1, 1–2, and 1–3 produce a system of weak bands whose centers are between 538.4 and 762.1 nm. Moreover, the O_2^{16} molecule has two noticeable bands with centers near 1.0674 and 1.2683 μm. Absorption bands of the second isotopic modification of oxygen occuring in the atmosphere, $O^{16}O^{18}$, are not only shifted with respect to the corresponding bands of O_2^{16}, but also contain a larger number of lines resulting from the decreased symmetry of this molecule.

In addition to characteristic electronic absorption bands, oxygen has diffusion bands whose origin arises from complexes of molecules $[O_2]_2$, described in [3.4]; however, their intensity is not great and they overlap with stronger absorption bands of ozone and water vapor.

As is seen from the above discussion, in the wavelength range from 0.2 to 20 μm, absorption by atmospheric gases is mainly due to vibration-rotation bands, the strongest of which are in the infrared spectral region. It is quite clear that very intensive absorption in the center of certain bands practically excludes the possibility of applying these regions for the purpose of laser monitoring of the atmosphere, irrespective of whether the wavelength of the laser radiation is near the central part of the absorption line of the gas, or is between strong lines of the central part of the absorption band. (In the latter case, however, the absorption coefficients are essentially smaller than their values at the centers of absorption lines; and absorption can reach 100% during the first centimeters or meters of laser propagation in the atmosphere due to the effects of

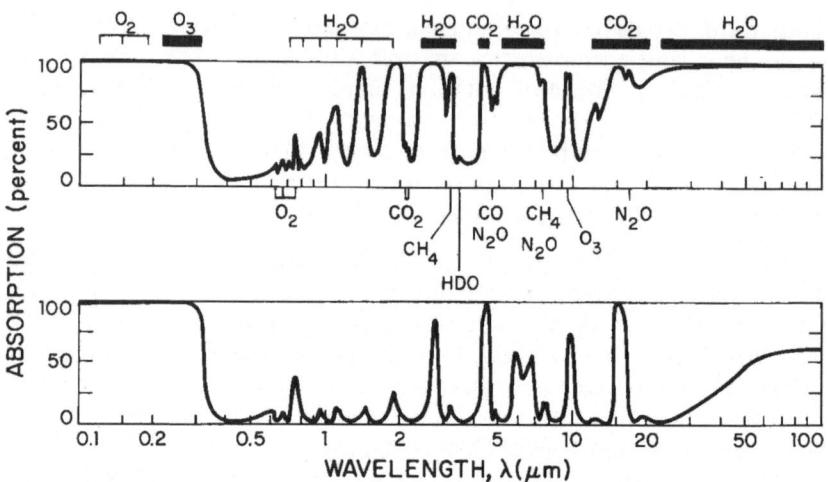

Fig. 3.1. General picture of absorption spectra of atmospheric gases. Upper curve: Absorption spectrum of solar radiation reaching the earth's surface. Lower curve: Absorption spectrum of solar radiation reaching an altitude of 11 km

wings of close, strong lines.) Within strong absorption bands, laser monitoring can be accomplished only in the upper layers of the atmosphere where the concentration of absorbing molecules is sufficiently small. Of course, if the pollutant to be monitored is causing the dominant absorption, and the technique of differential absorption is being used, then this wavelength region can be used in any event.

The most useful wavelengths for laser monitoring are generally those in intervals between absorption bands, or macrowindows of atmospheric transparency. Figure 3.1 gives a clear idea of these windows. Two low-resolution absorption spectra of solar radiation are shown. The upper curve characterizes the absorption spectrum of solar radiation reaching the earth's surface; the lower one was obtained under the same conditions, but at an altitude of 11 km. Positions of the centers of the basic absorption bands of atmospheric gases are also indicated.

Each of the absorption bands given in Fig. 3.1 is, as a rule, the result of superposition and overlapping of a whole series of different bands of the same gas, as well as of others. The same is true for the intervals between the bands, or transparent macrowindows, where combinations of various weak bands overlapping each other can be found, as well as peripheral regions of neighboring strong bands.

If an absorption spectrum of the atmosphere is recorded with high resolution (e.g., several hundredths of a cm^{-1}), each absorption band and each transparency window shown in Fig. 3.1 will consist of thousands of separate absorption lines. The strong lines belong to the band centers, while the transparency windows contain the weaker ones. Thus, within the atmospheric windows shown in Fig. 3.1 we have numerous transparency microwindows representing the intervals between neighboring lines.

In order to perform laser monitoring of the atmosphere, it is necessary to have quantitative information on absorption coefficients in the very atmospheric transparency microwindows located close to the wavelength of the laser radiation, along with information on absorption due to nearby spectral lines of atmospheric gases. We shall pay particular attention to this problem below.

3.1.4 Experiments

Before describing the results of experimental measurements of absorption of laser radiation in the atmosphere, we shall consider the accuracy with which the absorption line center wavelengths must be known in order to obtain reliable quantitative data on energy losses resulting from molecular absorption. We first deal with fixed-frequency lasers whose emission wavelengths are stable and well known. The widths of the laser lines are usually much smaller than the absorption linewidths of atmospheric gases. Let us estimate in what way the absorption cross section can vary with the difference frequency between laser line and absorption line center.

Table 3.1 presents calculations of the spectral dependence of the absorption cross section $\sigma(v)$ relative to the peak value $\sigma(v_0)$ for Lorentzian lines with halfwidths of 0.1 and 0.03 cm^{-1} atm^{-1} at pressures of 1 atm (ground level) and 0.1 atm (16 km altitude). We see that for the widest lines in the ground atmospheric layer the value of $\sigma(v)$ decreases by a factor of ten at a distance of 0.3 cm^{-1} from line center. With the narrow lines, such a decrease occurs already for $\Delta v = 0.1$ cm^{-1}. At an altitude of 16 km, $\sigma(v)$ decreases by a factor of ten for wide and narrow lines at distances of 0.03 and 0.01 cm^{-1} from the center, respectively. Furthermore, with laser radiation 0.1 cm^{-1} from line center for wide lines and 0.03 cm^{-1} for narrow lines, $\sigma(v)$ is one hundred times smaller than its value at line center for this altitude.

Thus, if monochromatic laser radiation overlaps the central part of an absorption line, then, for an accurate estimate of its absorption, the laser frequency and position of the absorption line center should be known very precisely. This precision must be hundredths of a cm^{-1} for gases in the ground layer of the atmosphere, and it increases considerably with altitude. Most of the atlases on absorption spectra of solar radiation in the atmosphere obtained until recently lack the precision necessary for

Table 3.1. Values of $\alpha(\nu)/\alpha(\nu_0)$ at various distances $\Delta\nu\,(=\nu-\nu_0)$ from absorption line center, for two values of atmospheric pressure (1 atm, 0.1 atm) and two line halfwidth pressure dependences ($0.1\ \mathrm{cm}^{-1}\,\mathrm{atm}^{-1}$, $0.03\ \mathrm{cm}^{-1}\,\mathrm{atm}^{-1}$). Data computed according to (3.17) for a Lorentzian line

$\Delta\nu\,[\mathrm{cm}^{-1}]$	$\sigma(\nu)/\sigma(\nu_0)$			
	$p=1$ atm $\gamma_L=0.1\ \mathrm{cm}^{-1}$	$p=1$ atm $\gamma_L=0.03\ \mathrm{cm}^{-1}$	$p=0.1$ atm $\gamma_L=0.01\ \mathrm{cm}^{-1}$	$p=0.1$ atm $\gamma_L=0.03\ \mathrm{cm}^{-1}$
0	1.000	1.000	1.000	1.000
0.005	0.998	0.973	0.800	0.265
0.01	0.990	0.900	0.500	0.083
0.02	0.962	0.692	0.200	0.022
0.03	0.917	0.500	0.100	0.010
0.04	0.862	0.360	0.059	0.006
0.05	0.800	0.265	0.038	0.004
0.1	0.500	0.083	0.010	0.001
0.2	0.200	0.022	0.0025	0.0002
0.3	0.100	0.010	0.0011	0.0001
0.4	0.059	0.006	0.0006	—
0.5	0.038	0.004	0.0004	—
1.0	0.010	0.001	0.0001	—
2.0	0.0025	0.0002	—	—

even rough estimates of radiation absorption of laser sources. As the laser radiation frequency moves further away from the absorption line center, requirements as to accuracy of determination of the center frequency are lowered. These conclusions are valid not only for water vapor, but also for other gases, since the halfwidths of their absorption lines are of the same order.

Proceeding to consideration of the results of experimental investigations of absorption of laser radiation by atmospheric gases, we begin with the common ruby laser[2], paying particular attention to it because it clearly illustrates specific features of the problems to be faced.

Investigations of ruby laser radiation absorption were made by a number of researchers [3.6–11] both in the natural atmosphere and under laboratory conditions. The data obtained on absorption coefficients during some of the first laser measurements, however, differed substantially. Soon thereafter, the reason for these differences became clear: the laser wavelength depends on the temperature of the ruby; and radiation can, therefore, interact with various regions of the atmospheric absorption spectrum. Subsequent measurements were made while

---

[2] For a modern treatment the reader is referred to W. KOECHNER: *Solid-State Laser Engineering* (Springer, New York, Heidelberg, Berlin 1976).

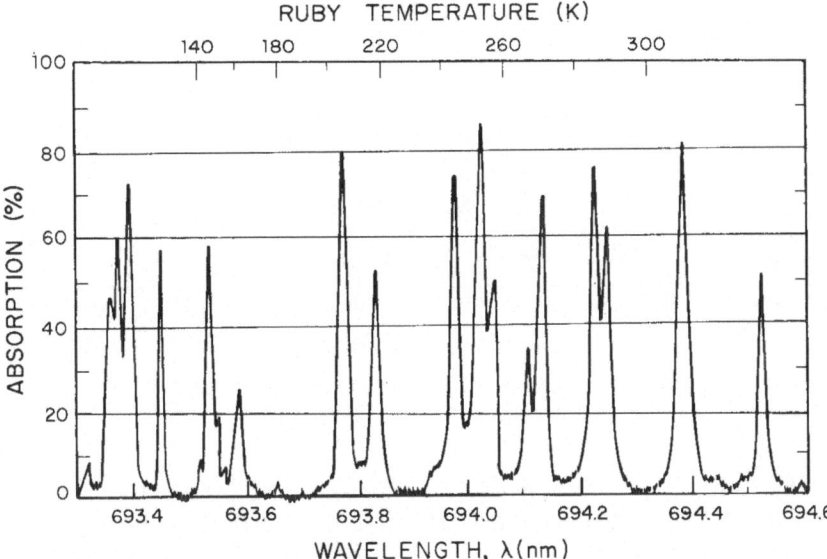

Fig. 3.2. Tunable ruby laser measurement of fine structure of the earth's atmospheric absorption spectrum around $\lambda = 0.69$ μm. (After LONG [3.57])

monitoring the laser wavelength; however, even in this case the results were not totally consistent. In the first place, measurements with one and the same wavelength gave different values for the absorption coefficient. These differences were not as great as in the first experiments, but they nevertheless exceeded measurement errors. Secondly, identical increases of the precipitated water layer in the atmosphere caused unequal variation of absorption coefficients for various wavelengths of ruby radiation.

For convenience in explaining these results, we consider the absorption spectrum of the earth's atmosphere obtained by LONG [3.57] with a high-resolution spectrometer, when the sun was used as radiation source. Figure 3.2 shows the spectral dependence of solar radiation absorption between 693.4 and 694.6 nm on a usual winter day with the sun 40° above the horizon. The spectrum was recorded on a spectrograph with a resolution of 300000 (0.002 nm). This figure also shows the scale of ruby temperatures and corresponding laser wavelengths.

As is seen from Fig. 3.2, absorption of solar radiation by the entire thickness of the atmosphere varies from nearly zero between the absorption lines, to 80% at the centers of the strongest lines. It is not difficult, therefore, to understand why, during the first measurements of absorption coefficients (made without monitoring the wavelength of the

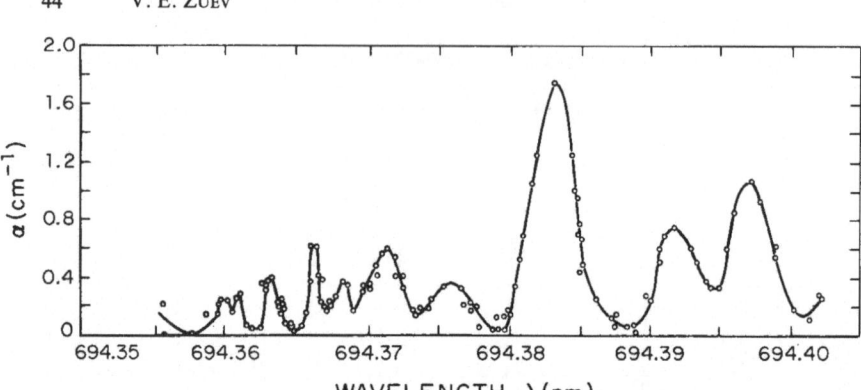

Fig. 3.3. Fine structure of water vapor absorption spectrum over $0.7\,cm^{-1}$ wide region, obtained with a laser spectrometer having very high resolution

ruby radiation), variations were observed under presumably identical conditions. Furthermore, if the halfwidths of the ruby laser lines for these measurements were not $0.1-0.2\,cm^{-1}$, but considerably narrower, the variations would be even larger.

Recent measurements have shown that the halfwidth of ruby laser radiation was close to $0.1\,cm^{-1}$. Thus, in all the measurements made, the data were obtained not on the absorption coefficients but on the absorption functions (see Subsect. 3.1.1); and the dependence of absorption function on absorbing mass (precipitated water layer) will be different depending on whether the laser radiation line is between the absorption lines or overlaps them.

One of the most sensitive methods of laser monitoring of humidity and concentration of other absorbing gases is the method of differential absorption using two wavelengths. For example, if we send two pulses of ruby laser radiation along one and the same path into the atmosphere after changing the wavelength in such a way that in one case it occurs at the line center of water vapor absorption (e.g. the line at 694.38 nm), and in the other one between the absorption lines; then from the lidar equation it is easy to obtain an algorithm for single-valued extraction of information on the humidity profile. However, in this case sufficiently exact data on humidity can be obtained only when the halfwidth, intensity and position of the center of the water vapor absorption line are well known. In this connection an attempt has been made [3.12–15] to obtain an absorption spectrum of the atmosphere in a narrow region of ruby laser radiation using a super-high resolution laser spectrometer (resolution $> 10^6$) constructed for this purpose.

Figure 3.3 shows the recording of an atmospheric absorption spectrum in a region $0.7\,cm^{-1}$ wide containing ten lines of water vapor. (In Fig. 3.2

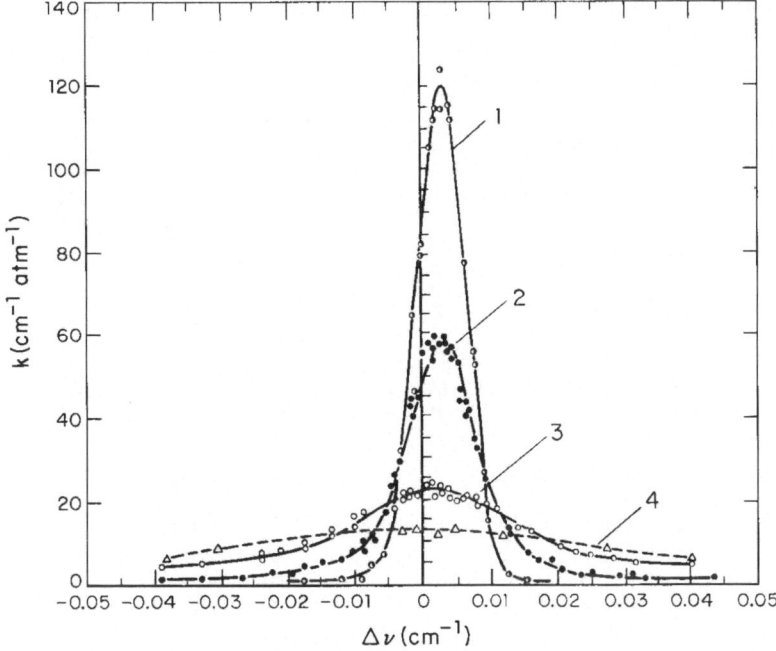

Fig. 3.4. Contours of methane spectral line at fixed CH_4 pressure p of 1 torr and various pressures of N_2 gas: 1–0 torr; 2–50 torr; 3–200 torr; 4–400 torr. $k[cm^{-1} atm^{-1}] = N\sigma/p$

only one line was recorded in this region.) Clearly, with increasing resolution of spectrometers we shall be able to register more and more absorption lines of atmospheric gases which should be taken into account when using monochromatic laser sources for monitoring.

An analysis of our results shows that in the region investigated the absorption coefficient due to water vapor varies between 0.17 and 0.01 km^{-1} at an absolute humidity of 10 g/m^3. It must be stressed here that we are concerned with true absorption coefficients calculated from Bouguer's law, which can be obtained only if the spectrometer used yielded a practically undistorted absorption spectrum.

As an example illustrating the results of experimental investigations of the absorption of gas laser radiation, we consider measurements using a He–Ne laser at 3.39 μm [3.16]. It is known that the line center of this laser is in the central part of a methane absorption line; hence, we are concerned here with a case in which an insignificant shift of the radiation line frequency may cause a large change in the absorption coefficient.

The purpose of our investigation was to obtain an undistorted contour of a methane absorption line under various conditions corresponding to different altitudes. These were created by appropriate

Table 3.2. Experimentally-measured absorption coefficients of various radiation lines of CO_2, CO, and DF lasers, scaled to mid-latitude ground-layer summer atmospheric conditions (14.3 torr H_2O, 294 K, 760 torr atmospheric pressure, 0.28 ppm N_2O, 1.6 ppm CH_4, 330 ppm CO_2). After LONG et al. [3.17]

Laser transition		Frequency [cm⁻¹]	Absorbing gas and absorption coefficient [km⁻¹]						
			CO_2	H_2O	HDO	CH_4	N_2O	N_2	Room air
CO_2	P(20)	944.194	0.08	0.215	x	x	x	x	0.295
	R(20)	975.931		1.3	x	x	x	x	
CO	6–5, P(14)	1957.050	x	4.9	x	x	x	x	4.9
	6–5, P(15)	1952.907	x	1.4	x	x	x	x	1.4
	6–5, P(16)	1948.729	x	2.6	x	x	x	x	2.6
	5–4, P(15)	1978.586	x	0.67	x	x	x	x	0.67
	5–4, P(16)	1974.374	x	0.78	x	x	x	x	0.78
	5–4, P(17)	1970.129	x	2.25	x	x	x	x	2.25
DF	3–2, P (6)	2594.25			0.0174		0.00227		
	P (7)	2570.51			0.00861		0.0373		
	P (8)	2546.42			0.00246		0.0214		
	2–1, P (6)	2680.17				0.00152	x		
	P (7)	2655.85				0.00113	x		
	P (8)	2631.06				0.00086	x		
	P(10)	2580.10					0.0453		
	P(11)	2553.97					0.0097		
	1–0, P(11)	2638.39				x	x		0.41

variations of temperature and pressure of a synthesized atmosphere. The measurements were made using a laser spectrometer consisting of a single-frequency, highly stabilized He–Ne laser, with frequency scanning by a magnetic field. Spectral resolution of the device was 10^{-3} cm⁻¹ or better.

Figure 3.4 shows the methane absorption line contours obtained at various pressures. The measurements were made in a narrow spectral range with the width of 0.08 cm⁻¹. As is seen from the figure, the center of the absorption line shifts with pressure. The value of this shift turned out to be 1.3×10^{-5} cm⁻¹ torr⁻¹. The value of absorption coefficient at line center for the ground layer of the atmosphere (pressure 1 atm, methane concentration 1.6 ppm) is 1.4 km⁻¹.

The case under consideration clearly illustrates the specific peculiarities of the problem of a quantitative estimate of laser radiation losses due to molecular absorption. It indicates that only spectral devices of super-high resolutions, such as the tunable lasers described in Chapter 6, permit us to obtain data on the absorption coefficients free of apparatus distortions.

As for the absorption functions of laser radiation, their values can be obtained by using the very lasers as radiation sources for appropriate measurements in the atmosphere. It should be kept in mind that these measurements must be made for various thicknesses of the atmosphere and must include a total interception of the laser beam by the receiver. Many examples of such data have been obtained by the author and his collaborators, and are described in detail in [3.1]. It should be noted here that these data are not universal. They are valid for quantitative estimates of radiation absorption only for the specific lasers used in the experiments.

Careful measurements of the absorption of various radiation lines of CO_2, CO, and DF lasers by atmospheric gases have recently been made by LONG and his collaborators [3.17]. The measurements were performed in a multipass cell with a total beam length of 730 m. The results are given in Table 3.2 where crosses (x) denote that no significant absorption for these lines can be expected. Blank spaces are not filled for lack of experimental data.

3.1.5 Theory

Bearing in mind the stringent demands mentioned above with respect to accurately knowing the positions, halfwidths, and intensities of molecular spectral lines in calculations of absorption coefficients and functions of atmospheric gases in the regions of laser emission spectra, all factors affecting the characteristics of corresponding absorption lines must be taken into account in any theoretical analysis. These characteristics are substantially influenced by interaction effects of vibrational and rotational molecular motion, which become more apparent with higher rotational quantum numbers of corresponding energy transitions of the molecule. Meanwhile, it is the absorption lines corresponding to large values of the rotational quantum number that, as a rule, get into the most interesting (in a practical sense) wavelength regions within the transparency windows of the atmosphere.

Methods for calculating absorption coefficients and absorption functions of various atmospheric gases, taking into account as many different effects of intermolecular and intramolecular interactions as possible, have been developed by a large group at the Institute of Atmospheric Optics of the Siberian Branch of the U.S.S.R. Academy of Sciences. In their work, the contour of the central part of a spectral line is assumed to be dispersing. The expression for the absorption cross section is

$$\sigma(v) = \sum_i S_i(T)\pi^{-1}\gamma_i(T, p)/[(v - v_{0i})^2 + \gamma_i^2(T, p)], \tag{3.28}$$

where summation is carried out over all lines contributing to the absorption with frequency v; S_i, γ_i, and v_{0i} are the intensity, halfwidth, and center frequency of the i-th line, and T and p are the temperature and pressure, respectively. The calculation procedures for $S_i(T)$, γ_i, and v_{0i} were described in [3.1] and in a series of subsequent publications [3.18–23].

Calculation of the continuous absorption due to wings of distant lines cannot be made when using the dispersion contour; however, an asymptotic theory has been developed [3.24–29] that qualitatively explains existing discrepancies between data of various experimental investigations and between the experimental results and calculations based on previous theories of spectral line contours.

Practical use of the methods described above requires application of sufficiently accurate data of parameters of a few lines in every absorption band. If such data are available, calculations can be made for all practically important lines of a given band. A fairly complete set of data on the absorption line parameters of atmospheric gases is given in tabular form by McClatchey et al. [3.30], wherein the positions of line centers have been taken from experiments, and the intensities and halfwidths calculated on the basis of definite assumptions on the character of intermolecular interactions and their influence on the parameters mentioned.

In conclusion, we note that despite definite progress toward a theoretical solution, we are still far from a sufficiently exact prediction of the absorption of monochromatic or quasi-monochromatic radiation in the atmosphere. The absorption of laser beams by various gases has so far been determined more reliably from experiments than from theory; however, current theoretical investigations and their intensive development will give rise to promising progress in the near future. We should stress that correct theoretical data on the absorption coefficients provide universal material necessary for solving any concrete problem requiring quantitative knowledge of the absorption of optical wave energy by atmospheric gases for any arbitrary models of the atmosphere, source radiation spectrum, and direction of wave propagation.

3.1.6 Limits of Applicability of Bouguer's Law to Absorption

In the definitions of absorption coefficient, spectral absorption (transmission), and absorption (transmission) function, given in Subsection 3.1.1, Bouguer's law was assumed to hold. This assumption is also often used in processing experimental data; and in this connection it is necessary to consider the applicability limits of this law, bearing in mind that real and apparent deviations may occur.

Let us write Bouguer's law in the following form for the case of a homogeneous absorbing medium [derived from (3.16) for monochromatic radiation]

$$I(v) = I_0(v) \exp[-N\sigma(v)L] .\tag{3.29}$$

In deriving (3.29) the absorption cross section was assumed to be independent of the incident radiation intensity and the concentration of absorbing molecules. Let us consider the validity of these assumptions.

The first assumption was subjected to a thorough experimental test by VAVILOV [3.31] who proved that, for electronic transitions with short lifetimes, $\sigma(v)$ is independent of $I_0(v)$ over nearly 20 orders of magnitude. For electronic transitions with long lifetimes, there is a dependence, and it was measured as a function of $I_0(v)$. In the case of vibrational-rotational transitions in molecular spectra of atmospheric gases, no check of the dependence of absorption coefficient on incident radiation intensity has, to our knowledge, been made. However, some conclusions are possible on the basis of a comparative analysis between predicted and measured absorption functions in narrow regions of the atmospheric spectrum. Calculations of the absorption function made without taking into consideration the dependence of $\sigma(v)$ on $I_0(v)$ turned out to be in satisfactory agreement with corresponding measurements using the sun as radiation source. Bearing in mind that the spectral power density of a laser can be many orders of magnitude greater than that of the sun, an effect due to the dependence of $\sigma(v)$ on $I_0(v)$ should be expected when using certain high-power lasers. As our analysis indicated [3.1], the spectroscopic saturation effect characterizing the absorption decrease in comparison with that predicted by Bouguer's law must take place with radiation power density $I_0(v)$ of $\gtrsim 10^7$ W/cm^2 propagated through the atmosphere. More definite conclusions will be made on the basis of special investigations now being conducted by the author's staff.

The invariance of absorption coefficient with concentration of absorbing gases indicates that every molecule absorbs radiation independently. As many investigations have shown, this assumption is valid for small concentrations. An increase in concentration of the absorbing gases and a higher concentration of foreign gases lead to amplification of intermolecular interaction effects, changing the values of $\sigma(v)$, and producing real deviations from Bouguer's law. The intermolecular interactions lead, in particular, to broadening of the spectral lines in a manner dependent on partial pressures of the absorbing and foreign gases, and on the temperature. This broadening, then, leads to a corresponding variation in $\sigma(v)$.

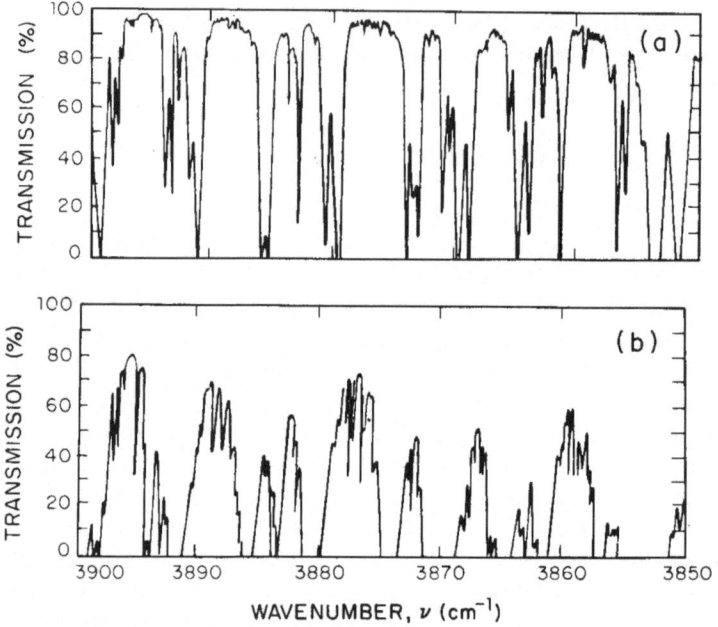

Fig. 3.5a and b. Water vapor absorption spectrum with full resolution in the region of 3850–3900 cm^{-1}. Pressure is 1 atm, and precipitated water layer thickness is 0.001 cm in (a), 0.01 cm in (b). (After Calfee [3.32])

The value of $\sigma(v)$ at any fixed altitude can be considered essentially independent of the concentration N of absorbing molecules because the pressures of foreign gases vary little with time and the partial pressures of the absorbing gases are small. However, values of absorption coefficient at various altitudes will be different since pressure, temperature, and partial pressures of absorbing gases do depend on altitude. The exact solution of this problem requires taking into account the dependence of $\sigma(v)$ on N for every given altitude, even if pressure variations at this altitude are negligible. Furthermore, self-broadening generally differs from foreign-gas broadening. On the basis of available data, this difference is largest for water vapor molecules, for in the central regions of H_2O absorption lines, self-broadening is greater than foreign-gas broadening (e.g., by N_2 or O_2) by a factor of 4 to 6. In the far wings of water vapor lines, this difference can amount to absorption changes by factors of ten or even hundreds. This effect cannot be ignored since the molecular concentration of water vapor in the atmosphere is of the order of 0.1–1% of the overall concentration of all molecules.

We now consider an *apparent* deviation from Bouguer's law resulting from the limited resolution of spectral instruments. We deal with cases

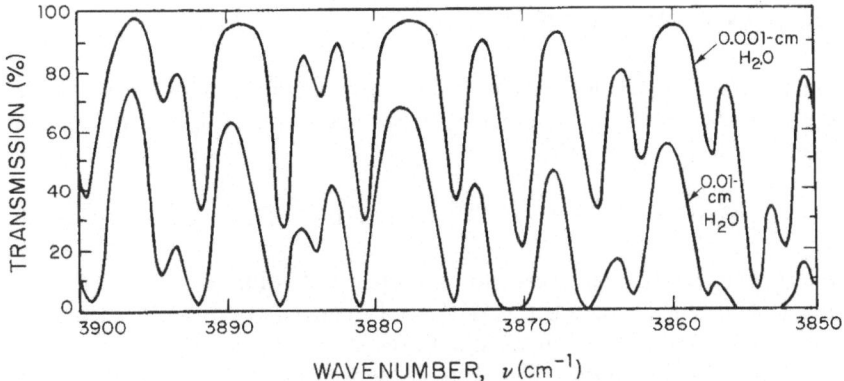

WAVENUMBER, $\nu \, (\text{cm}^{-1})$

Fig. 3.6. Water vapor absorption spectrum obtained by multiplying $T(\nu)$ of Fig. 3.5 by an apparatus function with spectral width $(2\gamma_0)$ of 2 cm^{-1}. (After CALFEE [3.32])

here for which the absorption coefficient cannot be considered constant when it is within the spectral region resolved by the instrument. Since, as mentioned before, the absorption line halfwidths of atmospheric gases in the ground layer are of the order of 0.01–0.1 cm^{-1}, in order to obtain a practically undistorted record of their contours the required resolution must be 0.001–0.01 cm^{-1}, or better. The poorer the resolution used in the experiment, the larger the distortions in the recorded spectrum. As a practical matter, therefore, we are not recording the spectral absorption (transmission), but the absorption (transmission) function distorted by the effect of the apparatus function of the spectral instrument. Figures 3.5 and 3.6 illustrate the effect of instrumental broadening on a fully-resolved spectrum. The measured spectral transmission $T'(\nu)$ is connected with the true one $T(\nu)$ and the apparatus function $U(\nu - \nu')$ by the equation

$$T'(\nu) = \int_{-\infty}^{\infty} U(\nu - \nu') T(\nu') d\nu' \, . \tag{3.30}$$

If the apparatus function is given by a (normalized) Gaussian distribution, then

$$T'(\nu) = (\ln 2/\pi)^{1/2} \Gamma^{-2} \int_{-\infty}^{\infty} \exp[-(\nu - \nu')^2 \ln 2/\gamma_0^2] T(\nu') d\nu' \, , \tag{3.31}$$

where γ_0 is the apparatus function halfwidth. Using the Fourier transform, we can write

$$T(\nu) = (1/2\pi) \int_{-\infty}^{\infty} [T'(\omega)/U(\omega)] \exp(-i\omega\nu) d\omega \, , \tag{3.32}$$

where $T'(\omega)$ and $U(\omega)$ are the transformations of $T(\nu)$ and $U(\nu)$.

3.2 Molecular Scattering

The theory for Rayleigh scattering from gas molecules yields the following expression for scattering coefficient (also known as the Rayleigh extinction coefficient)

$$\alpha_R(\lambda) = [8\pi^3(n^2 - 1)^2/3N_g\lambda^4][(6 + 3\delta)/(6 - 7\delta)], \qquad (3.33)$$

where N_g is the number of gas molecules per unit volume, n is the refractive index of the medium, and δ is the depolarization factor of the scattered radiation (equal to 0.035 according to some recent measurements [3.33]).

Detailed calculations of α_R were made for various wavelengths in the range from 0.2 to 20 μm [3.34]. On the basis of these results, it is easy to

Table 3.3. Molecular scattering coefficients α_R at $T = 15$ °C, $p = 1$ atm, and optical thickness τ_R of vertical layer of the atmosphere [3.34]

λ [μm]	α_R [km^{-1}]	τ_R	λ [μm]	α_R [km^{-1}]	τ_R
0.30	1.446×10^{-1}	1.2237	0.65	5.893×10^{-3}	0.0499
0.32	1.098×10^{-1}	0.9290	0.70	4.364×10^{-3}	0.0369
0.34	8.494×10^{-2}	0.7188	0.80	2.545×10^{-3}	0.0215
0.36	6.680×10^{-2}	0.5653	0.90	1.583×10^{-3}	0.0134
0.38	5.327×10^{-2}	0.4508	1.06	8.458×10^{-4}	0.0072
0.40	4.303×10^{-2}	0.3641	1.26	4.076×10^{-4}	0.0034
0.45	2.644×10^{-2}	0.2238	1.67	1.327×10^{-4}	0.0011
0.50	1.716×10^{-2}	0.1452	2.17	4.586×10^{-5}	0.0004
0.55	1.162×10^{-2}	0.0984	3.50	6.830×10^{-6}	0.0001
0.60	8.157×10^{-3}	0.0690	4.00	4.002×10^{-6}	0.0000

Table 3.4. Molecular scattering coefficients α_R for various altitudes Z and optical thicknesses τ_R of atmospheric layers ($Z \to \infty$) for $\lambda = 0.30$, 0.55, 1.06 μm [3.33]

Altitude	$\lambda = 0.30$ μm		$\lambda = 0.55$ μm		$\lambda = 1.06$ μm	
Z [km]	α_R [km^{-1}]	$\tau_R (Z \to \infty)$	α_R [km^{-1}]	$\tau_R (Z \to \infty)$	α_R [km^{-1}]	$\tau_R (Z \to \infty)$
0	1.446×10^{-1}	1.2237	1.162×10^{-2}	0.0984	8.458×10^{-4}	0.0072
5	8.693×10^{-2}	0.6538	6.988×10^{-3}	0.0526	5.085×10^{-4}	0.0038
10	4.881×10^{-2}	0.3212	3.924×10^{-3}	0.0258	2.855×10^{-4}	0.0019
15	2.999×10^{-2}	0.1471	1.848×10^{-3}	0.0118	1.345×10^{-4}	0.0009
20	1.049×10^{-2}	0.0672	8.436×10^{-4}	0.0054	6.138×10^{-5}	0.0004
30	2.173×10^{-3}	0.0146	1.747×10^{-4}	0.0012	1.271×10^{-5}	0.0001
40	4.716×10^{-4}	0.0035	3.791×10^{-5}	0.0003	2.758×10^{-6}	0.0000
50	1.212×10^{-4}	0.0010	9.743×10^{-6}	0.0001	7.089×10^{-7}	0.0000

determine the optical thickness τ_R [$=\ln(I_0/I)$] due to Rayleigh scattering in various geometrical thickness of the atmosphere. Tables 3.3 and 3.4 present data on α_R and τ_R for different wavelengths and geometrical thicknesses, which subsequently will be used for comparison with appropriate data on aerosol scattering and molecular absorption coefficients.

Molecular scattering has a single normalized scattering phase function. If the intensity of scattered radiation at an angle θ from the direction of propagation is denoted as $I(\theta)$, with $\theta=0$ denoting the forward direction and $\theta=180°$ the backward direction, then the molecular scattering phase function for unpolarized light takes the form

$$I_{g,s}(\theta)=(3/4)(1+\cos^2\theta)\,.$$

It is not difficult to notice that the molecular scattering phase function is symmetric relative to a plane drawn through the scattering volume, perpendicular to the direction of radiation propagation. In the forward and backward directions $I_{g,s}(\theta)$ takes a maximum; for $\theta=90°$ and $270°$, a minimum.

3.3 Particulate Scattering

3.3.1 Scattering Coefficient; Absorption Coefficient; Attenuation Coefficient

The coefficients of scattering, absorption, and attenuation per unit volume by airborne particles (or aerosols) are defined by the following equations:

$$\alpha_{p,s}(\lambda)=N_p\int_0^\infty \alpha_{p,s}(\lambda,a)f(a)da\,, \tag{3.34}$$

$$\alpha_{p,a}(\lambda)=N_p\int_0^\infty \alpha_{p,a}(\lambda,a)f(a)da\,, \tag{3.35}$$

$$\alpha_p(\lambda)=N_p\int_0^\infty \alpha_p(\lambda,a)f(a)da\,, \tag{3.36}$$

where N_p is the number density of scattering particles, $\alpha_{p,s}(\lambda,a)$, $\alpha_{p,a}(\lambda,a)$, and $\alpha_p(\lambda,a)$ are the coefficients of scattering, absorption, and total attenuation for the particle of radius a; and $f(a)$ is the particle size distribution function satisfying the condition

$$N_p'(a)da=N_pf(a)da\,, \tag{3.37}$$

where $N_p'(a)da$ is the number density of particles varying in radius from a to $a+da$.

The coefficients $\alpha_{p,s}(\lambda)$, $\alpha_{p,a}(\lambda)$ and $\alpha_p(\lambda)$ characterize energy losses of an optical wave propagation through an aerosol medium, as determined by scattering, absorption, and overall attenuation, representing a simple sum of scattering and absorption. Consequently, we may write

$$\alpha_p(\lambda) = \alpha_{p,s}(\lambda) + \alpha_{p,a}(\lambda) . \tag{3.38}$$

The attenuation of radiation propagating in an aerosol medium is described by the equation

$$dI(\lambda) = -\alpha_p(\lambda)I(\lambda, z)dz . \tag{3.39}$$

Integration of (3.39) gives the well-known expression for Bouguer's law in the case of an inhomogeneous medium:

$$I(\lambda) = I_0(\lambda) \exp\left[-\int_0^L \alpha_p(\lambda, z)dz\right] , \tag{3.40}$$

which, for a homogeneous aerosol medium, acquires the following form

$$I(\lambda) = I_0(\lambda) \exp\left[-\alpha_p(\lambda)L\right] . \tag{3.41}$$

The term $\alpha_p(\lambda)$ is called the linear extinction coefficient (units of cm^{-1} or km^{-1}), or the volume extinction coefficient (units of cm^2/cm^3).

3.3.2 Scattering by a Single Particle

The theory of light scattering by a spherical particle was developed by MIE in 1908 and described in detail in a well-known monograph by VAN DE HULST [3.35]. This theory yields the following expressions for efficiency factors of scattering, absorption, and total extinction by a single particle in terms of the corresponding coefficients

$$Q_s(x, m) = \sigma_s(a, \lambda, m)/\pi a^2 , \tag{3.42}$$

$$Q_a(x, m) = \sigma_a(a, \lambda, m)/\pi a^2 , \tag{3.43}$$

$$Q_E(x, m) = \sigma_E(a, \lambda, m)/\pi a^2 . \tag{3.44}$$

These efficiency factors of scattering, absorption, and total extinction are numerically equal to the ratio of the energy scattered, absorbed, and attenuated, respectively, by a particle to the energy incident onto its geometric cross-section πa^2. They are represented theoretically by infinite, slightly convergent series. The parameters x and m characterize the relative size and relative refraction coefficient of the particle

$$x = 2\pi a/\lambda, \tag{3.45}$$

$$m = m_1/m_2, \tag{3.46}$$

where m_1 and m_2 are the complex refractive indices of the particle and medium, respectively. In the atmosphere we may consider $m_2 = 1$, and $m_1 = n - i\kappa$, where n and κ are the real (index of refraction) and imaginary (index of absorption) parts of the complex refractive index of the particle material.

The functions Q_s, Q_a, and Q_E are tabulated in detail for a large set of parameters x and m encompassing the conditions of optical wave propagation in real aerosol systems such as clouds, fogs, hazes, and precipitations. The efficiency factors of scattering, absorption, and attenuation for particles of ellipsoidal and cylindrical shapes are given in [3.35]. Theoretical and experimental determinations of Q_s, Q_a, and Q_E for irregularly-shaped particles have proved to be extremely difficult, and corresponding data have not been obtained thus far.

3.3.3 Scattering by Clouds and Fogs

The author and his collaborators have made extensive calculations of volume attenuation coefficients of water clouds and fogs, whose particle (droplet) sizes are described by a gamma distribution. The particle concentrations and gamma-distribution parameters were varied to encompass values occurring in nature [3.36–39]; and along with the calculations, measurements were made of volume attenuation coefficients and fog microstructure parameters which allowed one to perform quantitative comparisons between experimental and calculated data. A detailed description of the results was given in [3.1]. Here we consider only some of the results of calculations made using our own data on the complex refractive index of liquid water [3.42–43]. Figure 3.7 shows one of these results.

Laser sounding of clouds and fogs requires knowledge of the relationships between such parameters as water content Q_W, particle concentration N_p, and meteorological visible range V_M. For a gamma distribution, these parameters are connected with one another by the following relations

$$N_p = 3.912\, q^2 \left[V_M F(0.5)\pi a^2 (q+2)(q+1) \right]^{-1}, \tag{3.47}$$

$$Q_W = 3.912\, (4a)(q+3)\varrho_m [3 V_M F(0.5)q]^{-1}, \tag{3.48}$$

where q and a are the gamma-distribution parameters [q is characteristic of the half-width of the distribution, and a is the most probable particle

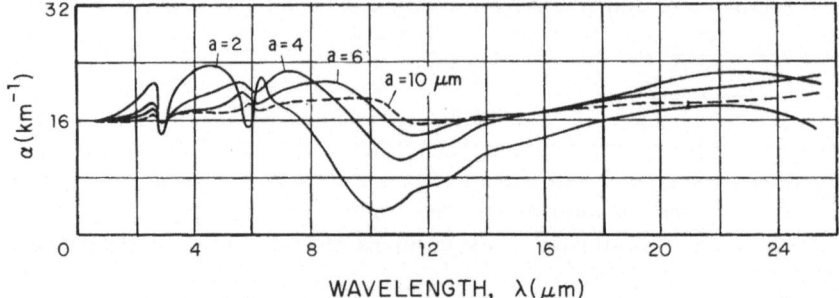

Fig. 3.7. Extinction coefficients of water clouds and fogs within wavelength range 0.5–25 μm for microstructural parameters a (most probable particle radius) of 2, 4, 6, and 10 μm, and q (characteristic of half-width of the distribution) of 2. Meteorological visibility, $V_M = 0.2$ km

radius in μm], $F(0.5)$ is the effective attenuation factor for wavelength $\lambda = 0.5$ μm, and ϱ_m is the water density. Table 3.5 shows the values of Q_W and N_p for some typical sets of microstructure parameters at $V_M = 200$ m, calculated according to (3.47) and (3.48). As we see, both the water content and (in particular) the particle concentration of clouds can vary significantly at one and the same visibility, depending on the microstructure parameters. According to data in the literature, in over 50% of the time visibility in clouds is within 100 to 300 meters, while the most probable values of cloud water content are in the range 0.1–0.3 g/m³ [3.1].

3.3.4 Scattering by Hazes

Detailed calculations of the coefficients $\alpha(\lambda)$ have also been made for hazes [3.44, 45]. A particle size distribution was assumed in accordance with Junge's empirical formula (cf. Sect. 4.2). The minimum (a_1) and

Table 3.5. Water content Q_W and particle concentration N_p of some water clouds and fogs. Parameter q characterizes the half-width of Γ-distribution of particle sizes, and a represents the most probable particle radius

Cloud type (Droplet size)	q	a [μm]	V_M [km]	Q_W [g/m³]	N [cm⁻³]
Small	1	2	0.2	0.031	971
	1	10		0.015	2050
Medium	6	2	0.2	0.194	28
	6	10		0.101	65
Large	10	2	0.2	0.324	10
	10	10		0.168	2.3

Table 3.6. Extinction coefficients $\alpha(\lambda)$ of hazes for most probable values of microstructural parameters: $a_1 = 0.05$ μm, $a_2 = 5$ μm, $\Lambda = 4$, $V_M = 10$ km

λ [μm]	α [km^{-1}]	λ [μm]	α [km^{-1}]	λ [μm]	α [km^{-1}]	λ [μm]	α [km^{-1}]
0.31	0.65	1.18	0.17	2.94	0.16	6.15	0.05
0.50	0.40	1.24	0.15	3.12	0.13	6.21	0.04
0.53	0.38	1.26	0.15	3.34	0.09	6.27	0.04
0.56	0.36	1.39	0.14	3.39	0.08	6.44	0.03
0.59	0.34	1.43	0.13	3.51	0.07	6.66	0.02
0.625	0.32	1.45	0.13	3.58	0.06	7.69	0.02
0.63	0.32	1.47	0.13	3.85	0.05	8.36	0.01
0.67	0.30	1.56	0.12	4.17	0.05	10.04	0.01
0.69	0.29	1.67	0.11	4.55	0.04	10.6	0.01
0.71	0.26	1.79	0.10	4.77	0.04	11.29	0.02
0.77	0.26	1.82	0.10	5.00	0.03	11.42	0.02
0.81	0.25	1.89	0.10	5.13	0.03	11.56	0.03
0.84	0.24	1.94	0.09	5.27	0.03	12.42	0.03
0.91	0.22	2.00	0.09	5.41	0.03	13.79	0.05
0.97	0.21	2.22	0.08	5.56	0.02	14.18	0.05
1.01	0.20	2.36	0.07	5.82	0.02	14.38	0.05
1.05	0.19	2.50	0.06	5.87	0.02	16.81	0.06
1.06	0.18	2.63	0.05	5.93	0.03	17.39	0.06
1.11	0.18	2.71	0.04	5.98	0.04	22.47	0.04
1.13	0.17	2.78	0.06	6.03	0.05	25.31	0.02
1.15	0.17	2.91	0.15	6.09	0.06		

maximum (a_2) particle sizes were taken to be: $a_1 = 0.01, 0.05, 0.1$ μm; and $a_2 = 1.0, 5.0, 10.0$ μm. The index in power Λ was assumed to be 2, 3, 4, 5. The calculations were made for spherical particles of water within a wavelength range from 0.3 to 25 μm. Table 3.6 presents the calculated results of $\alpha(\lambda)$ for the most probable values of the haze microstructure parameters. We note that the coefficient $\alpha(\lambda)$ for hazes decreases sharply in the wavelength region from 0.3 to 2.7 μm, with minimum values occurring between 8 and 11 μm. The difference between maximum and minimum values of $\alpha(\lambda)$ is about two orders of magnitude, while its absolute values differ from those of clouds and fogs by 1–3 orders of magnitude, depending on the wavelength.

The calculated results shown in Table 3.6 cannot claim to be an exact description of the real picture of attenuation by hazes, since the attenuation may consist of both two-layer particles and irregularly-shaped particles; moreover, Junge's formula gives only an approximate idea of the haze particle size distribution. Sufficient data are not yet available to permit corresponding calculations to be made for actual particle size distributions. There is also a lack of experimental data on the coefficients $\alpha(\lambda)$ for various types of hazes upon which to base construction of a statistically valid optical model for haze.

3.3.5 Scattering by Precipitations

Precipitation particles of spherical shape can be considered large ($x = 2\pi a/\lambda \gg 1$) in the wavelength range of interest, and the attenuation efficiency, $Q_E(x, n)$, is 2. In this case, the attenuation coefficient becomes

$$\alpha(\lambda) = N \int_0^\infty \pi a^2 Q_E(x, \hat{n}) f(a) da = 2G_p \qquad (3.49)$$

where G_p is the geometrical cross-section of the particles per unit volume. Rain drops can be considered spherical; hence the attenuation co-efficient does not depend on wavelength—it is defined only by the geometrical cross-section of particles per unit volume which, with given microstructure characteristics, is uniquely associated with water content.

According to data in the literature, the dependence of the coefficient $\alpha(\lambda)$ on microstructure parameters of rains and snowfalls is insignificant compared with its dependence on precipitation intensity. In the case of rains,

$$\alpha \approx 0.21 \, J^{0.74} \,, \qquad (3.50)$$

where J is the rainfall rate [mm/hr] and α is the attenuation coefficient in km^{-1}. The correlation coefficient between $\ln \alpha$ and $\ln J$ is 0.95 ± 0.01. If d is the water content of rain in g/m^3, we find that the correlation co-efficient between $\ln \alpha$ and $\ln d$ is even larger: 0.97 ± 0.01. The correlation coefficient between $\ln \alpha$ and $\ln J$ for snowfalls is also sufficiently large: 0.91 ± 0.02. Numerical values for the attenuation coefficients of precipi-tations can be determined without difficulty if their intensity is known.

3.3.6 Aerosol Scattering Phase Functions

The aerosol scattering phase function is not symmetric with respect to forward and backward scattering. Every ensemble of aerosol particles has its own scattering phase function. The degree of asymmetry is characterized by the asymmetry coefficient δ_p, which is the ratio of fluxes scattered by particles into the front and back hemispheres. Analysis shows that the forward-stretched scattering phase function increases with increasing size parameter $x = 2\pi a/\lambda$. Thus, considering the scattering phase functions of hazes, clouds, and precipitations in the visible, δ_p will generally increase in that order. For greatly forward-stretched scattering phase functions (e.g., for clouds, fogs, and, in particular, precipitations in the visible), the main part of the scattered radiation is concentrated in a very narrow range of angles close to 0° (forward direction). Thus, for example, the value $I(\theta)$ for various fogs can decrease by three to seven orders of magnitude with a change in scattering angle from 0.2° to 7°.

3.3.7 Limits of Applicability of Bouguer's Law to Scattering

All the material considered in this section is based on the validity of Bouguer's law, and it is important to analyze the limits of applicability of this law as it applies to laser sounding of the atmosphere. The lidar equation was obtained on the basis of a single-scattering theory based on the validity of Bouguer's law.

The main assumption made when deriving Bouguer's law for scattering media consists in stating that the particles scatter electromagnetic radiation independently of each other and, therefore, the effects of multiple scattering may be neglected. If we now turn to the scheme of laser sounding of an aerosol medium, neglect of the multiple scattering effects means that a recorded echo-signal of a sounding laser pulse must essentially exceed a noise signal due to the acts of the 2nd, 3rd, etc. scattering multiplicity in the direction of the receiving system. The noise signal due to multiple scattering effects depends on the optical thickness and optical properties (scattering phase function) of the scattering medium, the initial diameter and angle of divergence of the sounding pulse, the distance to the volume sounded, the viewing angle of the receiving system, and the radiation wavelength. Detailed data on the results of appropriate investigations are given in the literature [3.46–49]. Only some general conclusions are considered here.

For laser sounding of hazes or monitoring through hazes, the effects of multiple scattering can usually be neglected, whereas with sounding of clouds and fogs they can be ignored only for limited optical thicknesses whose values essentially depend, first of all, on the angle of divergence of the source and the viewing angle of the receiving system. Calculation algorithms developed by the author's staff have made it possible to predict the values of optical thickness for any aerosol medium for which it is possible to neglect multiple scattering effects, and if the attenuation coefficient, scattering phase function, and geometrical parameters of the sounding scheme are given [3.50, 51].

3.4 Total Attenuation: Wavelength Regions Efficient for Laser Sounding of the Atmosphere

3.4.1 Nonturbid Atmosphere

A nonturbid atmosphere is one which is free of aerosol particles. Strictly speaking, such states of the atmosphere do not occur since samples of even the clearest atmospheric air contain at least several small aerosol particles per cm^3. Nevertheless, in certain cases it is worthwhile speaking about nonturbid atmospheres. If the atmosphere is turbid by particles of 0.01 µm diameter or so, then these particles should be considered optically inactive in the visible, infrared, and beyond. For example, cloud particles are optically inactive with respect to radio waves. On the other hand, in regions distant from industrial centers, in mountainous areas, or after prolonged precipitation, the concentration of aerosol particles becomes so small that laser attenuation due to them can be neglected.

During nonturbid atmospheric conditions the total attenuation of laser radiation is defined by its molecular absorption and molecular scattering. A clear idea of the role of molecular scattering in attenuating optical waves can be obtained from a known formula connecting meteorological visibility V_M with the scattering coefficient. This expression is usually given for the wavelength of 0.55 µm corresponding to maximum sensitivity of the human eye

$$\alpha(0.55 \text{ µm}) = 3.912/V_M . \qquad (3.51)$$

This expression is derived assuming that the threshold of contrast sensitivity of the eye equals 0.02. If (3.51) is written for other wavelengths, and the atmosphere is assumed to be free of aerosol particles, then for $\lambda = 0.3$, 0.4, 0.5, 0.55, and 0.7 µm, the corresponding values of V_M are 27, 90, 230, 340, and 900 km.

The molecular scattering coefficients have been calculated with high accuracy for a standard model of the atmosphere; hence, within the applicability limits of this model, we may calculate the component of atmospheric transmission stipulated by molecular scattering for any laser wavelength. The total attenuation will, therefore, be determined by the unknown values of molecular absorption coefficients which vary within wide limits depending on the spectral absorption of each atmospheric gas at the laser radiation wavelength. At the centers of the strongest absorption lines the radiation is fully absorbed during the first few millimeters of travel. The intervals between these lines may permit the radiation to travel tens or hundreds of meters before total absorption occurs. The centers of weak lines located between strong bands have absorption coefficients of the order of $0.1–1$ km^{-1}. Finally, in the intervals between the weak absorption lines, the molecular absorption coefficients may be greater than, less than, or comparable with the molecular scattering coefficients, depending on the laser wavelength.

When calculating the total attenuation of laser radiation in a nonturbid atmosphere, we take into account the fact that the molecular scattering coefficients within the laser spectrum may, with great accuracy, be considered independent of wavelength; whereas the molecular absorption coefficients may vary even within the rather narrow spectral range of certain lasers. In such a case Bouguer's law is inapplicable and the transmission component due to molecular absorption can be determined only in terms of the absorption function.

When selecting effective wavelengths for laser sounding, the following contradiction must be faced: On the one hand, the higher the atmospheric transmission for the wavelength being used, the greater the atmospheric range a sounding pulse can penetrate; on the other hand, for a strong echo signal it is advantageous to have appreciable interaction of the laser pulse with the atmosphere, at the expense of absorption and scattering. Therefore, when the echo-signal wavelength coincides with the sounding pulse wavelength (e.g., when using molecular, aerosol, or resonance scattering), it must be selected in an appropriate optimal way. However, when we use spontaneous Raman scattering for laser monitoring, the wavelength of the laser radiation must be selected in such a way that both the sounding pulse and the echo-signal wavelength of Raman scattering are between weak absorption lines (microwindows of atmospheric transmission).

In a number of cases it is advisable to perform remote sensing when line centers of laser radiation and absorption of atmospheric gases coincide (e.g., when applying the phenomena of resonance molecular absorption, resonance scattering and resonance Raman scattering). To increase the maximum altitude for laser sounding, certain sets of absorption lines having various intensities can be used. Thus, e.g., while sounding the profiles of water vapor or any other atmospheric gas from outer space, use of the strongest absorption lines will yield appropriate data from the upper atmosphere. As intensity of the lines used decreases, corresponding information can be extracted for lower and lower layers of the atmosphere.

3.4.2 Hazes

In the case of hazes the total attenuation comprises three components. In addition to molecular absorption and scattering, aerosol scattering must be considered. For hazes the aerosol scattering coefficients vary between wide limits, depending on concentration, size, and chemical composition of the particles. At the same time, within the laser frequency spread, the aerosol scattering (attenuation) coefficients as well as the molecular scattering coefficients are essentially constant.

Meteorological visibility determined by the joint influence of aerosol and molecular scattering by hazes varies greatly, whereas when it is limited by molecular scattering, it can be considered practically constant. Furthermore, the molecular scattering coefficient is uniquely connected with wavelength; this cannot be said about the aerosol scattering coefficient. Hence in hazes a great variety of situations can be realized depending on the relative sizes of the aerosol attenuation coefficient and the molecular scattering coefficient. Some features of this variety are the following: Under the most usual meteorological conditions (visibility in the ground layer $\sim 10\,$km), the coefficients of aerosol attenuation and molecular scattering in the ultraviolet are the same order of magnitude. In the visible the aerosol attenuation is considerably stronger than the molecular scattering. Finally, in the infrared, molecular scattering can be neglected in comparison with aerosol scattering.

Most transparent spectral intervals for laser wavelengths are situated, on the one hand, at minimum values of coefficients of aerosol and molecular scattering, and, on the other hand, at the macro- and micro-windows of atmospheric transmission. Most of these intervals are found in the wavelength range 8–13 μm corresponding to the long-wave transmission window. It must be stressed that the use of this wavelength range for laser monitoring of the atmosphere using aerosol scattering has a definite advantage in comparison with the visible wavelength range, since the scattering phase function of aerosol particles here is less stretched forward; hence, more of the radiation is scattered backwards.

As for molecular absorption, its role has already been considered in the previous section. Here we only note that the absorption coefficients at the centers of the strongest lines are generally greater than the aerosol attenuation coefficients, whereas in the microwindows of atmospheric transmission the situation reverses with strong turbidities.

3.4.3 Clouds and Fogs

The attenuation of laser radiation by clouds and fogs is conditioned by the same three components as for hazes. That component connected with aerosol attenuation is dominating for all wavelengths except those near centers of the most intense absorption lines. For all kinds of clouds and fogs the coefficient of molecular scattering is negligible compared with the aerosol scattering coefficient, not only in the visible and infrared, but also in the ultraviolet.

In laser sounding of clouds it should, first of all, be taken into account that a sounding pulse penetrates through the atmospheric layer from the lidar system to a cloud, and then its echo-signal returns from the cloud to the lidar. It is important, therefore to minimize energy losses in the atmospheric layer between the lidar and the sounded cloud. The most preferable technique is to select a spectral interval between weak lines in the atmospheric transmission windows in the visible and infrared, since in this case the losses due to molecular scattering are usually negligible. In addition, such intervals have maximum transmission since the sum of the coefficients of continuous absorption and aerosol scattering is minimal. Analysis shows that these intervals are located in transmission windows with a wavelength range of about 2–12 μm. The total attenuation in the visible is greater on account of aerosol scattering from hazes.

As to the selection of wavelengths of laser pulses for most effective cloud sounding, the following should be noted. For large-droplet clouds the attenuation coefficient in the wavelength range encompassing the ultraviolet, visible and infrared (up to 25 μm) depends to a small extent on the wavelength; consequently the penetration depth of the sounding pulse into the cloud also depends slightly on wavelength. Nevertheless, the use of pulses of longer wavelength is preferable, at the expense of a less-forward-stretched scattering phase function; and in this case greater values of echo-signal should be expected.

For small-droplet clouds, the most effective wavelengths of sounding pulses are in the spectral interval from approximately 10 to 12 μm, where attenuation coefficients are minimal and the sounding pulses can penetrate through greater cloud thicknesses. In this wavelength range the attenuation coefficients of hazes are also minimal.

Taking the above into account, the conclusion may be drawn that the spectral interval from 10 to 12 μm is the most effective for the sounding of clouds and fogs if we pay no attention to associated purely technical problems, e.g., the availability of appropriate receivers for the mentioned wavelength range.

3.4.4 Precipitations

The attenuation coefficients of precipitations, like those of clouds, fogs and hazes, consist of three components, for which that associated with aerosol attenuation does not depend on wavelength in the wavelength range of interest to us. That is why the sounding pulses must have wavelengths for which the values of the other two attenuation components (molecular scattering and absorption) are minimal. Their selection does not differ from that considered in the previous section. As in the case of clouds and fogs, greater values of echo-signals are expected from pulses with longer wavelengths because of the scattering phase function effect.

3.5 Models of the Atmosphere

Quantitative predictions of the propagation of laser radiation through the atmosphere, and associated optical characteristics depend upon knowledge of parameters such as concentration of absorbing gases, total pressure, temperature, and concentration, chemical composition, size, spectra and shape of aerosol particles. Changes in these characteristics give rise to variations of attenuation coefficients and atmospheric transmission for a given wavelength. In this connection, it is important to have models of the atmosphere which have been statistically verified, i.e., data on the mean profiles of appropriate parameters as well as their repetition and confidence intervals. Development of such models is being intensively carried out at the present time by a number of scientific groups.

3.6 Deviations from Homogeneity

3.6.1 Effects of Atmospheric Turbulence

Laser pulses propagating through the atmosphere not only suffer energy losses caused by absorption and scattering phenomena, but are influenced by atmospheric turbulence as well. Turbulent fluctuations of the refractive index lead to distortion of the initial laser beam parameters (which are important for interpreting the results of laser sounding of the atmosphere), and lead to signal fluctuations which limit sensitivity of the long-path absorption technique. Thus, for example, the echo signal of a sounding laser pulse depends on coherence and space-time fluctuations of intensity and field phase within the propagating laser beam, the sizes of the latter and its random shifts. For long-path propagation, turbulence can cause the laser beam to break up and change direction, thereby causing partial or total loss of signal on the retroreflector or collection optics.

The decrease of coherence of a sounding laser beam is very important in the case of remote wind velocity measurements using Doppler methods. The degree of focusing of the radiation determines the minimum volume sounded, and random shifts of laser beam position restrict the stability of the sounded volume—both of these parameters are functions of turbulence.

Detailed consideration of the effects of atmospheric turbulence on parameters of infinite and spatially-limited optical waves is presented in monographs [3.1 and 52], and in a series of reviews (see, e.g. [3.53, 54]). Here we shall give only the most essential results important for laser monitoring techniques.

The *intensity fluctuations* of a laser beam propagating through the atmosphere are characterized by the variance of the logarithm of intensity. In the case of weak fluctuations, this is expressed by the following formula obtained from perturbation theory

$$\delta_{\ln I}^2 = 1.23 \, q_n C_n^2 K^{7/6} L^{11/6} , \tag{3.52}$$

where the coefficient q_n depends on the diffraction size of the radiating aperture and the beam divergence (and changes by not more than a factor of four), C_n^2 is the structural characteristic of fluctuations of the medium refractive index, $K \, (= 2\pi/\lambda)$ is the amplitude of the radiation wave vector, and L is the path length. This equation correctly describes the experimental data for cases in which $\delta_{\ln I}^2 \lesssim 0.6$. For $\delta_{\ln I}^2 > 0.6$, the measured value of variance does not obey (3.52), since at a certain point it reaches saturation (region of strong fluctuations). In this case the following approximate formula is obtained for the relative variance of intensity fluctuations in the Fresnel zone of diffraction of a collimated beam

$$\delta_I^2 = 1 + 0.87 \, (1.23 \, C_n^2 K^{7/6} L^{11/6})^{-2/5} . \tag{3.53}$$

The above equation describes the experimental data with an error of not more than 10–30% in the applicability range indicated.

In the case of a focused beam, the value δ_I^2 can be estimated by the equation

$$\delta_I^2 = 1 + 4 \, [D_s(2a_0)]^{-4/5} , \tag{3.54}$$

which agrees satisfactorily with experiment. $D_s(2a_0)$ is the variance of the phase difference over the diameter of the transmitting aperture.

The distribution of laser beam intensity in a plane perpendicular to the direction of propagation determines the *beam size*, and can be represented as

$$I(z, \varrho) = I(z, 0) \exp(-\varrho^2/a^2) , \tag{3.55}$$

where z and ϱ are the longitudinal and transverse coordinates, and a is the effective radius of the beam

$$a = a_0 \{(1 - z/R_p)^2 + Q_A^2 [1 + 0.47 \, D_s^{6/5}(2a_0)]\}^{1/2} , \tag{3.56}$$

where a_0 is the initial beam radius, R_p is the radius of the phase front curvature, and $Q_A \, (= K a_0^2/z)$ is the wave parameter of the aperture. This equation says that focusing of a laser beam (i.e., $z/R_p = 1$) is possible only when $Q_A [1 + 0.47 \, D_s^{6/5}(2a_0)]^{1/2} < 1$. This condition is clearly not fulfilled for very large phase variations over the diameter of the transmitting aperture [i.e., $D_s(2a_0) \gg 1$] when the beam size in a focus does not depend on the diameter of the transmitting aperture.

The variance of fast shifts of laser beam position due to *random refraction fluctuations* in the atmosphere can be estimated by the following equation

$$\delta_d^2 = 1.7\, C_n^2 z^2 (2a_0)^{-1/3} . \qquad (3.57)$$

This equation is valid near the diffraction zone of the beam ($Q_A \ll 1$) for small values of the parameter $D_s(2a_0)$. For arbitrary values of Q_A and $D_s(2a_0)$, (3.57) must be corrected.

Defocusing of the beam ($z/R_p \to \infty$) and increasing diffraction divergence result in a decrease in variance of its random shifts. If the initial beam size becomes comparable with the outside scale of turbulence (usually for radiating apertures having radii of 50 cm or more), δ_d^2 decreases several times compared with its value determined from (3.57).

3.6.2 Scintillation Due to Aerosols

The origin of laser beam intensity fluctuations in an aerosol atmosphere is connected with particle concentration variations, their mutual spacing, size distribution, shape, and orientation in space. It should be mentioned that intensity fluctuations in this case can increase with increasing particle concentration since the fields are additive, but the intensities are not; however, these fluctuations are not observed in most experiments involving incoherent sources and receivers for which space-time averaging is performed on the radiation intensity.

Solution of the general scintillation problem leads to very cumbersome equations for the intensity moments. We shall, therefore, confine ourselves to a corresponding qualitative consideration of intensity fluctuations of light scattered by a system of statistically independent particles in a random, inhomogeneous medium.

If the condition for a *system of statistically independent particles* may be realized, then the particles are, on the average, in the wave zone of one another. Spherical waves scattered by particles form an interference pattern which changes according to the particle distribution. As can be shown, the average light beam intensity in this case is defined by the transfer equation, and the statistical characteristics of the fluctuations are expressed in terms of certain integrals of this intensity [3.55].

When particles are in the near zone of one another, statistical characteristics of the intensity fluctuation are determined mainly by mutual shading or screening. As shown in [3.55], the intensity fluctuation variance δ_I^2 increases in this case with an increase in particle concentration.

This is connected with the fact that, with an increase in particle concentration, the average intensity decreases according to Bouguer's law, while δ_I^2 increases at the expense of comparatively rare "realizations" when the "screens" do not overlap the whole beam.

Light intensity fluctuations in a *random inhomogeneous scattering medium*, e.g., in an aerosol suspended in a turbulent ground layer of the atmosphere, are determined by solving the stochastic transfer equation in which the optical thickness and scattering coefficient are expressed in a random field of particle concentration. Solution of the equation for variance of intensity fluctuations can be numerically calculated using the Monte-Carlo method.

Experimental investigations of laser beam intensity fluctuations in an aerosol atmosphere are in the initial stage. Nevertheless, existing preliminary data indicate that the phenomena discussed above will have to be taken into account when interpreting the results of laser sounding and long-path monitoring of the atmosphere. In this connection, it should be stressed that further theoretical and experimental studies of these phenomena are greatly needed.

Acknowledgments. The author considers it a pleasant duty to thank Drs. A. G. BOROVOY and V. L. MIRONOV for their assistance in writing the last section of the chapter, and also N. P. MALINCHEVA, T. V. KUZNETSOVA, and M. KH. KURMAN for contributing greatly in the technical design of the manuscript.

References

3.1   V. YE. ZUYEV: *Propagation of Visible and Infrared Waves in the Atmosphere* (NASA TT F-707, 1972). Translation of *Rasprostraneniye Vidimykh i Infrakrasnykh Voln v Atmosfere* (Sovetskoye Radio Press, Moscow 1970)

3.2   H. A. LORENTZ: Proc. Amst. Acad. Soc. **8**, 591 (1906)

3.3   L. I. GALIBINA, V. E. ZUEV: Soviet Phys. J., No. 4, 69 (1962)

3.4   V. I. DIANOV-KLOKOV: Optika i Spektroskopiya **16**, 409 (1964)

3.5   V. E. ZUEV: Soviet Phys. J., No. 3, 138 (1967)

3.6   V. E. ZUEV: Soviet Phys. J., No. 10, 53 (1967)

3.7   V. E. ZUEV, V. V. POKASOV, YU. A. PKHALAGOV, A. V. SOSNIN, S. S. KHMELEVTSOV: Izv. Akad. Nauk, SSSR, Ser. Fizika Atmosfery i Okeana **4**, 63 (1968)

3.8   E. L. KERR, J. G. ATWOOD: Appl. Opt. **7**, 915 (1968)

3.9   V. A. BORISOV: Optiko-Mekhanicheskaya Promyshlennost (Optical-Mechanical Industry), No. 7, 11 (1970)

3.10 B. A. ANTIPOV, YU. N. PONOMAREV: Soviet Phys. J., No. 3, 145 (1972)

3.11 V. E. ZUEV, A. V. SOSNIN, S. S. KHMELEVTSOV: Zh. Prikladnoy Spektroskopii, No. 8 (1972)

3.12 V. E. ZUEV, V. P. LOPASOV, M. M. MAKOGON: Dokl. Akad. Nauk, SSSR **199**, 1041 (1971)

3.13 V. E. ZUEV, V. P. LOPASOV, M. M. MAKOGON: Appl. Opt. **10**, 2452 (1971)

3.14 V. E. ZUEV, V. P. LOPASOV, M. M. MAKOGON: Soviet Phys. J., No. 11, 135 (1971)

68   V. E. ZUEV

3.15 V.E.ZUEV: Vestnik Akad. Nauk, SSSR, No. 8, 18 (1972)
3.16 B.A.ANTIPOV, V.E.ZUEV, P.D.PYRSIKOVA, V.A.SAPOZHNIKOVA: Optika i Spektroskopiya **31**, 899 (1971)
3.17 R.K.LONG, F.S.MILLS, E.K.DAMON: "Molecular Absorption Studies Using Infrared Lasers (CO, DF, CO_2)". Paper presented at 2nd High-Resolution Molecular Spectroscopy Symposium, Novosibirsk, U.S.S.R. (11–13 September 1974)
3.18 O.K.VOYTSEKHOVSKAYA, V.E.ZUEV, I.I.IPPOLITOV, YU.S.MAKUSHKIN: Zh. Prikladnoy Spektroskopii **17**, 164 (1972)
3.19 O.K.VOYTSEKHOVSKAYA, I.I.IPPOLITOV, YU.S.MAKUSHKIN: Optika i Spektroskopiya **35**, 42 (1973)
3.20 O.K.VOYTSEKHOVSKAYA, I.I.IPPOLITOV, YU.S.MAKUSHKIN: Optika i Spektroskopiya **33**, 78 (1972)
3.21 YU.S.MAKUSHKIN, VL.G.TYUTEREV: Optika i Spektroskopiya **35**, 439 (1973)
3.22 V.N.BRYUKHANOV, YU.S.MAKUSHKIN: Optika i Spektroskopiya **34**, 56 (1973)
3.23 A.D.BYKOV, YU.S.MAKUSHKIN, M.P.CHERKASOV: Materialy II Vsesoyuznogo Simpoziuma po Rasprostraneniyu Lazernogo Izlucheniya v Atmosfere (Proc. 2nd All-Union Symposium on Laser Radiation Propagation in the Atmosphere), Tomsk (1973)
3.24 S.D.TVOROGOV, V.V.FOMIN: Optika i Spektroskopiya **30**, 413 (1971)
3.25 S.D.TVOROGOV, V.V.FOMIN: Optika i Spektroskopiya **31**, 1026 (1971)
3.26 L.I.NESMELOVA, S.D.TVOROGOV, V.V.FOMIN: Izv. Akad. Nauk, SSSR, FAO **9**, 1205 (1973)
3.27 L.I.NESMELOVA, S.D.TVOROGOV: Izv. Akad. Nauk, SSSR, FAO **9**, 1209 (1973)
3.28 V.V.FOMIN, S.D.TVOROGOV: Appl. Opt. **12**, 584 (1973)
3.29 V.E.ZUEV, S.D.TVOROGOV, V.V.FOMIN: "Some Problems of the Spectral Line Broadening Theory and Formation of the Absorption Band Contour". Prikladnaya Spektroskopiya. Collection of Review Reports of 17 All-Union Spectroscopy Symposium, Minsk (5–9 July, 1971) pp. 41–66
3.30 R.A.McCLATCHEY, W.S.BENEDICT, S.A.CLOUGH, D.E.BURCH, R.F.CALFEE, K. FOX, L.S.ROTHMAN, J.S.GARING: "Atmospheric Absorption Line Parameters Compilation". AFCRL-TR-73-0096, Environmental Research Papers, No. 434 (1973)
3.31 S.I.VAVILOV: "Mikrostruktura Sveta (Light Microstructure)". Izd. Akad. Nauk, SSSR (1950)
3.32 R.F.CALFEE: J. Quant. Spectr. Rad. Transfer. **6**, 221 (1966)
3.33 L.ELTERMAN: "Atmospheric Attenuation Model, 1964, in the Ultraviolet, Visible, and Infrared Regions for Altitudes to 50 km". Environmental Research Papers, N. 46, AFCRL, Cambridge, Mass. (1964)
3.34 R.PENNDORF: J. Opt. Soc. Am. **47**, 176 (1957)
3.35 H.C. VAN DE HULST: *Light Scattering by Small Particles* (John Wiley and Sons, Inc., New York; Chapman and Hall, Ltd., London 1957)
3.36 V.E.ZUEV, M.V.KABANOV, B.P.KOSHELEV, S.D.TVOROGOV, S.S.KHMELEVTSOV: Soviet Phys. J., No. 3, 92 (1964)
3.37 V.E.ZUEV, B.P.KOSHELEV, S.D.TVOROGOV, S.S.KHMELEVTSOV: Izv. Akad. Nauk, SSSR, Ser. Fizika Atmosfery i Okeana **1**, 509 (1965)
3.38 V.E.ZUEV, S.D.TVOROGOV: Soviet Phys. J., No. 2, 143 (1966)
3.39 V.E.ZUEV, V.V.SOKOLOV, S.D.TVOROGOV: Soviet Phys. J., No. 4, 73 (1971)
3.40 V.E.ZUEV, M.V.KABANOV, B.P.KOSHELEV, S.D.TVOROGOV, S.S.KHMELEVTSOV: Soviet Phys. J., No. 2, 90 (1964)
3.41 V.E.ZUEV, B.P.KOSHELEV, S.D.TVOROGOV, S.S.KHMELEVTSOV: Soviet Phys. J., No. 3, 121 (1966)
3.42 V.E.ZUEV, V.P.LOPASOV, V.K.SONCHIK: Izv. Akad. Nauk, SSSR, Ser. Fizika Atmosfery i Okeana **3**, 16 (1967)

3.43 V. E. ZUEV, V. K. SONCHIK: Izv. Akad. Nauk, SSSR, Ser. Fizika Atmosfery i Okeana **5**, 745 (1969)

3.44 V. E. ZUEV, V. V. SOKOLOV, S. D. TVOROGOV: Soviet Phys. J., No. 3, 7 (1966)

3.45 V. E. ZUEV, V. V. SOKOLOV, S. D. TVOROGOV: Soviet Phys. J., No. 1, 107 (1969)

3.46 V. E. ZUEV, M. V. KABANOV, B. A. SAVELEV: Soviet Phys. J., No. 5, 80 (1964)

3.47 V. E. ZUEV, M. V. KABANOV, B. A. SAVELEV: Dokl. Akad. Nauk, SSSR **175**, 327 (1967)

3.48 V. E. ZUEV, M. V. KABANOV, B. A. SAVELEV: Izv. Akad. Nauk, SSSR, Ser. Fizika Atmosfery i Okeana **3**, 724 (1967)

3.49 V. E. ZUEV, M. V. KABANOV, B. A. SAVELEV: Appl. Opt. **8**, 137 (1969)

3.50 YU. F. ARSHINOV, V. A. DONCHENKO, V. E. ZUEV, M. V. KABANOV, G. M. KREKOV, G. G. MATVIENKO, A. I. POPKOV, I. V. SAMOKHVALOV: Proc. Intern. Symp. on Radiation, Sendai, Japan (1972)

3.51 V. E. ZUEV, G. M. KREKOV, A. I. POPKOV: Izv. Akad. Nauk, SSSR, Ser. Fizika Atmosfery i Okeana **9**, 770 (1973)

3.52 V. I. TATARSKII: *Theory of Fluctuation Phenomena in Wave Propagation in a Turbulent Atmosphere* (Izd. Akad. Nauk, SSSR 1957)

3.53 S. S. KHMELEVTSOV: Appl. Opt. **12**, 2421 (1973)

3.54 P. S. LAWRENCE, J. W. STROHBEHN: Proc. IEEE **58**, 1523 (1970)

3.55 A. G. BOROVOY, M. V. KABANOV, B. A. SAVELEV: (to be published)

3.56 S. S. PENNER: *Quantitative Molecular Spectroscopy and Gas Emissivities* (Addison-Wesley, Reading, Mass. 1959)

3.57 R. K. LONG: "Atmospheric Absorption and Laser Radiation"; Ohio State University Engineering Publications, Bulletin 199 (Columbus, Ohio 1967)

The following English translations are available:

1) Soviet Phys. J. is a cover-to-cover translation of Izv. vuz Fizika, 1965, vol. 7+.
2) Optika i Spektroskopiya = Optics and Spectrosc., 1959, vol. 6+.
3) Izv. an SSSR, Ser. Fizika Atmosfery i Okeana (FAO) = Atmospheric and Oceanic Physics, 1965, No. 1+.
4) Optiko-Mekhanicheskaya Promyshlennost = Soviet J. of Optical Technology, 1966, vol. 33+.
5) Zh. Prikladnoy Spektroskopii = J. Appl. Spectrosc., 1965, vol. 2+.

4. Lidar Measurement of Particles and Gases by Elastic Backscattering and Differential Absorption

R. T. H. COLLIS and P. B. RUSSELL

With 22 Figures

Electromagnetic energy is scattered by gas molecules of the atmosphere and also by its particulate or droplet constituents. In the case of energy at the optical and near-optical wavelengths generated by lasers, such scattering is sufficient to permit application of the radar principle to observations of the atmosphere itself. Even in the visibly "clear" atmosphere backscattered signals from gases and suspended particles at ranges of several kilometers may readily be detected with laser "radars", or *lidars*, of modest performance. It is accordingly possible to detect the presence and location of particulate clouds or layers and, by tracking inhomogeneities in particle concentration, to determine atmospheric structure and motion. In addition, this capability of obtaining returns from distant atmospheric volumes provides, from a single location, an extended path along which the effects of atmospheric absorption may be determined. In this way, by using energy at specific wavelengths, resonant absorption may be used to make range-resolved measurements of the quantity of a particular absorptive gas along the path. In this chapter we consider various techniques of atmospheric probing, all based upon elastic backscattering, and in some cases also making specific use of the absorption of laser energy.

4.1 Background Information

It is interesting to note that the radar principle, which was first widely exploited at radio frequencies for the detection of targets such as aircraft or ships, and was only later applied to detecting hydrometeors in storm detection applications, was first applied to atmospheric probing at optical wavelengths. As early as the 1930's [4.1, 2] the concept of assessing the density and dust loading of the upper atmosphere by scattering techniques was explored by observing the beam of a vertically-pointing searchlight with a remotely located photodetector—an arrangement that would now be called a bistatic radar configuration. In pulsed light form, the conventional monostatic radar configuration was similarly applied to remote sensing of the upper atmosphere in the 1940's–50's [4.3, 4]. Indeed, in 1939, French meteorologists anticipated meteorological radar (and subsequent lidar or laser radar) in almost every respect in a pulsed-light cloud-base measuring system using a spark as a light source [4.5]. (This was several years before centrimetric wavelength developments made it possible to use radio frequency 'radar' for meteorological pur-poses.) In fact, familiarity with weather radar techniques provided both the initiative and the basic approaches to the earliest workers in the field of atmospheric probing using laser energy as a source of power.

As it happened, because of the nature of the first practical lasers with adequate power, their use in remote probing applications of the atmo-sphere followed easily and readily the pulsed radar methods which have become so well established in many meteorological applications. Thus in 1963, LIGDA of Stanford Research Institute, who was himself a pioneer of weather radar applications, turned almost naturally to the newly developed giant-pulse, Q-switched ruby laser as the transmitter of a "new" optical radar [4.6]. The developments of LIGDA and his co-workers, and an increasing number of other workers in the field, were thus con-siderably influenced by weather radar experience, although most of the basic principles involved had a sounder basis in optical technology and practice. At least one reason for this was the fact that weather radar provided a well-established background of practical operational applica-tions in atmospheric probing and it was only later that the more complete approach due to optical science was fully adopted in the emerging field of laser atmospheric probing.

In the present chapter we focus our detailed discussion on this basic pulsed radar technique of atmospheric probing with laser energy, since it provides a basis of understanding for virtually all forms of radar probing. As indicated in Chapter 2, laser energy scattered or re-emitted by atmospheric constituents may differ in wavelength (hence frequency) from the incident energy, and atmospheric information may be derived from these *inelastic* processes (see especially Chapter 5). In this chapter, however, we will be concerned with measurements of the *elastic* com-ponent of atmospheric backscattering, for which scattered and incident wavelengths are equal. Specifically, pulsed radar measurements at the transmitted frequency provide: 1) direct observation of atmospheric elastic backscattering from which it is possible to determine the presence and distributions of particles and hydrometeors (or, in certain condi-tions, atmospheric molecular density); 2) reference signals from distant scattering volumes from which, using differential absorption at multiple wavelengths, gaseous absorption may be measured.

The direct observation of backscatterers provides information which is either of significance in itself—such as the presence and location of smoke plumes, haze layers, cloud layers, etc.—or from which significant information may be inferred (as, for example, thermal stratification as revealed by haze layers in certain circumstances). Again, gaseous density (and hence temperature) at high altitudes may be deduced from back-scattering profiles, while in fog or haze, the variation of backscattered signals as a function of range may be analyzed to assess visibility.

Differential absorption measurements make it possible to determine the presence and concentration of atmospheric pollutant gases, or such natural atmospheric constituents as water vapor, by selecting appropriate wavelengths at which such absorption is specific.

4.2 Principles of Light Detection and Ranging with Lasers (LIDAR)

The basic radar principle of remote atmospheric probing by lasers is called LIDAR, from the acronym *Light Detection and Ranging*. This term, in analogy to RADAR (where *RA*dio is the active element), was in fact first applied to pulsed light techniques using conventional sources by MIDDLETON and SPILHAUS [4.7].

In the simplest form, lidar employs a laser simply as a source of pulsed energy of useful magnitude and suitably short duration. Typically [4.8–11] Q-switched ruby (wavelength $\lambda = 0.69\,\mu m$) or neodymium ($\lambda = 1.06\,\mu m$) laser systems are used and generate pulses having peak power measured in tens of megawatts and durations of 10–20 nano-seconds. Pulses with such energy (i.e., of the order 1 Joule) are directed in beams by suitable optical systems, an example of which is shown schematically in Fig. 4.1. Because the laser energy is virtually mono-chromatic and is highly coherent, such beams are highly collimated.

As the transmitted laser energy passes through the atmosphere, the gas molecules and particles or droplets encountered cause scattering. A small fraction of this energy is backscattered in the direction of the lidar system and is there available for detection. The scattering of energy *out* of the forward direction or propagation, or absorption by the gases and particles, reduces the intensity of the beam, which is said to be at-tenuated. Such attenuation applies to both the path *to* and *from* a distant backscattering region.

At the lidar, backscattered energy is collected in a suitable receiver by means of reflective or refractive optics (see, e.g., Fig. 4.1) and trans-ferred to a photodetector (commonly, a photo-multiplier). This produces an electrical signal, the intensity of which at any instant is proportional to the optical power received. Since light travels at known velocity, the

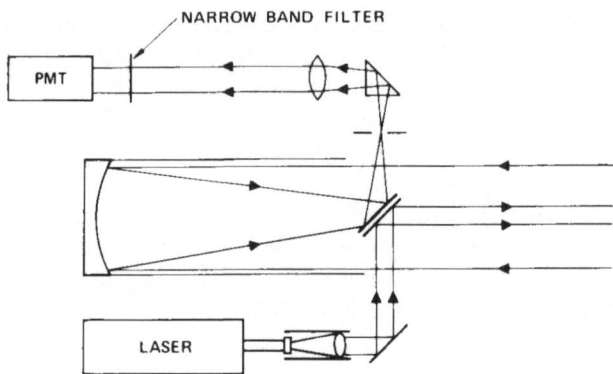

Fig. 4.1. Schematic diagram of fundamental lidar configuration, with coaxial transmitter/receiver geometry

range of the scattering volume producing the signal received at any instant can be uniquely determined from the time interval since transmission of the pulse. The magnitude of the received signal is determined by the backscattering properties of the atmosphere at successive ranges, and also by the two-way atmospheric attenuation. Atmospheric backscattering in turn depends upon the wavelength of the laser energy used, and the number, size, shape and refractive properties of the particles or droplets (or molecules) intercepted by the incident energy. Although, as discussed in detail in Subsection 4.3.1, backscattering from an assemblage of scatterers is a complicated phenomenon, in general terms backscattering increases with increasing concentrations of scatterers.

The electrical signal from the photo-detector thus contains information on the presence, range, and concentration of atmospheric scatterers and absorbers. Various forms of presenting and analyzing such signals are available. In the simplest form they may be presented on an oscilloscope in a coordinate system showing received signal intensity as a function of range. Since such signals are very transient (1 km of range is represented by an interval of time of 7 μs), it is necessary to photograph such oscilloscope displays to obtain adequate data presentation. An example of a signal recorded in this manner is shown in Fig. 4.2. (Alternatively the electrical signals may be recorded by magnetic tape, disc, or digital storage devices prior to presentation or further, slow-speed processing. These methods are described more completely in Subsection 4.2.2.)

Sequences of observations in a given direction, or in various scanning patterns, can be used to provide information on the atmospheric condition on an extended scale in space or time. Figure 4.3, for example,

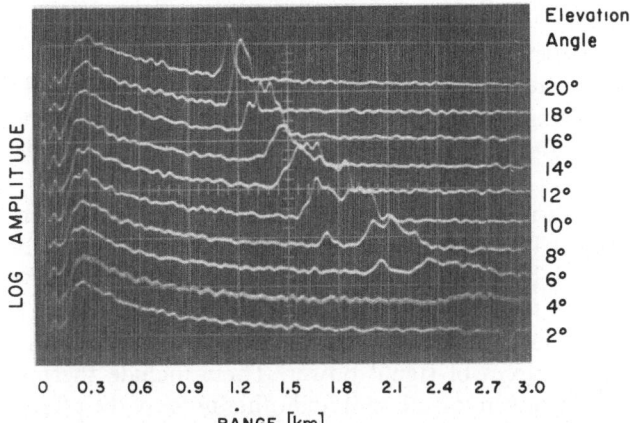

Fig. 4.2. "A Scope" (intensity vs. range) display of received lidar signals, obtained by photographing the oscilloscope screen. Traces show $1/R^2$ decrease of "clear air return." Increased signals from 1.0 to 2.5 km range are caused by smoke plume

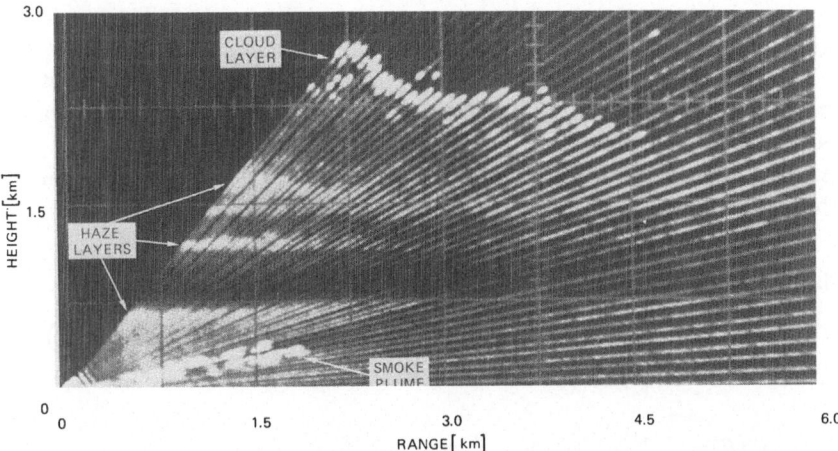

Fig. 4.3. Intensity-modulated "Picture" display obtained using magnetic disc recording and playback techniques. The vertical cross section delineates a smoke plume 1.5 km downwind of a 245 m power plant stack. Also readily evident is the top of the turbid mixed layer at 0.75 km altitude, stratified haze layers aloft at 1.1 and 1.5 km, and the base of a visible cloud at about 2 km altitude. Attenuation limits pulse penetration into the cloud

shows an intensity-modulated vertical cross section generated by making a series of observations while scanning in the vertical plane. The rate at which such information can be collected depends upon the pulse repetition rate of the laser system. Other forms of data presentation both in terms of temporal and spatial scanning geometries and in terms of data processing and display are illustrated in the examples of different types of applications shown in Section 4.5 below.

In many cases the intensity of the lidar signal as displayed may be readily interpreted in terms of atmospheric conditions by virtue of relative variations. The presence of haze and cloud layers, as shown in Fig. 4.3 is a typical example.

Quantitative evaluation of the lidar signal is a matter of some complexity that is discussed further below (Subsection 4.5.3). It should be noted, however, that in both qualitative displays and in quantitative evaluations the limiting factor to signal detectability (given adequate sensitivity of the detection and display devices) is the level of noise due to various extraneous sources of signal power. These include thermal noise of the electrical circuits involved, shot and other noise of the photodetector and, probably most importantly, noise caused by optical energy, normally of solar origin, present as background light, particularly of course, in daytime [4.12]. Because of the high degree of monochromaticity of laser energy, extraneous light can be excluded to a substantial degree by the use of a narrow band filter centered on the laser frequency.

Incidentally, because laser energy can be highly collimated, it is possible to direct all the transmitted energy in a narrow beam (typically with a divergence of the order of 2 or 3 milliradians). Accordingly, the field of view of the receiver can be restricted to this angle, thus minimizing background light entering the system.

4.2.1 The Lidar Equation

The basic lidar principles outlined above may be expressed formally in the single-scattering lidar equation

$$P_r(R) = P_0 \left(\frac{c\tau}{2}\right) \beta(R) A_r R^{-2} \exp\left[-2 \int_0^R \alpha(r) dr\right] \tag{4.1}$$

where P_r is the instantaneous received power at time t, P_0 is the transmitted power at time t_0, c is the velocity of light, τ is the pulse duration, β is the volume backscattering coefficient of the atmosphere, R is range [see (4.2) below], A_r is the effective receiver area, and α is the volume extinction coefficient of the atmosphere[1].

The determination of range R and the increment of range from which returns are received simultaneously is illustrated in Fig. 4.4. Since the

---

[1] This is defined as cross section per unit volume [cm^2/cm^3], and is numerically identical to the linear extinction coefficient used in laser transmission equations, which has units of cm^{-1}.

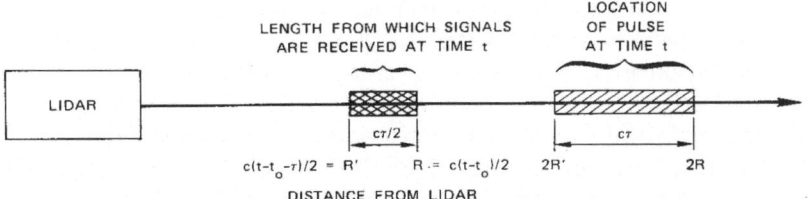

Fig. 4.4. Location and length of range increment from which lidar signals are simultaneously received at time t. Pulse has duration τ; Leading edge was transmitted at time t_0

transmitted pulse has a finite duration τ, it illuminates a finite geometrical length $c\tau$ of the atmosphere at any instant. However, since the received energy must travel a two-way path, the atmospheric length (or range increment) from which signals are received at any time t is just half this value. To illustrate this quantitatively, suppose that the leading edge (beginning) of the pulse is transmitted at time t_0. Then the maximum range from which energy is received at time t is given by half the distance that the leading edge could travel in the intervening time. That is,

$$R = c(t - t_0)/2 . \tag{4.2}$$

At the same instant t, additional energy is received from ranges illuminated by portions of the pulse transmitted after the leading edge. Since the trailing edge (end) of the pulse was transmitted at time $t_0 + \tau$, the minimum range from which energy is received at time t is given by

$$R' = c(t - t_0 - \tau)/2 . \tag{4.3}$$

Thus the distance

$$L_p = R - R' = c\tau/2 , \tag{4.4}$$

called the "effective pulse length", is the range interval from which signals are received at any instant. It is exactly half of the actual length instantaneously illuminated by the pulse because of the two-way path that all received energy must traverse.

The volume backscattering coefficient β is defined as the fractional amount of incident energy scattered per unit solid angle (steradian) in the backward direction per unit atmospheric length. (It thus has dimension $\ell^{-1}\, \mathrm{sr}^{-1}$.) Since, in this chapter, we are concerned only with elastic backscattering processes (for which received and transmitted wavelengths are equal), the symbol β will herein be taken as the elastic component of the volume backscattering coefficient.

The effective receiver area A_r enters (4.1) because it determines the solid angle A_r/R^2 subtended by the receiver at range R. (Provided that the angle of divergence of the transmitter beam does not exceed the angle of acceptance of the receiver, all the energy transmitted is available for backscattering at range R).

The expression

$$\exp\left[-2\int_0^R \alpha\,dr\right]$$

represents the fractional transmittance (T) of energy along the two-way path between the lidar and range R. This depends upon α, the volume extinction coefficient of the atmosphere, which represents the fraction by which the flux of energy in the direction of propagation is reduced per unit length. (It thus has dimension ℓ^{-1}.) It is due to both absorption α_a and dissipation of energy by scattering α_s. Strictly, this term applies in this simple form only when Bouguer's law is valid, that is, when essentially all scattered energy is permanently removed from the outgoing lidar beam. This simplification is generally true in fairly transparent atmospheres (cf. Subsections 3.1.6 and 3.3.7). In strongly scattering atmospheres—e.g., in cloud, fog, or thick haze—the fraction of scattered energy that remains in, or re-enters, the lidar beam can become significant. Multiple scattering corrections to the single-scattering equation (4.1) must then be made, as shown in Section 4.3, which also contains a full discussion of the relation of β and α to atmospheric properties.

4.2.2 Lidar Systems and Equipment

The basic form of lidar in its simplest configuration has been outlined above in Section 4.2 and illustrated schematically in Fig. 4.1. Extensions and variations of the basic concept follow the needs of any particular application or the exploitation of more sophisticated principles. The lidar systems assembled for such applications may be employed either in fixed locations or from moving vehicles or aircraft to map and monitor atmospheric targets over extended areas by scanning. A summary and overview of the various components available for such applications, and their integration into typical system configurations, are presented in this section. Detailed component specifications for several existing systems are provided in a number of published references [4.8–11, 13, 14, 18–20].

Lasers

The earliest lidars employed pulsed ruby or neodymium glass lasers of relatively high power (~ 1 Joule per pulse) as sources of nearly monochromatic and coherent energy [4.8–11]. While such ruby ($\lambda = 694.3$ nm) and neodymium glass ($\lambda = 1061$ nm) lasers are limited to a few pulses per second, YAG (yttrium aluminum garnet) lasers generate pulses at 1065 nm wavelength at rates of several hundred per second or more, at useful energies (0.1 J per pulse), thus offering considerable possibilities for high speed scanning. The lasers mentioned (ruby, neodymium glass, and YAG) use a crystal as the lasing medium. Laser energy can also be produced by junction diodes; and in particular, gallium arsenide ($\lambda_0 = 890$ nm) lasers produce several hundred pulses per second at power levels which, while relatively modest on a single pulse basis, amount to several kilowatts when averaged. By integrating returns from such lasers, performance adequate for cloud detection and fog measurements has been achieved [4.14].

The laser output wavelength is important for a number of reasons, firstly because the proportion of energy scattered by atmospheric constituents depends upon wavelength (as discussed in detail in Section 4.3) and secondly because of eye-safety considerations (discussed in Subsection 4.2.3). Generally speaking, scattering by atmospheric particles and gases is much stronger at visible and ultraviolet wavelengths than in the infrared, whereas the probability of eye damage is greatly reduced in the infrared. Laser output wavelength is especially critical for differential absorption applications and for the other wavelength-dependent techniques described in Chapters 2 and 5. Such applications usually require the capability of "tuning" or varying the output wavelength; and this has been achieved in a number of ways, including thermal tuning of solid state lasers. The wide range of present and prospective tunable lasers for lidar applications has been reviewed in a recent study by WRIGHT et al. [4.15]. The greatest wavelength range is generally available by using tunable liquid dye lasers and optical parametric oscillators [4.16, 17].

There is, of course, no reason why such tunable lasers cannot be used in simple backscattering applications, and indeed dye lasers have been. Because Q-switching techniques are not applicable to dye lasers, the pulses of energy generated are normally at least 1 μs in duration (thus limiting lidar range resolution to 150 m). For certain purposes, however, this is wholly tolerable, and practical and efficient low cost dye lasers would be immediately useful for many simple lidar applications [4.13].

Partly for reasons of eye-safety and partly to achieve lower cost systems than possible with high powered pulse lasers, low powered continuous wave (cw) gas lasers are also being applied in lidar applications. In this approach, range information is derived by modulating the cw energy with a characteristic frequency variation pattern, normally a sawtooth frequency modulation (fm). Such fm-cw systems operate at very low instantaneous power level—of the order of a fraction of a watt—and are thus completely eye-safe [4.18].

Detectors

Turning to the receivers of lidar systems, we find that the crucial problem lies in effective detection of the very low intensities of returned energy. At visible and shorter wavelengths ($\lambda \approx 200$ to 700 nm), the multi-stage photomultiplier is most efficient and is used almost exclusively. Although marginally effective, it is also commonly used with neodymium systems ($\lambda = 1060$ nm). At IR wavelengths (i.e., above about $\lambda \approx 800$ nm), where the materials used in the photocathodes of photomultipliers cease to be effective, photoconductive solid-state detectors (particularly those that can be used in the avalanche mode) have been used successfully. They are far less satisfactory, (relatively speaking), however, than the photomultipliers used at shorter wavelength. The lack of effective detectors of longer wavelength energy seriously hampers the use of such energy in lidar developments to date.

Optical Components

The optical aspects of lidar systems are relatively straightforward and follow well established optical practice. The use of reflective optics is generally preferred where possible, although refractive lenses are widely used, with appropriate coatings for increased transmission of the monochromatic laser energy. The exclusion of unwanted background light energy by the use of narrow-band filters is commonly limited to some 0.5 to 1 nm on either side of the central wavelength (narrower filters sometimes lead to difficulty with solid state lasers since the wavelength of their emissions tends to vary as a result of thermal effects.) The references cited below contain detailed information on the optical components used in various actual lidar systems.

Data Acquisition, Display, and Processing

The methods of recording and presenting received signal data are an important aspect of lidar measurements and should be mentioned in somewhat more detail than in Section 4.2 above. Although the photography (particularly, Polaroid photography) of such signals, as displayed on an oscilloscope, remains in common use, more sophisticated techniques of data presentation are now commonplace and are significantly enhancing the utility of lidar data. The use of magnetic discs to record the return from each pulse, and provide a reiterative read-out for the display of lidar signals on a cathode ray tube (CRT) display was an early example of such techniques [4.11]. Magnetic disc recording is capable of providing considerable flexibility in the form of presentation on a virtually real-time basis as well as providing for more permanent storage of the data. (An example of one type of intensity-modulated "pictorial" display provided in this manner has already been shown in Fig. 4.3, and other examples are given in Section 4.6.) The magnetic discs used, however, have a bandwidth limited to approximately 4 MHz and thus cannot fully accomodate the range resolution inherent in the signals of typical pulsed ruby and neodymium lidars (which need approximately 50 MHz bandwidth to match the 20 ns pulses used.)

Very weak signals detected by photomultipliers take the form of series of photoelectron pulses rather than a continuous current. (The intensity of the signal in these cases is given by the rate of generation of such pulses.) Such signals can readily be processed directly in digital form by pulse-counting techniques, and a number of systems employing such methods have been used [e.g., 4.9, 10]. More recently the analog signals of higher intensity have been converted to digital form and processed accordingly. A very effective combination of digital processing and analog display has been achieved by UTHE and ALLEN [4.19] who, while recording data in digital form, with real time computation and printout of quantitative products from the raw data, retain direct graphical presentation of data in the form of digitally driven, intensity-modulated CRT displays. As illustrated in Fig. 4.5, the system is assembled from commercially available components, including a transient signal digitizer, minicomputer, magnetic tape storage unit, TV monitor, and teletype keyboard control. Examples of data acquired by this system are presented in Section 4.6.

System Integration

Optimum integration of the components described above into complete lidar systems is determined by the application for which the system is intended. Corresponding to the wide range of lidar applications (see Section 4.6), there is hence great variety in the types of systems that have been developed. Detailed specifications of many of these systems have been published in the literature [e.g., 4.8–11, 13, 14, 18–20, 108] and will not be repeated here. To provide concrete examples of what is possible, however, the characteristics of three quite different systems are described below and listed in Table 4.1.

Fig. 4.5. An advanced lidar digital data acquisition, processing, and display system. (From top to bottom: Dual DECTAPE, TV monitor, PDP-11 mini-computer, Biomation transient digitizer, and teletype with thumb switches on lower right)

Figure 4.6 shows a photograph of a portion of the SRI Mark IX lidar system. This system, which was designed primarily for lower atmospheric (below 15 km) studies of elastic backscattering, employs a ruby laser in the optical configuration shown in Fig. 4.1 to permit truly co-axial transmitter/receiver geometry. The integrated transmitter/receiver assembly is mounted on a pedestal that provides for automatic scanning in both elevation and azimuthal angle. Received signals can be simultaneously recorded using both the digital system described above (pictured in Fig. 4.5) and an analog magnetic disc system. (The video disc recorder is under the table in Fig. 4.6, and the data maniuplation and display electronics are mounted in the rack on the right.) The entire system is mounted in a truck, complete with power-generation facilities, to permit operation at remote sites and while in motion along highways.

Fig. 4.6. Interior view of the SRI Mark IX truck-mounted lidar system, showing pedestal mounted transmit/receive unit and analog recording and display electronics

Table 4.1. Specifications of three research lidar systems

	SRI mark IX mobile system [4.11, 19]	Univ. of West Indies upper atmospheric system [4.10]	NCAR-Wisconsin compact dye system [4.13, 20]
Transmitter			
Wavelength	694 nm	694 nm	585 nm
Laser	Ruby (1.0 × 7.6 cm rod)	Ruby (1.9 × 22.9 cm rod)	Dye (Rhodamine 6 G)
Beamwidth	<3/4 mrad	0.28 mrad	~2 mrad
Optics	Gallilean beam expander plus mirror coaxial with receiver telescope	Convex lens and parabolic mirror	(see Fig. 4.7)
Pulse energy	1.0 J	7–10 J	0.25 J
Pulse length	30 ns	10 µs	<1 µs
Pulse repetition frequency	60 min^{-1}	10 min^{-1}	60 min^{-1}
Receiver			
Optics	Newtonian reflector	Newtonian mosaic reflector (36 individual mirrors)	Fresnel lens with diaphragm and collimating lens
Area	~180 cm^2	16 m^2	~1100 cm^2
Field of view	1.0 to 5.5 mrad	0.78 mrad	~3 mrad
Predetection filter passband width	1 nm	2 nm	1 nm
Detector	RCA 7265 PMT	4 EMI 9558 PMTs	EMI 9658 PMT

Table 4.1. (continued)

	SRI mark IX mobile system [4.11, 17]	Univ. of West Indies upper atmospheric system [4.10]	NCAR-Wisconsin compact dye system [4.13, 18]
Data storage processing and display	1. Analog system: logarithmic amplification; selectable gain $(1/R^2)$ compensation; magnetic disc storage; intensity modulated or standard oscilloscope displays, cartesian or polar coordinates, direct or from disc storage 2. Digital system: transient signal digitizer; minicomputer; magnetic tape storage, TV monitor; teletype keyboard control	Pulse counting in each of three altitude ranges, each connected to a separate detector	Gain-switching amplifier; transient signal digitizer; paper tape storage
Operating mode	Ground based, fixed or mobile. Automatic scanning possible.	Ground-based, fixed	Aircraft or ground-based

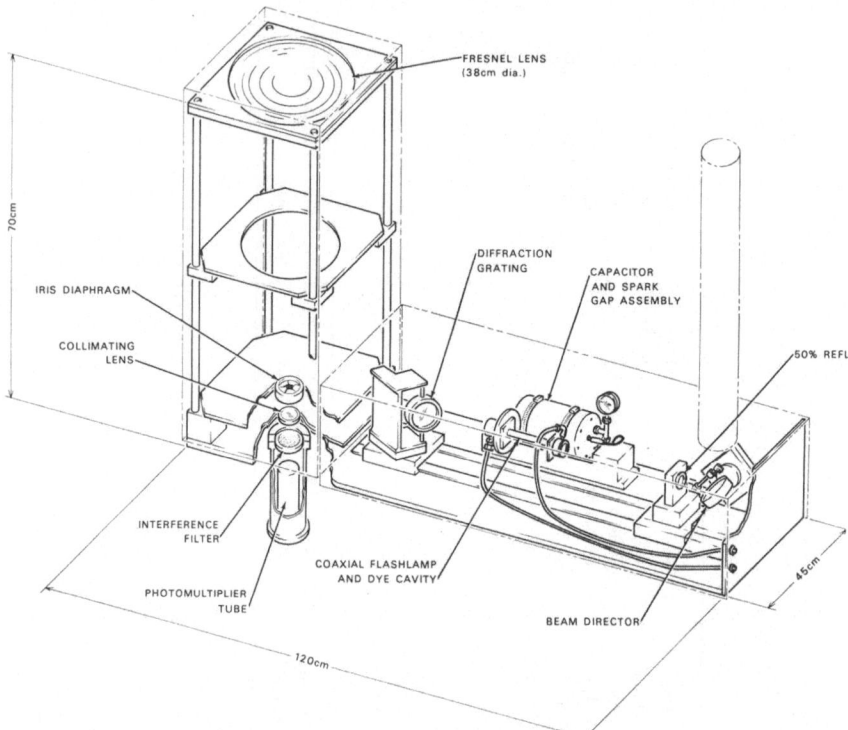

Fig. 4.7. Schematic drawing showing the major elements of the transmit/receive unit of a compact dye laser radar. Approximate dimensions of the unit are shown; the unit weighs about 60 kg [4.13]

A more detailed description of a similar system is given by Allen and Evans [4.11].

A different type of system has been described by Kent et al. [4.10]. It is designed for ground-based studies of the upper atmosphere (~ 20 to 100 km), and thus emphasizes maximum sensitivity, rather than mobility or aiming capabilities. To achieve high sensitivity it uses a very high power (> 7 J/pulse) ruby laser and a receiving mirror system of area 16 m^2 constructed as a mosaic of 36 individual mirrors, each 0.7 m in diameter. Data acquisition is by pulse-counting, using a combination of several photomultipliers to increase system dynamic range.

A third type of system, utilizing a dye laser, was constructed by Grams and Wyman [4.13], and is schematically illustrated in Fig. 4.7. Use of the dye laser and Fresnel-lens receiving optics resulted in a system that was relatively compact, light-weight, reliable, and low in cost. The system was mounted in a research aircraft, permitting rapid measurement of widespread atmospheric volumes. The data acquisition system used a gain-switching amplifier in conjunction with a commercially-available transient signal digitizer [4.20].

Examples of measurements made with all three of the above systems are presented in Section 4.6.

4.2.3 Eye Safety

Of primary concern with regard to operator and public safety in lidar studies is the risk of damage to the retina of the eye. As described in detail by WOLBARSHT [4.21], the probability of retinal damage depends on pulse energy, wavelength, and exposure time. Whereas extremely high concentrations of laser energy of any wavelength can cause lesions to skin and the corneal surface, such concentrations are not normally used in lidar applications. On the other hand, significant retinal damage could result from the typical pulse energies (~ 1 J) and durations used by many visible-wavelength lidars, even at ranges of tens of kilometers. Moreover, the danger is increased by atmospheric turbulence effects, which can cause intermittent local focusing of energy in the beam. At wavelengths above the visible, successive defocussing of laser energy by the eye reduces the risk of lesion to the sensitive retina, and at wavelengths above about 1.4 µm the material of the eye effectively and harmlessly absorbs the energy before it reaches the retina. At the present time, therefore, eye-safe lidars must either operate at low intensity in the visible or near visible range or operate at much longer wavelengths. (Erbium lasers, generating energy at 1.54 µm are currently being used for this reason.) Specific eye safety criteria for laser operation have been promulgated by the American National Standards Institute, and are described in [4.22].

4.3 Atmospheric Interactions Determining the Received Lidar Signal

The atmosphere determines the received elastically-scattered lidar signal through two factors appearing in (4.1): the elastic volume back-scattering coefficient β, and the attenuation (or extinction) coefficient α. Conversely, it follows that lidar measurements at the transmitted wavelength λ_0 can provide information on both the elastic backscattering and the attenuating characteristics of the atmosphere, provided these two effects can somehow be separated (see Section 4.5). The relation of β

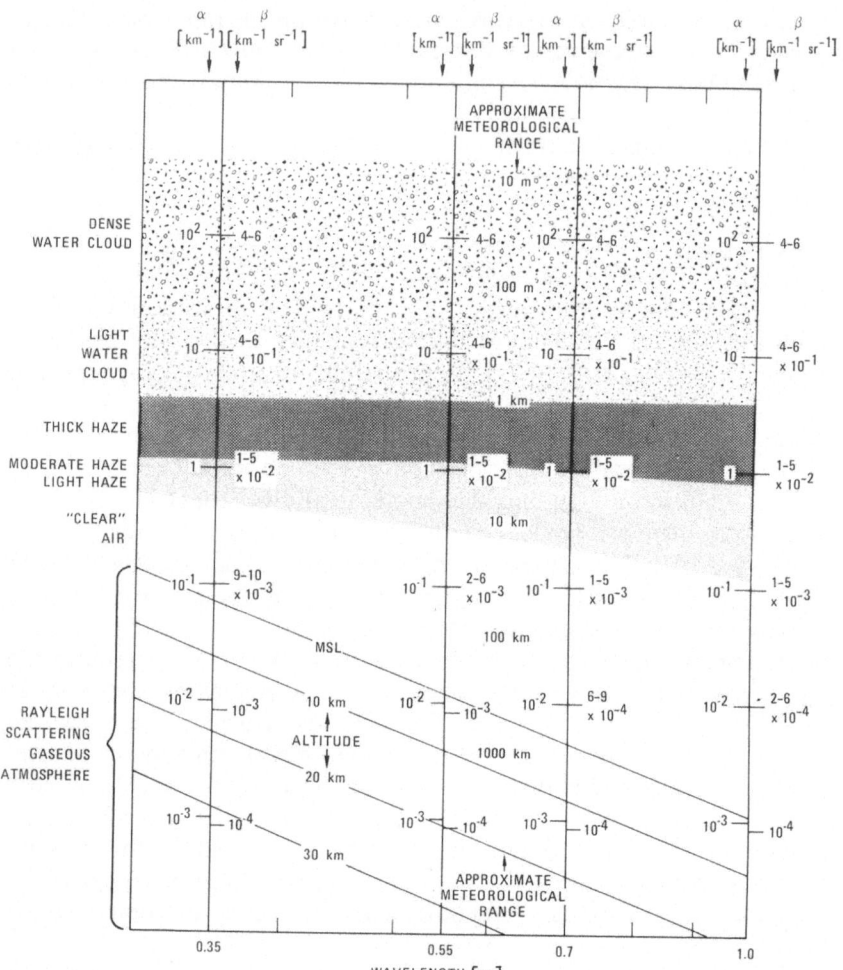

Fig. 4.8. Approximate variation of attenuation and backscattering coefficients with wavelength and atmospheric condition. For haze backscattering coefficients, larger values (for a given attenuation) apply to spherical particles of real refractive particles, partially absorptive particles, or particles with real refractive index $n \approx 1.33$ (e.g., water). [4.26, 33, 39–42, 45–48, 50–54, 67, 68, 107, 113, 115]. For cirrus clouds see [4.125]

and α to atmospheric composition and state is explored in Subsections 4.3.1 and 4.3.2 below. First, however, we present Fig. 4.8, which shows typical values of β and α for a range of wavelengths and atmospheric conditions. These values, along with the parameters of a particular lidar (cf. Table 4.1), may be substituted into the lidar equation (4.1) to obtain performance estimates under various conditions.

Strictly speaking, the functional dependence given by (4.1) of P_r on β and α is valid only under conditions where each received photon has been scattered but once, (i.e., multiple scattering does not occur). While atmospheric conditions frequently satisfy this requirement to a high degree, practical situations do occur in which multiple scattering can have important effects on the measured lidar signal. These effects in modifying the received lidar signal from the value predicted by the conventional single-scattering lidar equation (4.1) are discussed in Subsection 4.3.3.

4.3.1 Elastic Backscattering

The volume backscattering coefficient is a sum of contributions from both the gaseous and particulate phases of the atmosphere, i.e.,

$$\beta = \beta_g + \beta_p , \tag{4.5}$$

where the subscript g stands for gases and p for particles. These contributions are discussed below.

Backscattering from Gas Molecules

Since the size of gas molecules is very small compared to all laser wavelengths, elastic backscattering by atmospheric gases, except at the wavelengths of absorption lines, is described by the Rayleigh scattering approximation [4.23–25]. Thus,

$$\beta_g = \beta_R = N_g d\sigma_R(\pi)/d\Omega , \tag{4.6}$$

where N_g is the number of gas molecules per unit volume and $d\sigma_R(\pi)/d\Omega$ is the differential Rayleigh scattering cross section, at scattering angle $\theta = \pi$, per "average" gas molecule. For the mixture of atmospheric gases which occurs below about 100 km,

$$d\sigma_R(\pi)/d\Omega = 5.45[\lambda(\mu m)/0.55]^{-4} \times 10^{-28} \text{ cm}^2 \text{ sr}^{-1} . \tag{4.7}$$

This equation displays the characteristic λ^{-4} wavelength-dependence of Rayleigh scattering, and neglects an additional very slight wavelength dependence due to dispersion of the refractive index of air. (Taking this dispersion into account changes the exponent of (4.7) to -4.09 [4.26].) At sea level, where the molecular number density $N_g = 2.55 \times 10^{19} \text{ cm}^{-3}$, (4.6) and (4.7) yield

$$\beta_g = \beta_R = 1.39[\lambda(\mu m)/0.55]^{-4} \times 10^{-8} \text{ cm}^{-1} \text{ sr}^{-1} . \tag{4.8}$$

This volume backscattering coefficient is large enough to produce measurable signals in lidars of very modest capability (see also Subsections 4.2.2 and 4.5.1, and [4.27, 28]).

As will be seen in Subsection 4.3.2, the Rayleigh backscattering coefficient is a constant multiple of the Rayleigh attenuation coefficient, with

$$\beta_R = \frac{1.5}{4\pi} \alpha_R = 0.119\alpha_R , \tag{4.9}$$

so that a measurement of either β_R or α_R may immediately be converted to the other. Moreover, the simple relation of (4.6) shows that a measurement of β_R can be used to provide a measure of atmospheric density N_g. In the troposphere, where backscattering by particles is frequently comparable to, or larger than, that by gases, this type of inference is not possible; but it has proven useful in the upper atmosphere, as shown in Subsection 4.6.1.

Resonance Scattering. At the wavelengths of absorption lines of specific gases resonance scattering may occur, producing an increase in the elastic backscattering coefficient by several orders of magnitude [4.29]. In a resonant scattering process, the characteristic time between photon incidence and re-emission is several orders of magnitude longer than for the Rayleigh scattering process, and is in fact somewhat longer than the time between molecular collisions in the troposphere. As a result, collisions cause non-radiative de-excitation of resonantly absorbing molecules, so that the effective resonant backscattering cross section is too small to be observed in the troposphere. In the upper atmosphere ($\gtrsim 50$ km), however, where molecular collisions are less frequent, resonant (elastic) scattering is not "quenched". Resonant backscattering has in fact been used to observe concentrations of atomic sodium and potassium at altitudes between 80 and 110 km, as described in Subsection 4.6.1.

Backscattering from Particles

As mentioned above, at altitudes below a few kilometers, typical concentrations of airborne particles result in a particulate backscattering coefficient β_p that frequently exceeds the gaseous component β_g. (See also Subsection 4.3.2.) In this section, we will use the term "particle" to describe any liquid or solid object that does not provide its own means of locomotion in the atmosphere. Thus we exclude, for example, airplanes and insects, but include haze and smoke particles, fog and cloud droplets and crystals, and even such larger hydrometeors as raindrops and snow flakes. Atmospheric particles range in linear dimension from about 0.001 μm to 10 mm or larger, with ion clusters and the very small "Aitken" nuclei at the smaller end and hydrometeors at the larger [4.25, 30, 31].

Typically, particles with linear dimension smaller than a few microns can remain suspended in the atmosphere for long periods of time, on the order of days. These particles, together with the gas molecules of the atmosphere, thus form a colloidal suspension, or, in the language of the colloid chemist, an *aerosol*. Frequently atmospheric scientists refer to the suspended particles themselves as "aerosols", but this terminology can lead to confusion in discussions of light scattering, when one desires to separate conceptually the effects of particles and gas molecules. Hence we will in this chapter reserve the term "aerosol" to describe gas molecules and suspended particles collectively, and refer to suspended particles as "aerosol particles".

Atmospheric aerosol particles vary widely in source and composition as well as in size. Anthropogenic processes contribute such materials as smoke fly ash, mining and milling dusts, and liquid droplets produced photochemically from gaseous automotive combustion products. On the other hand, volcanic dust, meteoritic dust, spores and seeds, sea salt, and soil dust are of natural origin, as of course are most water droplets and ice particles. (See also Subsection 4.6.1.)

As in the case of gaseous molecules, light scattering by particles that are small compared to the wavelength of light (i.e., radius $a \lesssim 0.05\,\lambda$) is described by the Rayleigh scattering approximation. In the atmosphere, however, the optically significant particles are usually too large to be described as Rayleigh scatterers. The light scattering properties of these larger particles are difficult, if not impossible, to describe exactly because of the great natural variability in particle shape, composition, and size distribution.

For particles of arbitrary size and shape, and inhomogeneous composition, an analytical solution to the electromagnetic scattering problem is not available. Nevertheless, many atmospheric particles (e.g., fog and cloud drops, and droplets of photochemical smog) are essentially homogeneous spheres. In addition, useful information on the light scattering properties of other particles may often be derived by treating them as equivalent homogeneous spheres. Hence, the backscattering properties of homogeneous spherical particles provide an important starting point for a more complete description of backscattering by the general atmospheric aerosol. These properties are described below, followed by a discussion of departures from these properties caused by irregular particle shape and composition.

Backscattering by Polydispersions of Homogeneous Spheres. The volume backscattering coefficient of a suspension of homogeneous spheres of varying size (i.e., a "spherical polydispersion"), when illuminated by light of wavelength λ is given by

$$\beta_\mathrm{p}(\lambda) = \int_0^\infty \sigma_\mathrm{B}(a, \lambda, m) N'_\mathrm{p}(a) da, \tag{4.10}$$

where $\sigma_\mathrm{B}(a, \lambda, m)$ is the backscattering cross section of a particle of radius a and refractive index m, and $N'_\mathrm{p}(a) da$ is the number of particles per unit volume with radius between a and $a + da$. We introduce the symbol σ_B here for convenience; in our previous notation we have

$$\sigma_\mathrm{B} = d\sigma(\pi)/d\Omega. \tag{4.11}$$

Moreover, the particle size distribution $N'_\mathrm{p}(a)$ is related to the total number of particles N_p by

$$N_\mathrm{p} = \int_0^\infty N'_\mathrm{p}(a) da. \tag{4.12}$$

The refractive index m is in general a complex number,

$$m = n - i\kappa, \tag{4.13}$$

where the imaginary part κ is a measure of particle absorptivity.

The dependence of the backscattering cross section σ_B on a, λ, and m was studied by a number of scientists around the end of the nineteenth century, as described by Kerker [4.24]. Today, scattering by homogeneous spherical particles of arbitrary size (and especially with $a/\lambda \gtrsim 0.1$) is referred to as "Mie scattering", after Mie [4.32] who published a solution in 1908. As shown in detail in a number of excellent references [e.g., 4.23, 24, 33, 34], the backscattering cross section may be written as

$$\sigma_\mathrm{B}(a, \lambda, m) = \pi a^2 Q_\mathrm{B}(x, m), \tag{4.14}$$

where Q_B is the "backscattering efficiency", or the ratio of backscattering cross section to geometrical cross section πa^2. The backscattering efficiency depends on particle radius a and light wavelength λ only through their ratio; or, more conventionally, the "size parameter" x, defined as

$$x \equiv 2\pi a/\lambda. \tag{4.15}$$

The dependence of the backscattering efficiency Q_B on x and m is rather complicated, being expressed as an infinite series of combinations

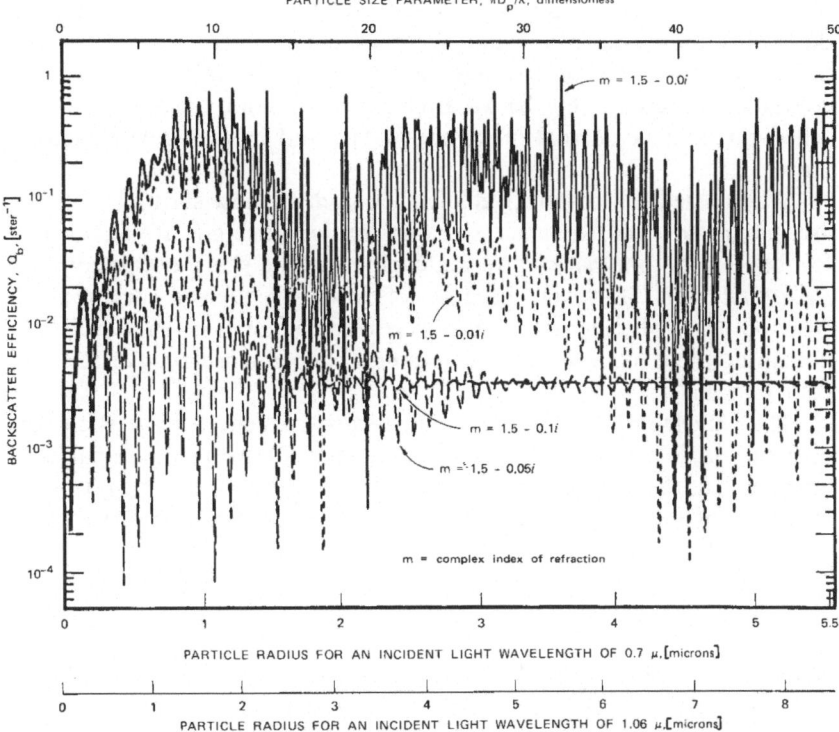

Fig. 4.9. Dependence of mie backscattering efficiency on particle size parameter x and complex refractive index m. ($x = \pi D_p / \lambda$, where D_p is particle diameter)

of Ricatti-Bessel functions of x and mx (see, for example, [4.24]). Numerical results for particular values of x and m can be obtained only by calculations that are too laborious to be done "by hand", but a number of efficient computer programs [e.g., 4.35] are now available that make Mie scattering results readily attainable. For a number of model size distributions $N'_p(a)$ [e.g., 4.33, 36–39], computational results for $\beta_p(\lambda)$ obtained by numerically evaluating the integral (4.10) have been tabulated and plotted in several published references [e.g., 4.33, 39–42].

Examples of the functional dependence of the Mie backscattering efficiency on size parameter x are plotted in Fig. 4.9 for several values of the complex refractive index m. The extremely oscillatory behavior is in marked contrast to the smooth x^4 dependence of the Rayleigh backscattering efficiency, which is obtained as the $x \ll 1$ limit of Mie scattering. The high frequency oscillations shown in Fig. 4.9 have been explained as the result of "surface waves" that travel along the interface between the sphere and the surrounding medium [4.43]. As the imaginary part of the refractive index (and therefore particle absorption) increases, the magnitude of both these oscillations and the backscattering efficiency decreases, at least for the range of κ values shown. (As particle absorption or size continue to increase, the backscatter efficiency approaches the Fresnel reflection limit at normal incidence [4.44], and increases with increasing κ.)

When Mie backscattering efficiencies, such as those shown in Fig. 4.9, are integrated over a particle size distribution [using (4.10) and (4.14)] to obtain a volume backscattering coefficient, the sharp oscillations are smoothed out. However, because of lower frequency variations in $Q_B(x, m)$, the volume backscattering coefficient for a given wavelength and integrated particle volume retains a dependence on particle size distribution.

The wavelength dependence of β_p is also influenced by particle size distribution, in a manner that in general does not have a simple analytical form, such as the λ^{-4} obtained in the case of Rayleigh scattering [cf. (4.7)]. However, if all optically significant particles have a size distribution given by the power-law or "Junge" model:

$$N'_p(a) = (\text{const})\, a^{-(A+1)}, \tag{4.16}$$

where A is a dimensionless fitting parameter[2], then it can be shown [4.23, 31] that the wavelength dependence of β_p is particularly simple, given by

$$\beta_p(\lambda) = (\text{const})'\lambda^{-\xi}, \tag{4.17}$$

where

$$\xi = A - 2. \tag{4.18}$$

For atmospheric hazes that roughly satisfy (4.16), typical values of the exponent fall in the range $2.5 \lesssim A \lesssim 4$ [4.26, 31, 39], and hence, via (4.18), $0.5 \lesssim \xi \lesssim 2$ (see also Subsect. 4.3.1). Thus, in general, the wavelength dependence of particulate backscattering by hazes is less pronounced than that of Rayleigh backscattering. Backscattering of visible light by fog and cloud droplets is virtually independent of wavelength because the large size parameters ($25 \lesssim x \lesssim 400$) fall nearly in the geometric optics region where the extinction efficiency $Q_E(x, m) \approx 2$ and the backscatter-to-extinction ratio (for polydisperse water spheres) has very weak wavelength dependence (e.g. [4.33]). The general wavelength dependence

---

[2] The symbol A is used here, rather than β (cf. Junge [4.31]) to avoid possible confusion with the backscattering coefficient.

of Rayleigh, aerosol particle, and water droplet backscattering is illustrated in an approximate manner by Fig. 4.8.

In further contrast to the Rayleigh scattering case [cf. [4.9]], the ratio of Mie backscattering to Mie extinction is also a function of particle size distribution and refractive index. Results computed for a number of "typical" aerosol size distributions and refractive indices have been published in the literature [4.33, 39–42]. They generally fall in the range

$$0.01 \, \text{sr}^{-1} \lesssim \beta_p/\alpha_p \lesssim 0.10 \, \text{sr}^{-1}, \tag{4.19}$$

and are thus smaller than the value $\beta_R/\alpha_R = 1.5/4\pi = 0.12 \, \text{sr}^{-1}$ obtained for Rayleigh scattering. This results from the increased proportion of light scattered into near-forward angles ($\theta < \pi/2$) with increasing particle size (see Chapt. 3).

A final characteristic of Mie backscattering (i.e., by spheres) which is of use in lidar investigations concerns the polarization properties of the scattering process. If the incident light is linearly polarized, the backscattered light retains the linear polarization of the incident light; moreover, the backscattering cross section is independent of the incident light polarization. These results follow as a direct consequence of the spherical symmetry of the Mie scatterers.

Particle Shape Effects. It should be recognized that elastic backscattering by real atmospheric particles of arbitrary shape generally differs in several important respects from the predictions of Mie theory. This is especially the case where particles are markedly non-spherical—e.g., elongated crystals, or, in the case of large particles, have plane surfaces (since these produce specular reflection).

Backscatter-to-Extinction Ratio. Laboratory measurements [4.45] of the angular distribution of light scattered by irregular, randomly-oriented silica plates have indicated that the backscatter-to-extinction ratio β_p/α_p could be an order of magnitude or more smaller than the values predicted by Mie theory for the same refractive index and an equivalent size distribution. This reduction in backscattering probably results from destruction of the "surface waves" discussed above, which require a smooth, spherical surface on which to propagate. Other laboratory measurements [4.46, 47] of light scattering by ice crystals have shown a similar reduction in backscattering; these measurements are in accord with the results of calculations [4.48] that approximate the ice crystals as infinitely long circular cylinders, a shape for which an analytical solution can be obtained [4.49, 50]. Atmospheric measurements of the backscatter-to-extinction ratios of haze and smoke polydispersions also indicate a reduction with respect to the Mie predictions for equivalent size distributions of homogeneous spheres with purely real refractive indices. Some of the values obtained in this manner are:

$$\beta_p/\alpha_p = \left\{ \begin{array}{l} 0.012 \pm 0.002 \\ 0.02 \\ 0.03 \\ 0.02 - 0.04 \end{array} \right\} \mathrm{sr}^{-1} \quad \begin{array}{l} [4.51] \\ [4.52] \\ [4.53] \\ [4.54] \end{array} \qquad (4.20)$$

These values are not as small as those obtained with silica plates and ice crystals, probably because of the fact that typical atmospheric haze and smoke particles are not as aspherical as those extreme examples.

Particle Absorption Effects. It should also be borne in mind that some of the reduction in backscatter of the observed atmospheric particles, as compared to the Mie values given by (4.19), is probably due to a small but non-zero absorbing (i.e., imaginary) component in their refractive index (cf. Fig. 4.9). Recent measurements [4.55–58] of the refractive index (for visible light) of "average" tropospheric, non-urban atmospheric haze and dust particles have shown the value

$$m = 1.5 - 0.007\,i \qquad (4.21)$$

to be fairly typical, with an uncertainty or variability of about 5% in n and a factor of about two in κ. Urban aerosol particles, which may contain significant amounts of carbon, tend to have larger values of κ [4.59–62]. In humid atmospheres, where particles may accumulate liquid water coatings, particle scattering properties may approach those of pure water, which has

$$m = 1.33 \qquad (4.22)$$

for visible wavelengths. In the infrared region, absorption bands of liquid water and numerous other substances can cause κ to increase significantly for a wide range of atmospheric particles.

Polarization. The polarization properties of aspherical particles also differ markedly from those of spheres. In general, the polarization of light backscattered by aspherical particles can differ from that of the incident light. Moreover, if the aspherical particles are aligned in some fashion, the size of the volume backscatter coefficient β_p can depend on the directions of polarization and incidence with respect to the alignment direction or plane. These effects are especially evident in light scattering by ice crystals, as demonstrated by both laboratory experiments

[4.63–65] and calculations [4.48–50, 65]. As discussed in Subsection 4.6.1, these effects are of considerable utility in the remote determination of cloud droplet phase.

Effects of Variable Composition. A real atmospheric polydispersion may consist of a mixture of homogeneous particles, with composition differing from particle to particle, or the particles themselves may be inhomogeneous. The effects of the first type of polydispersion inhomogeneity on light scattering have been discussed by BERGSTROM [4.66], who emphasized that the light scattering properties of such a mixture cannot, in general, be predicted from a single "average" refractive index. The effects of the second type of inhomogeneity are extremely difficult to treat theoretically, except for the case of the layered sphere [4.24]. Other (more typical) intra-particle inhomogeneities are best handled through empirical approximations.

4.3.2 Attenuation

The atmosphere attenuates a laser pulse through scattering and absorption by both gases and particles. Thus the extinction (attenuation) coefficient can be written as the sum of four terms

$$\alpha = \alpha_{g,s} + \alpha_{p,s} + \alpha_{g,a} + \alpha_{p,a}, \tag{4.23}$$

where the subscripts s, a, g, and p stand, respectively, for scattering, absorption, gases, and particles. These four terms may be grouped in pairs as

$$\alpha_g = \alpha_{g,s} + \alpha_{g,a}, \tag{4.24}$$

$$\alpha_p = \alpha_{p,s} + \alpha_{p,a}. \tag{4.25}$$

The gaseous and particulate extinction coefficients α_g and α_p are described in more detail in the sections below. A more extensive treatment is given in Chapter 3.

Attenuation Due to Gas Molecules

The attenuation coefficient $\alpha_{g,s}$ due to scattering by gas molecules has both an elastic and an inelastic component, but the elastic component is always dominant (by several orders of magnitude at most wavelengths), so the inelastic component is typically neglected. Since, as mentioned earlier, the size of gas molecules is small compared to the wavelengths of laser radiation, elastic scattering by gases is described by the Rayleigh scattering cross section σ_R, i.e.,

$$\alpha_{g,s} = \alpha_R \equiv N_g \sigma_R, \tag{4.26}$$

where N_g is the number density of gas molecules in the atmospheric volume of interest. For light of wavelength λ, the atmospheric Rayleigh scattering cross section at altitudes below about 100 km is given by

$$\sigma_R = 4.56(\lambda[\mu m]/0.55)^{-4} \times 10^{-27} \text{ cm}^2. \tag{4.27}$$

[see also (4.7) and accompanying discussion.] Extensive tabulations of α_R based on several model atmospheres were given by ELTERMAN [4.26] and McCLATCHEY et al. [4.67].

The gaseous absorption coefficient $\alpha_{g,a}$ is a very strong function of wavelength, becoming the dominant component of the extinction coefficient α in the vicinity of absorption lines and bands of the various atmospheric gases. These strong absorption features occur most frequently in the ultraviolet ($\lambda < 300$ nm) and infrared ($\lambda \gtrsim 900$ nm) regions of the spectrum, where they can severely limit the effective range of laser remote sensing measurements. On the other hand, these spectral features may be exploited in the differential absorption technique to provide remote measurements of specific gaseous constituents, as described in Section 4.4. More detailed descriptions of atmospheric gaseous absorption are given in Chapters 3 and 6.

Attenuation Due to Particles

At mid-visible wavelengths ($\lambda \approx 550$ nm), where gaseous absorption is small compared to scattering, (4.26) and (4.27) give a sea-level ($N_g = 2.55 \times 10^{19} \text{ cm}^{-3}$) Rayleigh scattering extinction coefficient of

$$\alpha_{g,s} = \alpha_R = 0.0116 \text{ km}^{-1}. \tag{4.28}$$

One measurement of atmospheric visibility, the "meteorological range" V_M, is given by the Koschmieder relation [4.68, 69] (see also Sect. 4.5)

$$V_M \equiv 3.91/\alpha. \tag{4.29}$$

Thus it can be seen that for a purely Rayleigh scattering atmosphere ($\alpha = \alpha_R$), the sea-level visibility would exceed 250 km! Since, even under exceptionally clear conditions the sea-level visibility rarely exceeds 50 km, it is evident that the purely gaseous scattering component of the atmosphere normally makes a very small contribution to the actual attenuation of mid-visible light. The reduction of visibility below the Rayleigh scattering value is due to the presence in the atmosphere of various solid and liquid particles, as described in Subsection 4.3.1.

The effects of particle shape on the volume extinction coefficient α_p are not as great as those on the backscattering coefficient β_p. Hence the extinction properties of atmospheric particles are quite frequently and reliably described in terms of equivalent spheres. In this formulation, the volume extinction coefficient for light of wavelength λ is given by

$$\alpha_p(\lambda) = \int \sigma_E(a, \lambda, m) N'_p(a) da,\tag{4.30}$$

where $\sigma_E(a, \lambda, m)$ is the extinction cross section for a particle of radius a and refractive index m, and $N'_p(a)$ is again the number density of particles per unit radius interval. We use the symbol σ_E to parallel the development of backscattering in Subsection 4.3.1. In our previous notation,

$$\sigma_E = \sigma_s + \sigma_a = \sigma_{p,s} + \sigma_{p,a},\tag{4.31}$$

which emphasizes that the extinction cross section includes contributions from both scattering and absorption. The absorption cross section vanishes for particles with a purely real refractive index.

For homogeneous spherical particles the dependence of the extinction cross section on a, λ, and m is given by Mie's scattering theory. Again, it can be written in terms of scattering and absorbing efficiencies, which are functions only of the particle size parameter x, see (4.15), and m

$$\sigma_s(a, \lambda, m) = \pi a^2 Q_s(x, m),\tag{4.32}$$

$$\sigma_a(a, \lambda, m) = \pi a^2 Q_a(x, m).\tag{4.33}$$

The infinite-series expressions for Q_s, Q_a, and Q_E $(=Q_s+Q_a)$ can be evaluated by use of the computer programs mentioned in Subsection 4.3.1, and numerous results are tabulated in the references cited therein.

The wavelength dependence of the particulate extinction coefficient α_p is a function of the particle size distribution and complex refractive index used in evaluating the integral (4.30). In general, this dependence does not have a simple analytical form. However, in analogy to the case for backscattering, if all the optically significant particles are distributed according to the power-law or "Junge" model of (4.16), then

$$\alpha_p(\lambda) = (const)'' \lambda^{-\xi}\tag{4.34}$$

and

$$\xi = \Lambda - 2, \tag{4.35}$$

where again $-(\Lambda + 1)$ is the exponent of the particle size distribution. As a note of caution in applying (4.34) and (4.35), we point out that the optically significant size range is a function of wavelength, so that, for example, (4.34) may not be applicable to ultraviolet light when it is to visible.

When the particle size distribution differs from a power law or is unknown, empirical expressions may be used to predict the wavelength dependence of particulate extinction. A frequently-quoted empirical relation [4.70] for visually clear air is

$$\alpha_p \approx \alpha_{p,s} = (3.91/V_M)\,(0.55/\lambda[\mu m])^q, \tag{4.36}$$

where

$$q = 0.585\,(V_M[km])^{1/3} \quad \text{for} \quad V_M \leqq 6\,km,$$

$$q = 1.3 \quad \text{for "average seeing conditions"},$$

and V_M is the atmospheric visibility. [We note the similarity of (4.36) to (4.29) when $\lambda = 0.55\,\mu m$ and the visibility V_M represents the meteorological range]. The limitations in applying (4.36) have recently been discussed by Woodman [4.71], who cites experimentally-observed values of q falling in the range 0.12 to 2.3. In addition, a number of investigators [4.36, 72–74] have actually observed negative values for q over the visible spectrum. Woodman concluded that (4.36) can provide useful estimates in the visible, but that in the infrared ($\lambda = 2$ to $10\,\mu m$) it differs significantly from the models of Elterman [4.26] and McClatchy [4.67], and, on the average, probably underestimates actual atmospheric particulate attenuation.

Since fog and cloud droplets are typically larger than one micron in diameter, their scattering cross section for UV, visible, and near-IR light is close to the geometric optics cross section. As a result, scattering of such light by fogs and clouds tends to be independent of wavelength.

4.3.3 Multiple Scattering

The two-way attenuation term $\exp\left(-2\int[\alpha_s + \alpha_a]dr\right)$ of the single-scattering lidar equation (4.1) is written under the assumption that all scattered photons except those scattered in the direct backward ($\theta = \pi$) direction are permanently removed from the transmitted and received lidar beams. In actual fact, however, a fraction of the scattered photons travel in near-forward ($\theta \approx 0$) directions and so never leave the lidar beam; in addition, some photons that are scattered out of the lidar beam may later be scattered back in. Thus the actually received lidar signal P_r includes photons that have been scattered more than once. As a result it somewhat exceeds the value given by (4.1); equivalently, the effective value of the scattering extinction coefficient α_s is somewhat smaller than the single-scattering values discussed in Subsection 4.3.2.

In addition to increasing the received power, multiple scattering can also affect its polarization. The paths of received multiply-scattered photons can lie on many different planes, and thus can transfer the plane of the electric vector. As a result a multiply-scattered received lidar signal is partially depolarized, even when all of the scatterers are spherical.

The effects of multiple scattering on the received lidar signal depend on the atmospheric scattering and absorption coefficients α_s and α_a, the size of scattering particles, the time after pulse transmission (i.e., the "effective range" R), and the widths of the transmitted and received lidar beams. In practical application, the effects of multiple scattering may be neglected in visually clear (i.e., non-cloudy) atmospheres where the one-way optical thickness $\int \alpha dR$ is less than about 1, provided that lidar beam widths are restricted to (typical) values of several milliradians or less. (See Table 4.1.) The effects tend to be much more important in fog, cloud, and rain, both because of the much larger probability of scattering (larger α_s) and the larger scattering particles, which focus a much larger fraction of scattered photons in near-forward directions.

The theoretical treatment of multiple scattering in pulsed lidar applications is complex. However, by taking a numerical simulation, or Monte Carlo, approach, predictions of received power for several cloud models and sets of lidar beam widths have been obtained by PLASS and KATTAWAR [4.75], and GOLUBITSKIY et al. [4.76]. LIOU and SCHOTLAND [4.77], and ELORANTA [4.78] have both developed analytical approaches to the somewhat restricted problem of double scattering. Unfortunately, when their analytical equations are applied to the same cloud models, their predictions for doubly-scattered power and depolarization differ by factors of about 10 and 4, respectively. ELORANTA's predictions for both quantities exceed those of LIOU and SCHOTLAND. When compared to the Monte Carlo results of PLASS and KATTAWAR, ELORANTA's results agree at small penetration depths of the cloud, where higher orders of scattering are not yet important.

As shown in Subsection 4.6.1, multiple scattering effects in fog, cloud, and rain are readily observable experimentally, and tend to exceed the theoretical estimates. Depolarizations of as large as 40% have been attributed to multiple scattering effects.

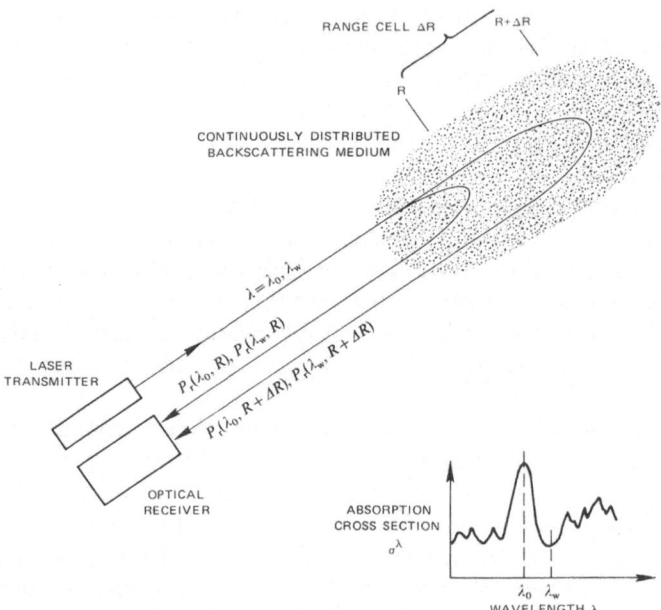

Fig. 4.10. The differential absorption lidar technique

4.4 The Differential Absorption Lidar Technique

The dependence of the received lidar signal [cf. (4.1)] on the two-way attenuation term, $\exp(-2\int \alpha dr)$, can be exploited to provide a range-resolved measurement of specific gaseous constituents at appreciable ranges and with potentially high sensitivity. As illustrated in Fig. 4.10, the method relies on the very strong wavelength dependence of the gaseous absorption coefficient $\alpha_{g,a}$ in the vicinity of a resonance absorption line, together with the readily measurable elastic backscattering return from atmospheric particles and gases. Using a tunable laser, near-simultaneous measurements are made over the same path at wavelengths ($\lambda = \lambda_0, \lambda_w$) within and outside of an appropriate absorption line of the gas being measured. The resulting signals are then compared to isolate absorption by the gas of interest from other extinction components in the atmosphere. The method was pioneered by SCHOTLAND [4.80–83], who called it Differential Absorption of Scattered Energy, or DASE. Various other investigators have used the terminology Differential Absorption Lidar (DIAL) [4.84, 85], Differential Absorption and Scattering (DAS) [4.86, 87], or simply differential absorption [4.88–91].

The technique may also be used to obtain path-integrated gaseous species concentrations by measuring the signal from a fixed remote target, rather than the range-resolved atmospheric return. In this type of measurement a significant reduction in the required transmitted power may be achieved. If the remote target is a topographical feature (e.g., a hill, a tower, a lake surface), the convenience of a single-ended (monostatic) experimental system is retained. Still higher sensitivity may be obtained through the use of specially-designed remote retroreflectors; this two-ended (bistatic) type of long-path measurement is described in Chapter 6.

4.4.1 Principles of Operation

To illustrate the principle of the differential absorption technique, we begin by rewriting the atmospheric extinction coefficient at wavelength λ as the sum of two terms

$$\alpha(\lambda) = \alpha_A(\lambda) + \alpha_G(\lambda) \tag{4.37}$$

where α_G is the extinction coefficient due to absorption by the gas G of interest, and α_A is the extinction coefficient due to scattering and absorption by all other constituents (particulate and gaseous) of the atmosphere [cf. (4.23)–(4.25)]. For a gas concentration[3] N and absorption cross section σ we have

$$\alpha_G(\lambda) = N\sigma(\lambda) . \tag{4.38}$$

An expression for $N(R)$ averaged over the range cell ΔR can be obtained by substituting (4.37) and (4.38) into (4.1), and then forming the difference of the logarithms of $P_r(\lambda, R)$ and $P_r(\lambda, R+\Delta R)$ for both a wavelength λ_0 within an absorption line of gas G and a wavelength λ_w on the "wing" of the line (i.e., outside of the main absorption peak)[4]. In

---

[3] N is a generalized symbol for concentration. In actual computations it may have dimensions of either number per unit volume $[\ell^{-3}]$ or mass per unit volume $[m\ell^{-3}]$, or it may be a dimensionless mixing ratio, so long as σ has the appropriate dimensions to give α_G dimensions of inverse length $[\ell^{-1}]$. If N has dimension $[\ell^{-3}]$, σ is an absorption cross section (dimension $[\ell^2]$); if N has dimension $[m\ell^{-3}]$, σ is a mass absorption coefficient (dimension $[m^{-1}\ell^2]$); if N is a dimensionless mixing ratio, σ has dimension $[\ell^{-1}]$.

[4] For a number of reasons (e.g., avoidance of interfering absorption by other gases, and optimum range achievement), the wavelengths λ_0 and λ_w may not fall at the idealized "peak" and "valley" positions schematized in Fig. 4.10.

this manner (following [4.83]), one obtains

$$\ln P_r(\lambda_0, R) - \ln P_r(\lambda_0, R + \Delta R) = \ln \beta(\lambda_0, R) - \ln \beta(\lambda_0, R + \Delta R)$$
$$+ 2\Delta R[\bar{\alpha}_A(\lambda_0) + \bar{N}\sigma(\lambda_0) + R^{-1}], \tag{4.39}$$

$$\ln P_r(\lambda_w, R) - \ln P_r(\lambda_w, R + \Delta R) = \ln \beta(\lambda_w, R) - \ln \beta(\lambda_w, R + \Delta R)$$
$$+ 2\Delta R[\bar{\alpha}_A(\lambda_w) + \bar{N}\sigma(\lambda_w) + R^{-1}], \tag{4.40}$$

which can be solved to yield

$$\bar{N}(R) = \frac{1}{2\sigma\Delta R} \left[\ln \left(\frac{P_r(\lambda_0, R)}{P_r(\lambda_0, R + \Delta R)} \right) - \ln \left(\frac{P_r(\lambda_w, R)}{P_r(\lambda_w, R + \Delta R)} \right) + B' + T' \right], \tag{4.41}$$

where

$$T' = -2[\bar{\alpha}_A(\lambda_0, R) - \bar{\alpha}_A(\lambda_w, R)]\Delta R, \tag{4.42}$$

$$B' = \ln [\beta(\lambda_0, R + \Delta R)/\beta(\lambda_0, R)] - \ln [\beta(\lambda_w, R + \Delta R)/\beta(\lambda_w, R)], \tag{4.43}$$

$$\sigma = \sigma(\lambda_0) - \sigma(\lambda_w), \tag{4.44}$$

and a bar over a symbol indicates the corresponding quantity averaged over the range ΔR.

Thus $\bar{N}(R)$ can be determined from a two-wavelength lidar measurement using (4.41), provided the differential transmission and backscattering terms T and B, and the differential absorption cross section σ, are known. In principle, σ can be determined in advance from a laboratory measurement or published data. Moreover, if the spectral dependence of α_A and β is weak in the wavelength region $\lambda = \lambda_0$ to λ_w, and if the line and wing measurements are made nearly simultaneously (to avoid significant temporal changes in α_A and β), B and T can be taken to be effectively zero. The validity of these and other assumptions is discussed in Subsection 4.4.3 below. First, however, we investigate the potential measurement sensitivity implied by (4.41).

4.4.2 Measurement Sensitivity

It follows from (4.41), assuming $B' = T' = 0$, that the minimum detectable concentration N_{min} of gas G in a differential absorption experiment is given by

$$N_{min} = \frac{\Delta^* \ln}{2\sigma\Delta R}, \tag{4.45}$$

where $\Delta*\ln$ is the minimum value of the logarithmic difference in (4.41) that can be accurately measured. For practical purposes of estimation, (4.45) may be rewritten in three forms, each corresponding to a different set of dimensions for describing N and σ. These are

$$N_{\min}[\text{cm}^{-3}] = 5 \times 10^{-3} \frac{\Delta*\ln}{\sigma[\text{cm}^2]\Delta R[\text{m}]}, \tag{4.46}$$

$$N_{\min}[\text{g cm}^{-3}] = 5 \times 10^{-3} \frac{\Delta*\ln}{\sigma[\text{cm}^2 \text{ g}^{-1}]\Delta R[\text{m}]}, \tag{4.47}$$

$$N_{\min}[\text{ppm atm}] = \frac{5 \times \Delta*\ln}{\sigma[\text{cm}^{-1} \text{ atm}^{-1}]\Delta R[\text{km}]}. \tag{4.48}$$

When σ is in the form shown in (4.48), it is commonly called the differential absorption coefficient, k, which may be obtained from $\sigma[\text{cm}^2]$ using

$$k[\text{cm}^{-1} \text{ atm}^{-1}] = N_0[\text{cm}^{-3} \text{ atm}^{-1}]\sigma[\text{cm}^2], \tag{4.49}$$

where N_0, the air molecular number density per atmosphere [atm], is equal to $2.55 \times 10^{19} \text{ cm}^{-3} \text{ atm}^{-1}$. The units of the minimum detectable concentration are parts per million-atmospheres [ppm atm].

Given typical detection and digitization equipment, a reasonable value for $\Delta*\ln$ is about 0.02 [4.84, 87–90]. This value, when combined with the differential absorption coefficients of several atmospheric constituents and pollutants, results in the minimum detectable concentrations shown in Table 4.2. The values for $N_{\min}$ given in Table 4.2 are number mixing ratios in units of ppm-atm, and assume a range increment ΔR of 100 m. Equivalently, all of these numbers may be multiplied by 100 and expressed in ppm-atm-meters, which, given the form of (4.45), is a more fundamental measure of sensitivity in a differential absorption experiment.

The form of (4.45) for the minimum detectable concentration is in marked contrast to corresponding expressions for minimum detectable concentration in experiments that measure the concentration of a material through its contribution to the backscattering coefficient β. In such experiments, as for example those employing Raman scattering or fluorescence (Chapt. 2 and 5), resonance scattering, or non-resonant elastic scattering to measure the concentration of backscattering particles and gases, the minimum detectable concentration is an increasing function of the range R between the scattering constituent and the lidar. The combined effects of receiver solid angle reduction and beam attenuation cause the minimum detectable concentration in such experiments

Table 4.2. Minimum detectable concentrations of various atmospheric gases using the differential absorption lidar technique; see (4.48, 49) for definitions

Species	Wavelength λ [μm]	Differential absorption coefficient k [cm^{-1} atm^{-1}]	Reference	Minimum detectable concentration per 100-m range cell[a] N_{min} [ppm atm]	Typical atmospheric concentrations	
					Urban [ppm]	Rural [ppm]
H$_2$O vapor	0.6944	0.00035	[4.83]	3000	4000–30000	4000–30000
CO	2.3	0.4	[4.92]	2.5	1–100 (15 typical)	0.3–1.0
CO	4.74	10	[4.93]	0.1		
NO$_2$	0.45	7.2	[4.94, 102]	0.4	0.01–0.3	0.001–0.005
SO$_2$	0.30	26	[4.95, 103]	0.04	0.01–0.3	0.01
	7.4	16	[4.96]	0.06		
	8.88	1	[4.93]	1		
C$_6$H$_6$	0.25	33	[4.97]	0.03		
O$_3$	0.29	12	[4.98]	0.08	0.04–0.5	0.03–0.06
O$_3$	9.48	10.8	[4.96]	0.1		
NH$_3$	10.7	30	[4.96]	0.03		0.005–0.02
NO	0.226	7	[4.99]	0.14	0.01–0.3	0.001–0.005
NO	5.2	4	[4.100]	0.25		
NO	5.31	10	[4.93]	0.1		
NO	5.5	1.2	[4.96]	0.8		
CH$_4$	3.39	15	[4.101]	0.07		

[a] Computed from (4.45) assuming Δ^*ln $= 0.02$ and $\Delta R = 0.1$ km.

to increase at least as rapidly as R^2, cf. (4.1). In a differential absorption measurement, on the other hand, minimum detectable concentration is independent of range R, so long as the backscattered signal from all atmospheric particles and gases at both on-line and off-line wavelengths is large enough to provide a measurable difference term, cf. (4.41), which exceeds Δ^*ln.[5] The range over which this condition holds true depends on the transmitter and receiver characteristics of the lidar, as well as the transmitted wavelength (UV, visible, or IR) and the atmospheric particulate content. However, even for particle-free air at sea level, several dye lidar systems (see, e.g., Table 4.1) exist for which the received signals at visible and UV wavelengths meet this criterion out to ranges of several km or more. At longer wavelengths (where many useful absorption lines occur), the decreased atmospheric backscattering coefficient β greatly reduces the possible range for spatially resolved measurements,

---

[5] At longer ranges, a measurement may still be made, but the decrease in signal-to-noise ratio (S/N) causes Δ^* ln to increase with range. For more detailed discussion of the resultant range dependence of N_{min}, see [4.84, 88–90].

but use of topographical targets permits path-integrated measurements in these cases.

The minimum detectable concentrations listed in Table 4.2 indicate that the differential absorption technique can be used to measure the concentration of several atmospheric gases, at ambient urban levels, in regions which are separated by several km from the sensing lidar. Such a capability makes differential absorption potentially the most sensitive of the single-ended schemes for laser detection of trace gases at tropospheric pressures (cf. Chapts. 2 and 5). A more detailed comparison of the detection sensitivities of the various laser methods is beyond the scope of this chapter. However, several such comparisons [4.87–90] have recently appeared in the literature, and have concluded that range-resolved differential absorption measurements of such gases as CO and NO_2 are 10^4 to 10^5 times as sensitive as Raman scattering measurements, and approximately 10 times as sensitive as fluorescence backscattering measurements. This increase in sensitivity occurs as a combined result of the very large absorption (relative to Raman scattering) cross sections of common gases and the readily-observable elastic backscattering return from atmospheric particles and gases (or topographical targets), both of which are exploited in the differential absorption method.

4.4.3 Potential Errors

The discussion of sensitivity presented in Subsection 4.4.2 is idealized because of the occurrence in any actual measurement of errors and uncertainties, not all of which are made explicit in (4.41). Recently, these potential errors have been investigated in some detail by SCHOTLAND [4.83] and WRIGHT [4.85]. In this section we present a synopsis of their findings.

SCHOTLAND evaluated uncertainties in the measured concentration arising from uncertainties in the differential backscatter and transmission terms B' and T', the absorption coefficient σ, and the measured power P_r. He first showed that for a typical gaseous absorption line with a frequency difference $\Delta v \lesssim 0.3 \text{ cm}^{-1}$ between line center and wing, the wavelength dependence of atmospheric backscattering β and extinction α (due to gaseous scattering and particulate scattering and absorption) is weak enough (less rapid than λ^{-4}) that in exactly simultaneous measurements, the differences $\beta(\lambda_0) - \beta(\lambda_w)$ and $\alpha(\lambda_0) - \alpha(\lambda_w)$ are indeed negligible (relative differences $\leq 10^{-4}$), provided absorption by other gases is essentially constant for $\lambda_0 \leq \lambda \leq \lambda_w$ (see below). More important than the spectral variability of β and α is their temporal variability due to natural fluctuations in atmospheric particulate content. SCHOTLAND showed that if a relative accuracy in N of 10% or better is desired (for a single $\lambda_0 - \lambda_w$ pair of shots) the time interval between line and wing shots should be kept on the order of a millisecond or less. If this condition is satisfied, SCHOTLAND concluded that the terms B' and T' in (4.41) may indeed be taken to be zero.

SCHOTLAND obtained expressions for the uncertainty in N due to uncertainties in σ and P_r, and applied them to the ground-based measurement of a vertical water vapor profile using a ruby lidar tuned on and off the absorption line at $\lambda = 694.38$ nm (see also Subsect. 4.6.2). He showed that in this case uncertainties in σ due to uncertainties in the temperature and pressure of the atmospheric volume being probed have negligible effect on the accuracy of N (relative error $\sim 1\%$ for a single measurement pair). However, uncertainties in σ due to uncertainty in the ruby laser output wavelength are a significant source of error, ranging from 3% at sea level to 6% at an altitude of 3 km (where the absorption line is narrower). In fact, at altitudes of 2 km or below, this error is dominant in SCHOTLAND's simulation. At an latitude of 3 km, on the other hand, the received on-line lidar signal $P_r(\lambda_0)$ for the assumed lidar system becomes small enough in comparison to sky background light that power-measurement errors begin to dominate, producing a relative error in N of 9% even when the variance is deduced by averaging 25 measurement pairs.

WRIGHT [4.85] has presented an error analysis that includes the terms considered by SCHOTLAND, and also considers the effects of interfering absorption by other gases and signal digitization errors. In addition, WRIGHT's analysis includes explicit terms for several different types of power measurement and signal detection errors. This general analysis shows that uncertainties arising from a given source, which may be insignificant in one particular type of measurement, may indeed become the dominant source of error in another type of measurement. Thus, for example, the error analysis for an experiment using visible wavelengths may be completely different from that for a measurement made in the infrared. Interference effects, which occur when a material other than the gas of interest has a significant wavelength dependence in the region $\lambda = \lambda_0$ to λ_w, tend to occur more frequently in the infrared and ultraviolet than in the visible. In many practical cases interference effects may indeed be significant [4.84]; they may be circumvented by using more than one absorption line of the gas of interest (i.e., more than two wavelengths), but this understandably complicates the analysis and requires more elaborate equipment. Finally, digitization and other signal processing errors are a fundamental aspect of any practical measurement, and may in fact be the dominant source of error by setting a lower limit to the attainable $\Delta^*\ln$ in (4.45).

Experimental applications of the differential absorption technique to the actual measurement of atmospheric constituents are reviewed in Subsection 4.6.2.

4.5 Interpretation of Single-Wavelength Lidar Measurements in Terms of Atmospheric Significance

In the preceding section we have seen how a particular type of two-wavelength lidar measurement may be processed to provide a result that is of direct atmospheric significance—i.e., the range-resolved concentration of a specific gaseous constituent. This example of the processing and interpreting of lidar signals differs in two major respects from the general procedure of extracting atmospheric information from single-wavelength measurements of aerosol (i.e., gas plus particle) backscattering. In the first respect, single-wavelength back-scattering measurements can often be directly interpreted in a semi-quantitative fashion without the necessity of performing any digital or appreciable analog signal manipulation (e.g., obtaining signal ratios and logarithmic differences, as required in differential absorption studies). On the other hand, the quantitative extraction of optical and physical aerosol densities from single-wavelength lidar measurements is in general quite difficult and rarely leads to the unambiguous results for physical concentration possible in differential absorption measurements. The reasons for and consequences of these differences are discussed in the following sections.

4.5.1 Semi-Quantitative Mapping and Monitoring of Particulate Material

The range and direction corresponding to a particular feature in a received lidar signal can be determined with high precision using time-of-flight considerations (Subsect. 4.2.1). Thus, for example, the positions of atmospheric regions producing high signal returns (bright areas) in Fig. 4.3 can be quantitatively located, simply by inspection of the signal display. On the other hand, as can be seen from the lidar equation (4.1), these bright areas result from the combined effects of two atmospheric processes: backscattering and attenuation. Moreover, as shown in Section 4.3, each of these processes depends in a typically complicated way on the number, shape, composition, and size distribution of scatterers present. Unambiguous determination of any one of these quantities from the lidar signal, with the other quantities unknown, is hence not possible. Thus, whereas quantitative locations of the haze layers shown in Fig. 4.3 are readily derivable from the lidar display, their nature (e.g., particle number or mass density) is not. (The manipulations required to derive this information, and the conditions when this is possible, are described in Subsect. 4.5.2.)

Nevertheless, the semi-quantitative information available directly from lidar displays such as Fig. 4.3 is of considerable value in a number of atmospheric study areas. Used in this fashion, lidar can provide better understanding of the natural atmosphere, as well as map and monitor the intrusion of pollutants.

In meteorology the measurement of cloud base, cloud layer thickness and height of the tops of tenuous clouds is a good example of such direct use. Again, the observation of haze layers provides ready grounds for inferring the thermal stratification in many circumstances. In connection with air pollution control, lidar observations of this type, by determining the depth of the mixed layer, can help in estimating the dilution volumes available beneath elevated inversions (see Subsect. 4.6.1).

Successive observations can reveal the displacement or changes in structure of natural or man-made atmospheric features; for example in showing the motion of clouds of particles or smoke plumes, or in revealing the changes in convective conditions. In certain cases the shapes of structural features are indicative of dynamic influences. Wave motion in cloud layers or between air layers of different turbidity is readily recognizable. Again, convective plumes may be apparent in otherwise stratified haze, indicating the thermal effects of surface heating. Successive vertical and horizontal cross sections through clouds of particles are of great value in research studies of the transport and diffusion of pollutant emissions, especially of smoke stack plumes.

4.5.2 Solution of the Lidar Equation

The first step in deriving quantitative information on the makeup of atmospheric regions from lidar signals is to solve the lidar equation (4.1) for one or both of the optical factors, backscattering β and attenuation α. In certain applications, such as remotely monitoring the attenuation coefficient of hazes or airport fogs, this optical information is of direct physical significance. In many other applications, however, the further step of converting optical quantities to such physical quantities as particle number or mass concentration is also required. This section primarily describes methods of solving the lidar equation (4.1) for β and α, and Subsection 4.5.3 deals with the feasibility and accuracy of optical-to-physical conversions.

Under certain atmospheric conditions the two-way opitcal thickness encountered by the lidar pulse is sufficiently small or well-defined that the transmission $\exp(-2\int_0^R \alpha dr)$ in (4.1) can be approximated as unity or a particular function based on an atmospheric model. This procedure effectively reduces the lidar equation to having one unknown, and has proven fruitful, e.g., in lidar measurements of the stratospheric aerosol (Subsect. 4.6.1). In the general case, however, where both β and α are effectively unknown, a relationship between them must be assumed (or derived from auxiliary measurements) before solution of the lidar equation is possible. In addition, because of the general difficulty of calibrating lidar systems, some type of externally derived boundary value must usually be employed.

Based on the above considerations, a number of solution techniques to evaluate the single-scattering lidar equation (4.1) for quantitative purposes have been described in the literature [4.104, 105]. The approach followed by JOHNSON and UTHE [4.106], and DAVIS [4.107] is typical.

The received signal $P_r(R)$, in logarithmic form, is range-normalized and corrected for instrumentation transfer anomalies. System constants (e.g., receiver area) and pulse-to-pulse variations in lidar performance (e.g., P_0) are eliminated by dividing the profile by its (range-normalized) value at a reference range R_0, typically a point where β is measured or can be assumed to be constant. The resulting "S-values", defined as

$$S(R) \equiv 10 \log \frac{P_r(R)R^2}{P_r(R_0)R_0^2} \tag{4.50}$$

evaluate, in relative terms, only the atmospheric dependent factors of the lidar equation, since, via (4.1),

$$S(R) = 10 \log \frac{\beta(R)T^2(R)}{\beta(R_0)T^2(R_0)}, \tag{4.51}$$

where $T = \exp(-\int \alpha(r)dR)$ is the one-way path transmittance.

Expressed in differential form, (4.51) becomes

$$dS/dR = 4.34\beta^{-1}(d\beta/dR) - 8.7\alpha, \tag{4.52}$$

and from this, given

    i) an assumption or data on the relationship between α and β, and

    ii) a boundary value of an appropriate parameter,

one can derive evaluations of the optical parameters, or, given additional relationships, certain physical parameters. By a linearization transformation, (4.52) can be expressed in a general form from which solutions for various aerosol parameters (e.g. β, α, particle number, or mass concentration) may be derived, according to the input parameters used.

Adopting the general symbol Φ for the aerosol parameter, the solution is

$$\Phi(R) = \frac{\exp[C_1 S(R)]}{\Phi^{-1}(R_0) - C_2 \int_{R_0}^{R} \exp[C_1 S(r)] dr}. \tag{4.53}$$

Specific examples of Φ, the corresponding values of C_1 and C_2, and the assumed relationship of aerosol quantities are listed in Table 4.3.

As noted in Subsection 4.3.3, in the case of turbid atmospheres, certainly for example in fog or cloud, multiple scattering occurs and the solutions proposed above are invalid. In such cases more sophisticated formulations of the lidar equation must be used [4.75–78], although useful evaluations of lidar observations in fog have been made by Viezee et al. [4.109] using a semi-empirical approach. However, in less turbid atmospheres and certainly in what is commonly thought of as "clear" air, the assumption of single scattering appears to be wholly acceptable for lidar data, where we are concerned with evaluations of backscattering within a narrow beam (of the order of 0.01°; see e.g., Table 4.1).

4.5.3 Deriving Aerosol Physical Quantities from Optical Measurements

Except for the special circumstances noted in the bottom rows of Table 4.3, the solutions to the lidar equation presented in the previous section yield data solely on the optical characteristics (β and α) of the atmosphere. In the case of visibility in fog or the attenuating effects of haze, the optical parameter transmissivity is of direct significance. In other cases, however, it is necessary to take a further step. This entails relating the observed *optical* factors to *physical* factors of appropriate concern. Thus a measurement of the volume backscattering coefficient of a smoke plume has little practical significance of itself in air pollution monitoring. Of far greater importance is information on the mass concentration, or number concentration if particle size distribution is known.

Derivation of such physical parameters from lidar measurements requires first, the separation of particulate from gaseous optical characteristics, and second, knowledge of the appropriate optical-to-physical parameter ratios (e.g., backscatter-to-number, extinction-to-mass, etc.). Whereas in general the separation of gaseous and particulate optical properties in solving the lidar equation is complicated [4.105], frequently situations arise where one or the other component may be realistically neglected. In the upper atmosphere ($\gtrsim 50$ km), for example, particulate

Table 4.3. Some lidar equation solution possibilities for given input parameters; see (3.53)

Solution for $(\Phi=)$	Basic relationship measured or assumed	C_1	C_2
β	$\beta = k_1 \alpha$	1/4.34	$2/k_1$
α	$\dfrac{d \ln \beta}{d \ln \alpha} = k_2$	1/4.34	$2/k_2$
N_p (number) (concentration)	Relative size distribution, $\tilde{N}(a)$ invariant with range R	1/4.34	$2 \int_0^\infty \pi a^2 Q_E \tilde{N}(a) da$
$[N'(a) = N_p \tilde{N}(a); \int_0^\infty \tilde{N}(a) da = 1; a = $ particle radius; $Q_E = $ Mie extinction efficiency factor]			
ϱ_m (mass concentration)	α/ϱ_m and β/ϱ_m are invariant with range R	1/4.34	$2\alpha/\varrho_m$ [4.106]

backscattering is typically negligible compared to gaseous backscattering [4.190] (see Subsect. 4.6.1.), whereas in urban atmospheres the reverse is frequently true. As shown in Section 4.3, the optical-to-physical ratios for a purely gaseous medium are constants, cf. (4.6.7, 9, 26), permitting ready inference of gas density from a backscattering measurement. For the particulate phase of an aerosol or cloud, on the other hand, the ratios of backscatter-to-mass, extinction-to-number, etc., are, in general, complex and variable functions of particle composition, shape, and size distributions.

In spite of the complexity and the number of unknowns, however, it is frequently possible to derive approximate physical information from lidar measurements, and since this information is obtained remotely, it is often of great value. To provide some idea of the degree of uncertainty inherent in such optically-derived physical information, Table 4.4 (by no means exhaustive) summarizes the results of a number of investigations in this area. Listed are the changes in backscatter-to-number, backscatter-to-mass, and backscatter-to-extinction ratios that result from exemplary changes in particle size distribution, refractive index, and shape. With the exception of the bottom row, all results were obtained from theoretical calculations for homogeneous or layered spheres, of the type described in Subsection 4.3.1. Since the ratios in columns ii and iii of Table 4.4 are divided into measured backscattering coefficients to obtain values for particle number and mass, their variability is a measure of the uncertainty in these derived physical quantities. The variability of the ratios in columns ii through iv is also an indication of how well satisfied are the criteria required for the lidar equation solutions listed in Table 4.3.

As Table 4.4 shows, considerable caution must be applied in drawing inferences regarding other optical or physical parameters from lidar measurements of backscattering coefficient *on the basis of theoretical assumption alone.* Although not excessive, the dependence of relationships between the various parameters on specific characteristics of an aerosol is significant, certainly where Mie theory (i.e., for homogeneous spheres) applies. However, there are strong grounds for believing that in practice, probably because of the non-sphericity of the particles involved and/or the complexity of their refractive properties, Mie theory is inappropriate for specifying the scattering characteristics of some *natural* aerosols in simple terms.

Table 4.4. Typical variations in the interrelation of optical and physical parameters of atmospheric aerosol particles (theoretically derived or indirectly inferred). N_p, M_p, and α_p are the particle number density, mass density, and extinction coefficient, respectively β_p is the backscattering cross section

(i)	(ii)	(iii)	(iv)	(v)	(vi) Notes
Particulate characteristic and change therein	Resulting change in			Reference	$N_p = \int_0^\infty N_p'(a)da$ Based on computations for homogeneous spherical particles unless otherwise noted
	β_p/N_p	β_p/M_p	β_p/α_p		
Size Distribution $\varLambda=3.5, a_1=0.275\ \mu m$				[4.111]	Based on Junge's distribution:
to $\varLambda=4.0, a_1=0.3\ \mu m$	$\times 0.7$				$\hat{N}(a)=Ca^{-(\nu+1)}, a_1 \leqq a \leqq a_2$
to $\varLambda=3.0, a_1=0.1\ \mu m$	$\times 1.6$				where $\hat{N}(a)da=$ number of par-
to $\varLambda=4.0, a_1=0.5\ \mu m$	$\times 0.5$				ticles with radii between a and
to $\varLambda=2.0, a_1=0.03\ \mu m$	$\times 3.3$				$a+da$ ($\lambda=0.694\ \mu m$, $a_2=3.3\ \mu m$, $m=1.5$)
Haze Model L				[4.33, 112]	Based on Deirmendjian's haze
to Haze Model H	$\times 0.3$				models [4.33] ($\lambda=0.694\ \mu m$, $m=1.33$)
$\varLambda=2.5, a_1=0.04\ \mu m$, $a_2=10\ \mu m$ to $\varLambda=4.0$, $a_1=0.08\ \mu m, a_2=3\ \mu m$ (extreme range)			$\times 1.6$	[4.41]	Based on Junge's distribution ($\lambda=0.694\ \mu m$, $m=1.5$)
$\varLambda=2.0$ to $\varLambda=3.75$ ($m=1.5$) (widest range for $m=1.5$)			$\times 1.6$	[4.113]	Based on Junge's distribution ($\lambda=0.694\ \mu m$)
to $\varLambda=3$ ($m=1.33$) (widest range for $m=1.5$)			$\times 1.3$		
$\varLambda=3, a_1=0.08\ \mu m$, $a_2=3.0\ \mu m$, to $\varLambda=4$, $a_1=0.04\ \mu m$, $a_2=10.0\ \mu m$ (extreme range)		$\times 0.5$		[4.40, 112]	Based on Junge's distribution ($\lambda=0.694\ \mu m$)

Table 4.4. (continued)

(i)	(ii)	(iii)	(iv)	(v)	(vi) Notes
Particulate characteristic and change therein	Resulting change in			Reference	$N_\mathrm{p} = \int_0^\infty \hat{N}(a)da$ Based on computations for homogeneous spherical particles unless otherwise noted
	$\beta_\mathrm{p}/N_\mathrm{p}$	$\beta_\mathrm{p}/M_\mathrm{p}$	$\beta_\mathrm{p}/\alpha_\mathrm{p}$		
Haze Model L to Haze Model H when $m = 1.43$ when $m = 1.33$		$\times 0.7$ $\times 1.0$		[4.33, 112]	Based on Deirmendjian's haze models [4.33] ($\lambda = 0.694\ \mu\mathrm{m}$)
Refractive Index $m = 1.5 - 0i$ to $m = 1.33 - 0i$	$\times 0.3$			[4.111]	($\lambda = 0.694\ \mu\mathrm{m}$) See note in Row 1 above
$m = 1.5 - 0i$ to $\kappa = 0.01$ (reasonable range) to $\kappa = 0.1$ (extreme range)	$\times 0.3$ $\times 0.1$			[4.58, 114]	($\lambda = 0.694\ \mu\mathrm{m}$, $n = 1.525$) empirical size distribution
$m = 1.7 - 1.84 i$ to $m = 1.33 - 0i$ for Haze Model L for Haze Model H	$\times 0.7$ $\times 0.3$				Based on Deirmendjian's haze models [4.33] ($\lambda = 0.694\ \mu\mathrm{m}$)
$\kappa = 0$ to $\kappa = 0.025$ $m = 1.6 - 0i$ to $m = 1.33 - 0i$			$\times 0.4$ $\times 0.3$	[4.41]	Based on Junge's distribution ($\lambda = 0.694\ \mu\mathrm{m}$) Effect on $\beta_\mathrm{p}/\alpha_\mathrm{s,\,p}$
$m = 1.5 - 0i$ to $m = 1.33 - 0i$			$\times 0.4$	[4.113]	Based on Junge's distribution ($\lambda = 0.694\ \mu\mathrm{m}$) Effect on $\beta_\mathrm{p}/\alpha_\mathrm{s,\,p}$
$m = 1.33 - 0i$ to $m = 1.54 - 0i$ for Haze model H		$\times 0.3$		[4.33, 112]	Based on Deirmendjian's haze models [4.33]
Shape Homogeneous sphere to "onion shaped artifact"			$\times 0.7$	[4.41]	Based on Junge's distribution ($\lambda = 0.694\ \mu\mathrm{m}$)
Sphere to irregular plate			$\times 0.1$ see note	[4.45]	Laboratory measurement. Differences of as large an order of magnitude noted at near backscatter angle, but measurements not extended to 180°. ($\lambda = 0.486$ and $0.546\ \mu\mathrm{m}$). Authors caution against generalization from their limited results but stress importance of shape.

Table 4.5. Selected determinations of the interrelation of optical and physical parameters of atmospheric particles (experimentally derived)

Results	Reference	Notes
Backscatter/extinction (β_p/α_p)		
0.012 ± 0.004	[4.51]	Comparison of ruby ($\lambda = 0.6943$ μm) lidar and nephelometer measurements in urban atmosphere with variable relative humidity below 70%
0.03	[4.107, 115]	For cirrus cloud, derived from ground and airborne ruby lidar observations
0.02–0.04	[4.54]	Both β and σ derived from lidar observations of boundary layer urban aerosol (ruby lidar)
Backscatter/Number or mass concentration		
Assessment of mass concentration of fly ash in smoke stack plume-good agreement with estimates based upon quite independent data	[4.106]	Fly ash material of known refractive index, size distribution and density (ruby lidar)
Series of comparisons of lidar observations of β_p and σ_p (at $\lambda = 0.6943$ μm and $\lambda = 1.06$ μm) with known concentrations of virtually mono-disperse aerosols in test chamber, gave good agreement with Mie theory predictions	[4.116]	Fly ash material of known refractive index and density, ruby and neodymium lidars (comparison also made for broad band light)
Lidar-observed backscatter profiles consistently related to profiles of particle concentration independently obtained by *in-situ* sampling	[4.117, 118]	Observation made over sea below 3 km
Comparison of lidar-observed backscatter profiles of stratospheric layers show close correspondence, with height and relative magnitude, to particle count profiles derived by *in-situ* balloon sampling	[4.119, 120]	NCAR and NASA/Langley ruby lidars

The inapplicability of Mie theory in these cases is manifest in two important ways. First, predictions of the relationship between the optical and physical characteristics, derived by Mie theory, can be inaccurate. Second, empirically derived relationships are evidently more consistent and less dependent upon critical values of individual parameters than would be expected on the basis of Mie computations. Some examples of these considerations are given in Table 4.5. This leads to two important conclusions: 1) that lidar observations of natural aerosols are likely to be less affected by minor changes in the detailed characteristics of an aerosol than is suggested by theoretical considerations; 2) that independently derived information can be used most effectively to provide useful and consistent interpretations of lidar data.

In concluding this section, it should be noted that there is little immediate prospect of obtaining, by remote single-wavelength lidar observations, such information as particle size distribution or composition. Multiple wavelength or bistatic (multiple angle) lidar observations potentially could provide information on these quantities, but demonstration of successful measurements remains an elusive goal [4.110].

4.6 Examples of Lidar Measurements

Laser technology itself is only in its second decade, and the first atmospheric probing techniques using laser energy date from the earliest years of that technology. Indeed, lidar was among the first practical applications of laser energy, and within the first five years or so lidar had made real and valuable contributions to a number of atmospheric investigations—ranging from tracking clouds of insecticide sprayed from aircraft over forested valleys [4.123] to the study of plume rise from smoke stacks [4.106, 124].

But while the application of lasers in a wide variety of practical tasks—in measurement and alignment in cutting, drilling and machining proceeded rapidly and has led to a multi-million dollar industry, the early promise of lidar has not yet been fulfilled.

Early researchers were quick to note, explore, and demonstrate many roles that lidar could play as an atmospheric probe. For the most part such demonstrations could not be carried much beyond an initial indication of the viability and promise of the concepts investigated, simply because of the limited performance of early lidar devices. Chief among such limitations were, and are, the limited data rates available (pulse repetition rates have only recently exceeded 1 pulse per second or so), and the limited power available, certainly at wavelengths that are eye-safe. To these inherent difficulties must be added that of effectively evaluating lidar observations. The latter point leads to the problems of processing and presenting lidar data either quantitatively or qualitatively. While well within the available technology, the digital and analog devices suitable for the purpose are costly, and by no means minor, adjuncts to the basic lidar system, which in any case is itself not trivial in cost or complexity. All these factors, especially when considered in light of the requirements, placed lidar at a disadvantage compared with existing devices for many operational purposes; for example for cloud base measurement or even the remote evaluation of smoke plume density, for which human observer estimates are currently used in enforcement practice. Lidar has thus not yet found use on a routine operational basis for such applications. On the other hand, in the area of atmospheric research, and in studies and investigations of atmospheric particulate pollutants, lidar has become an effective and important tool. Although still relatively undeveloped technologically in terms of its obvious potential, lidar's accomplishment to date is not inconsiderable and is certainly wide ranging in its scope. Certainly, of all the applications of lasers in atmospheric probing, lidar has established the most advanced position to date.

In the following section we illustrate a number of such applications and also show lidar could eventually be used routinely on an operational basis in meteorology and air pollution control. These examples, which are by no means exhaustive, have mainly been drawn, as a matter of convenience, from the work of the Stanford Research Institute (SRI) groups.

4.6.1 Elastic Backscatter Applications

Observations of Clouds and Fog

"*Ceilometry*". The measurement of cloud base height or 'ceiling' by lidar is very straightforward, particularly when the lower surface of the cloud is well defined [4.126, 127]. In this case, the rapid increase of signal that marks the backscattered return from the cloud base can be readily distinguished and used to operate a timing counter, from which height can be read out in digital form. This form of data presentation (which is that of laser range finders) is economical and practical and has been used in the commercial lidar ceilometers which, alone of lidar devices, have been marketed for operational use. The ceilometer produced by the ASEA company of Sweden is a simple pulsed ruby system very similar to the basic system described in Section 3.2. By contrast, equipment offered by Sperry-Rand in the USA employs a high-pulse-rate gallium-arsenide laser and integration techniques to achieve adequate average signal levels.

While such systems do well with well-defined cloud bases (as in fact do conventional optical ceilometer systems) they do not fully exploit lidar's capability for providing complete information on ragged or diffuse cloud bases, with patches of cloud below. Such conditions are characteristic of much low cloud, especially when visibility at the surface is reduced by mist or fog that merges with the cloud layer aloft. In fact, with graphical data presentation, lidar can provide unique information on such conditions. For practical purposes, however, specifically in airfield operations, the concept of "cloud base" or "ceiling", though long used, is hardly meaningful in many circumstances [4.128]. Thus, the critical factor in landing an aircraft is the height from which the pilot, looking along a slant path, can acquire visual reference. The measurement of low cloud base height is therefore closely related to evaluation of "visibility", as discussed below.

Cloud Observations in General. The observation of clouds in general and the measurement of layer thickness or cloud top height (when attenuation is not too great), is also readily accomplished by lidar and has obvious application in meterological research. Even very tenuous cirrus clouds, invisible to the eye, can readily be detected, their structure and shape mapped, and, within useful limits, their water content evaluated.

This capability is well demonstrated by observations of tenuous cirrus cloud made by UTHE and ALLEN [4.19] using the SRI Mark IX ruby lidar (described in Section 4.2.2). Apart from illustrating lidar's capability for making such measurements, these observations provide an example of the use of the system's digital data acquisition, processing, and display resources (see Fig. 4.5).

As used for these observations, the digital data handling system automatically performs the following operations:

i) Corrects the lidar trace for the inverse range-squared dependence and instrument response functions.

ii) Computes cloud densities over an altitude interval determined by input switch data. The computation assumes a clear air density at the lower input altitude, a mean crystal radius, and a ratio between the back-scatter and extinction coefficients that includes a multiple scattering correction [4.125].

iii) Approximately corrects for cloud attenuation on the basis of observed cloud thickness and relative density as determined by the maximum cloud-to-clear air ratio.

iv) Outputs the computational results to a teletype.

v) Plots the complete lidar signature on a digitally-driven CRT display.
(These operations may be performed in real time on the data resulting from each lidar firing with a lidar fire rate of 6/min and a density computation on over 100 data points, i.e., for a cloud 3 km thick.)

The form of data presentation on the CRT display may be selected from various alternatives. These include the basic A-scope presentation (see Fig. 4.2) where signal intensity is shown as amplitude versus range,

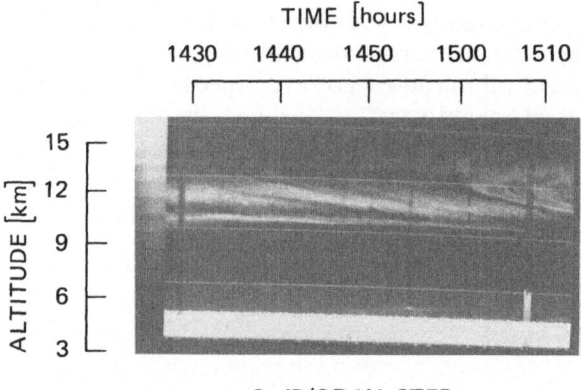

Fig. 4.11. Intensity-modulated analog display of cirrus cloud structure recorded with the digital lidar data system shown in Fig. 4.5

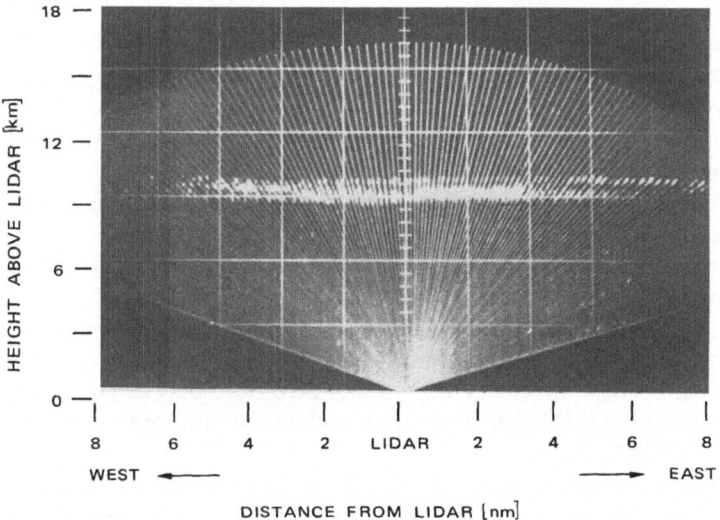

Fig. 4.12. Intensity-modulated display of cloud structure, revealing wave forms

or the "Z-screen" plot in which two-dimensional intensity-modulated displays represent range on one axis and time in the other, as successive lidar observations contribute to the data displayed. (With the lidar pointing vertically the latter presentation provides a height/time cross section.)

In the following example, lidar observations were made of cirrus cloud layers at a site in the tropics. Figure 4.11 shows a "Z-screen" plot of an extensive cirrus cloud layer system extending from about 9 to 12 km altitude. The CRT display in this case was programmed to present data in 16 gray-scale steps, each representing a 3 db increment of corrected lidar signal intensity. (The gray-scale steps are presented for reference purposes at the left-hand side of the screen, and are fully distinguishable at the console. The illustration as printed here suffers from numerous losses in photographic and printing processes involved.)

The structure revealed in the cross section shows the main features of the cloud layers, but the complicating effect of the time dimension should be noted. In addition to the graphical presentation, data can be presented in numerical form by teletype printout. Like the graphical data, these data are available in essentially real time. In the experiment illustrated in Fig. 4.11 the lidar was being fired every 3 seconds. The tabulated teletype data gave, for each shot, times of lidar firings, the height of maximum cloud density, the cloud-to-clear air density ratio in dB, the inferred crystal concentration and ice water content at the height of maximum cloud density, and in addition, an index related to the vertically integrated liquid water content.

Wave Clouds. The capability of lidar to observe cirrus structure in this way has revealed wave patterns of much interest in connection with mountain lee wave studies [4.129, 130]. As shown in Fig. 4.12, the amplitude and length of such standing waves are readily shown by the cross section patterns of the clouds. Successive cross sections of this type show how the wave mechanisms change as the wind flow and temperature conditions vary. Because of the limited data rate available when these observations were made, the lidar cross sections take several minutes to observe. Since the standing wave is a relativelyy persistent feature on this scale the observations are generally valid representations of the wave structure, although the possibility of distortions introduced by the method of observation should be borne in mind. With higher data rates, higher resolution in space and time would extend the technique to study dynamic features of waves of all types in the atmosphere—or for that matter, other significant atmospheric motions. (See *Observations of Atmospheric Motion* below.)

Fog. In the case of clouds "on the surface" or fog, lidar observations are restricted, in the same way as vision, by severe attenuation imposed by the suspended hydrometeors comprising the fog.

The evaluation of this attenuation is, of course, inherent in the measurement of visibility, which will be discussed in more detail below. However, the capability of lidar to monitor variations in fog density, even although only over limited distances, has been of value in research into fog dissipation techniques.

Collis et al. [4.131] showed how the density of fog on an airfield changed in the area of chemical seeding from an overflying aircraft. Subsequently, in much more comprehensive studies to provide clearings for landing aircraft by directing heat plumes from an array of ground generators [4.132], lidar-observed vertical cross sections revealed the extent and nature of the volume cleared by the heat. These observations provided both qualitative data and also quantitive data in the form of fields of extinction coefficient values derived by techniques used to evaluate visibility, as described below.

In both cases the two-dimensional lidar cross-sectional data supplemented and extended in situ measurements made on an instrumented tower. For example, the lidar readily revealed the relationship between the cleared volume and the tower measurements when, for example, the heated plume did not pass directly over the tower—or, as sometimes happened, the heated plume broke into several branches.

Visibility. Quantitative measurements of atmospheric turbidity are significant for the assessment of air quality or the description of meteorological conditions. In restricted visibility, such measurements have immediate operational importance in such activities as aircraft landings or marine navigation.

The parameter "visibility" used in both these contexts is a surprisingly complex concept and one that has been frequently misunderstood. Defined for meteorological purposes as the distance at which certain types of objects may barely be seen by a human observer by day, or the distance at which lights of specified intensity may barely be perceived at night, the term visibility involves physiological and psychological factors, in addition to the transparency and illumination of the intervening atmosphere. This transparency is, in fact, the basic factor of concern for many purposes, and will be recognized as the transmittance $T(\equiv \exp \int_0^{-R} \alpha dr)$ of the discussion Section 4.2. Visibility, for meteorological purposes, is often assessed by visual observation of known land marks or lights. This approach directly addresses the transparency of the total path over a line of sight, but is also affected by changes in ambient lighting conditions (sun position, etc.).

Alternatively, the transparency of the atmosphere over a short path may be measured by a "transmissometer," and various other instruments are used to make *local* (point) measurements of the atmosphere's scattering and absorbing properties. Inferences of general visibility conditions made on the basis of such local measurements are only valid in relatively homogeneous atmospheres. Because lidar offers flexibility in sampling the atmosphere in direction and range (to the limit of penetration in dense atmospheres) it offers advantages for assessing visibility

over extended and remote paths, especially above the surface. As noted by
MIDDLETON [4.7], the measurement of the volume extinction coefficient
α, is a preferred approach to making instrumental determinations of
visibility. *Meteorological range* V_M, i.e., the visibility of black skyline
objects in uniform daytime illumination (for an eye barely able to
distinguish objects with a 2 % difference in apparent brightness) is related
to α as follows

$$V_M = 3.9/\alpha,\tag{4.54}$$

as defined earlier by (4.29).

The visual range of lights at night, V_L, is a function of the intensity
I of the light and E_t, the threshold illuminance detectable by the eye,
according to the transcendental equation

$$V_L = (1/\alpha)\left[\ln\left(I/E_t\right) - 2\ln V_L\right].\tag{4.55}$$

The evaluation of α as a function of range by lidar observations has
already been noted in Section 4.1, and the difficulties of so doing in
highly turbid atmospheres have been stressed. A significant problem in
fogs lies in the complicating effects of multiple scattering.

In a homogeneous atmosphere, and where single scattering can be
assumed, lidar observations can be readily evaluated by the so-called
'slope' method, since $dS/dR = -8.7\alpha$, cf. (4.52). While this is appropriate
in relatively clear atmospheres ($V_M \geqq 3$ km or so) [4.133], the relationship
is modified by multiple scattering effects. In a series of studies and
experiments to develop practical methods of assessing slant visibility
for aircraft landing operations by surface lidar observations, VIEZEE
et al. [4.132] were able to obtain useful results by using this method,
applying empirically-derived factors to allow for effects such as
multiple scattering. The simple "slope" method, however, ceases to be
applicable when the atmosphere is inhomogeneous, and particularly
when the density of scatters *increases* with range.

In these cases, despite the practical difficulties in the case of signals
from dense fog, it is necessary to attempt a solution of the lidar
equation along the lines discussed in Section 4.5. VIEZEE et al. [4.132]
reported limited success in this very difficult problem.

Composition of Clouds (Polarization). The determination of the presence
of non-spherical scatterers by measuring the depolarization ratio of
lidar returns (cf. Subsect. 4.3.1) has immediate value in the observation
of clouds, where the presence of such scatterers indicates that ice crystals
must be present. This technique is especially suitable for distinguishing

between ice crystal clouds and water droplet clouds, particularly at medium altitudes. Field measurements of backscattered depolarization from clouds have been reported by SCHOTLAND et al. [4.64], PAL and CARSWELL [4.134], and UTHE et al. [4.135]. Depolarization ratios of 30 to 80 percent were observed from ice clouds, in accord with the theoretical and laboratory results described in Subsection 4.3.1. PAL and CARSWELL emphasized that for quantitative studies of this type, more than one receiver channel is required to permit simultaneous measurement of two or more polarization components. This requirement results from the typically rapid changes in the viewed portion of a cloud, which would confuse sequential measurements.

As indicated by the discussion of Subsection 4.3.1, a portion of the depolarization measured in backscattering from clouds may be due to multiple scattering effects, which can transfer the plane of vibration of the incident electric vector. These effects have been observed by PAL and CARSWELL [4.134], producing nonzero depolarization in backscattering from pure water clouds. As pointed out by PAL and CARSWELL, effects of multiple scattering and particle shape on measured depolarization may be partially separated by noting the dependence of depolarization on cloud penetration depth. Specifically, depolarization due to multiple scattering is zero at cloud base and increases with penetration depth, while depolarization due to particle nonsphericity assumes a large non-zero value at cloud base, and changes only as a result of changing particle characteristics (shape, size, and orientation).

Observations in Visually "Clear" Air

Mixing Depth, Strata, and Waves. In the same way that lidar may reveal even tenuous cloud layers by observing backscattering from ice crystals or water droplets, the suspended particles present in relatively clear air may be detected. Such particles may reveal depth of the mixed layer at the earth's surface, as discussed below, or show the vertical extent of haze layers from which the presence of inversions may be inferred. Although such observations offer a unique method of evaluating atmospheric turbidity (especially remotely) their most spectacular capability is in revealing features and structures of the clear atmosphere. The latter were among the first applications of the lidar technique in the troposphere to be reported [4.136]. An example showing later lidar measurements of such structure is illustrated by the intensity-modulated "picture" display of Fig. 4.13 [4.137]. It shows a height/time sequence of the atmospheric condition at a location in downtown St. Louis, Missouri, during the course of a typical summer day. In this form of presentation, the cross section shows the variation with time of received lidar "S-function"

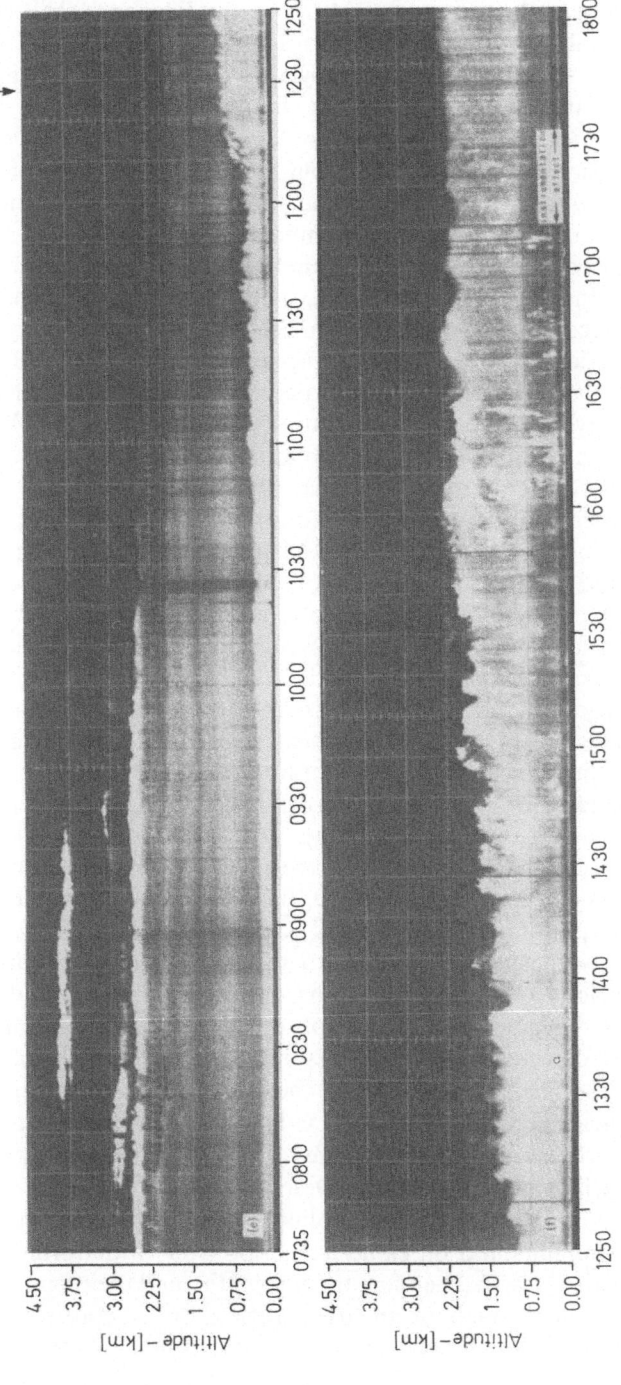

Fig. 4.13. Height/time cross section of the aerosol structure over St. Louis, Mo., on 13 August 1971 as observed by the SRI/EPA Mark VIII lidar system [4.137]

[cf. (4.50)] along a vertical line of sight above the lidar. At 0730 local time, the section shows a fairly homogeneous haze layer between 0.2 and 2.3 km, with some stratification, capped by a thin layer of stratus cloud that had formed at its upper surface. As the morning progresses, this cloud layer dissipates under the influence of solar heating (or "burns off," in the jargon of the weather forecaster), but the layered structure of the haze remains (1030 local time). Meanwhile, the denser layer of particles that has formerly been confined to the immediate surface layer is seen to grow to a depth of 500 m or so, and around 1200 noon begins to reveal cumulus shapes indicative of convection. The heads of these thermal "bubbles" are accompanied by deformation of the overlying layers, showing the vertical extent of the uprising motion; and by about 1400 local time, the bubbles have penetrated the stable layers and the deep convective layer is the dominant feature. At its upper level (~ 2 km), some condensation occurs and a visible cloud forms (as evidenced by the more intense echoes—confirmed by visual observations); while at the surface, a clearance is apparent as visibility improves. (Some injections of particles are apparent, probably smoke plumes.) By the end of the observation period (1700–1800 local time), the more uniform appearance of the haze layer suggests a reduction of convective activity and the achievement of a fairly well mixed haze condition. These observations, during which the surface visibility varied in the range 5–10 km in urban haze, were made during a study of the effect of cities on surrounding climate. They well illustrate the role lidar can play in such studies, even on a semi-quantitative basis (where mass or number concentration of the particles is considered only in relative terms, but layer heights are provided in precise and continuous fashion). They also show how lidar observations can delineate the depth of the mixed layer and establish the height of inversions, which restrict the dilution volume available for the dispersal of pollutants. Collis and Uthe [4.117], discuss this example further, with special reference to the problems of measuring atmospheric turbidity or the concentration of particles in air pollution studies. They also describe airborne lidar observations of dust concentrations which were detected at a height of some 2 km in the atmosphere over the Caribbean Sea, in the course of the Barbados Oceanographic and Meteorological Experiment (BOMEX) in 1969. This layer is interpreted as being the stream of dust carried across the Atlantic by the northeast trade winds from the Sahara Desert.

In addition to its importance to air pollution, the depth of the mixed layer has a determining effect on cloud formation over cities during the afternoon convective period. Therefore mixing depths, and their temporal and spatial variation throughout an urban region, are of considerable interest to studies of urban weather modification, especially when meas-

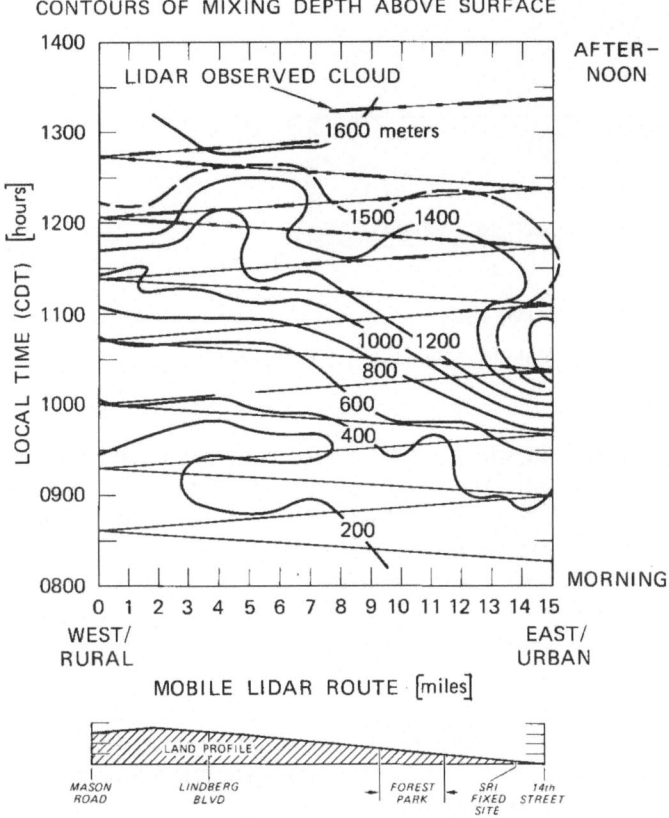

CONTOURS OF MIXING DEPTH ABOVE SURFACE

Fig. 4.14. Spatial and temporal variation of mixing depth in the St. Louis Region, as observed by mobile lidar. Thin diagonal lines trace the lidar route in space and time. Dark bands along the route lines mark times and locations when clouds were above the lidar van

ured in conjunction with cloud base heights and formation frequencies. Mobile lidar (either truck- or aircraft-borne) can provide this information with high spatial and temporal resolution. Figure 4.14, for example, shows a time-space contour diagram of mixing depth and cloud occurrence on a summer day in the vicinity of St. Louis, Missouri [4.138]. These data were obtained as part of the METROMEX study of urban weather effects by firing the Mark IX lidar (Fig. 4.6; Table 4.1) vertically as it was being driven along highways connecting downtown and outlying regions. The contours show a definite doming of the mixed layer over the city, and that the diurnal maximum in both mixing depth and cloud formation occurred over the city before over the outlying area.

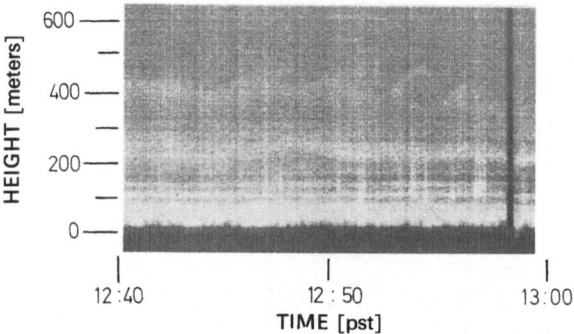

Fig. 4.15. Intensity-modulated display of clear-air aerosol structure, revealing wave forms [4.139]

Paralleling the way in which cirrus clouds observed by lidar reveal waves, particulate layers in the lower atmosphere frequently manifest wave forms. A particularly interesting and elegant series of lidar observations of such wave motion at the interface of the marine layer and the overlying air was made coincidentally with FM-CW radar observations as reported by Noonkester et al. [4.139]. Figure 4.15 taken from that paper, shows overturning, breaking waves, as observed in height-time cross sections obtained by lidar.

Again, the limited data rate of lidars so far used in such studies has restricted the scope of such observations. Higher data rate systems could provide greater time resolution and enable the dynamic features of the clear air to be studied in hitherto unknown detail.

Recently, acoustic radars, or *sodars* (for Sound Detection and Ranging) have begun to play an increasingly useful role in remote sensing studies of the atmospheric boundary layer. By measuring the scattering of sound pulses from small-scale temperature structure ("thermal turbulence"), such monostatic acoustic sounders can (under proper conditions) monitor both mixing depth and wave structures. Simultaneous boundary layer observations by both acoustic sounder and lidar were conducted by Russell et al. [4.211] in St. Louis to evaluate their relative merits for air pollution and weather modification research. By analysis of intensity-modulated displays of both sounder and lidar data (similar to Fig. 4.13), they found the summertime diurnal development could usefully be divided into three periods. The first was early forenoon, when aerosol particles had accumulated under a radiation inversion; here the lidar and the acoustic sounder performed about equally well in defining the depth of the mixed layer. The second was a strong convection period in which thermal plumes rose far above the early forenoon mixing depth; the lidar depicted the full vertical extent of these plumes by using surface-generated aerosols as tracers (cf. Fig. 4.13), but the sounder detected only the lower regions, where thermal turbulence was strong. The third was the period of re-formation of the inversion, which the sounder detected immediately but which the lidar detected only after a spatial discontinuity of aerosol content had developed. The sounder had a similar advantage in detecting subsidence inversions aloft. Wave structures observed by both sensors were usually well correlated. The authors caution against extrapolating these results to locations and seasons differing from St. Louis summer, since subsequent studies in the San Francisco Bay Area have given results that differ in significant ways. Moreover, higher-powered acoustic sounders have subsequently been demonstrated to be capable of monitoring the top of the mixed layer even during times of active convection.

Particulate Pollutants. Pollutant particles are important from an environmental health standpoint because they frequently serve as carriers for many toxic trace species (see, for example, [4.140] and Chapt. 2). This is especially true of the smaller (equivalent aerodynamic diameter $\lesssim 1$ μm) particles, which most readily deposit in the lung [4.141] and are most difficult to control at the source. Aerosol particles are also important in determining regional and global climates, both through their role in cloud and precipitation formation [4.142], and through their direct influence on atmospheric transfer of solar radiation [4.62, 142–146]. This latter influence also makes airborne particles the primary determinants of atmospheric visual range, as noted in Subsection 4.3.2.

For these reasons, the evaluation of the concentration of particulate pollutants is an important requirement in air pollution studies and monitoring operations. Measurement of the turbidity of the ambient atmosphere, particularly as it varies with height, is an obvious application of lidar, and has been discussed in Section 4.2 above and in the preceeding paragraphs.

Two other applications are of especial concern in air pollution control. These are the evaluation of the opacity of smoke stack emissions and the tracking of smoke and similar effluents. For air pollution control enforcement purposes there is a need for an objective method of remotely determining smoke opacity to replace the present visual judgements of trained "smoke inspectors." The obvious approach to this is to make lidar measurements of the backscattering coefficients of such smoke plumes. As discussed by COLLIS and UTHE [4.117], in which they describe UTHE's experiments with suspensions of flyash in a laboratory aerosol chamber, good determinations of mass concentration or plume opacity may be derived from backscatter measurements of this material. Further work is needed before these results may be applied to aerosols of different shape and refractive index. These experiments, however, also confirmed a close correspondence between attenuation measurements made at both ruby and neodymium wavelengths ($\lambda = 0.69$ μm and $\lambda = 1.06$ μm, respectively) and with a white light transmissometer. This corroboration is relevant to another method of remotely measuring plume opacity by lidar. This is accomplished by comparing the magnitude of the lidar signals from the ambient air along the path through the plume at points on the *near* and *far* side of the plume [4.148, 9]. Although the feasibility of these approaches has been initially demonstrated, their practical utilization awaits the development of simple, low-cost and eye safe equipment. Research is proceeding on the application of FM-CW techniques for this purpose [4.18].

The determination of mass concentration of plumes from coal-burning power stations was also successfully demonstrated by Johnson and Uthe [4.106], who used auxiliary information on particle size distribution obtained by concurrent direct sampling. The main thrust of their program, however, was the investigation of plume rise from tall stacks (of the order 260 m). In this application [4.106, 124, 150] the capability of lidar for observing the position of effluent plumes (in three dimensions), even when their density was too tenuous to be readily visible, has proved helpful in verifying plume-rise prediction models, and has shown additional effects due to atmospheric turbulence and vertical wind shear.

The dispersal of such particulate emissions as clouds of insecticide sprayed from an aircraft, or dust general by underground nuclear or other explosions has also been observed by lidar, again in conditions in which visual or photographic observations were impossible. In the first case, Collis [4.123] showed how clouds of low volume insecticide spray applied from an aircraft were distributed over forested valleys under the influence of natural air drainage. In the case of explosion-generated dust clouds [4.151], one instance is noteworthy in being probably the first use of lidar in an aircraft. With a neodymium lidar pulsing at a rate approaching once per minute, a cloud of dust was tracked over a distance of some 9 km for a period of 20 min, and its envelope delineated, despite being invisible to the eye.

Atmospheric Turbidity and Energy Transfer. Apart from its importance to public health and esthetics, the turbidity of the polluted atmosphere is of significance in determining the transfer of radiant energy between the sun, the earth, and outer space. Many authors (e.g. [4.143]) have recently expressed concern that effects of aerosol pollution on this radiant energy exchange could significantly modify the climate on regional and global scales. To assess this possibility in a quantitative

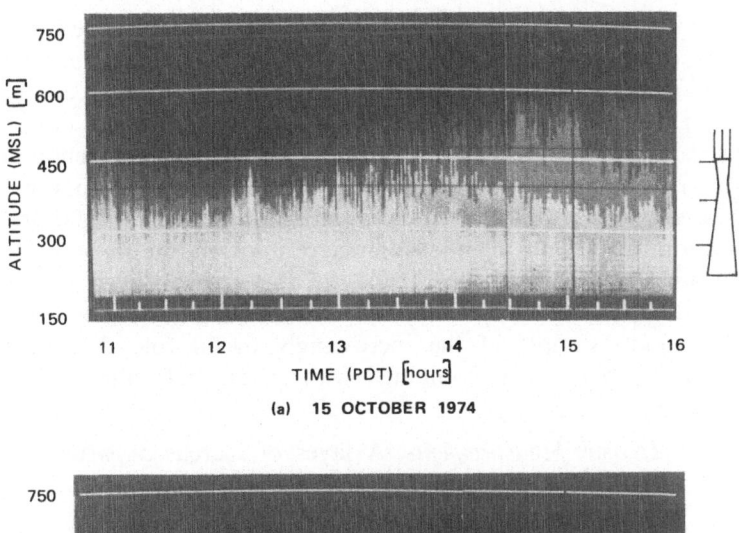

(a) 15 OCTOBER 1974

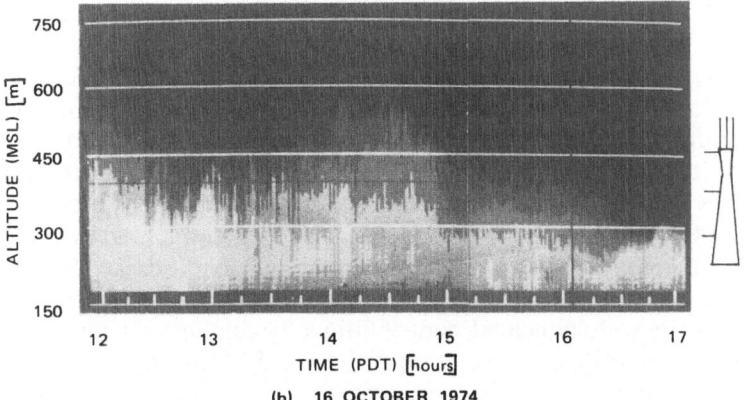

(b) 16 OCTOBER 1974

Fig. 4.16a and b. Intensity-modulated analog display of aerosol structure recorded with the digital lidar data system shown in Fig. 4.5 [4.152]

fashion, numerous mathematical models [e.g., 4.144–6] have been developed; but application and testing of the (sometimes conflicting) models have been hampered by a lack of suitable experimental data on the optical properties and radiative effects of particulate pollution.

To provide some of these required data, RUSSELL and UTHE [4.152] undertook a study in San Francisco that included coordinated measurements with lidar and solar radiation sensors, some of which were mounted on a tower to provide vertical resolution. Figure 4.16 shows an example of the lidar data obtained in that study, using the SRI Mark IX digital data system described in Subsection 4.2.2. The intensity-modulated lidar display graphically depicts the behavior of the dense, surface-based haze layer with respect to the tower. The three tick marks on the tower symbols indicate the levels at which continuous measurements of both upward and downward solar irradiance were made. As can be seen, the local marine inversion confined the dense smog layer below the uppermost sensors during the observation period, but the vertical extent and optical density of particulate pollution between the lower sensors and the sun varied considerably. Future comparisons between the lidar and solar radiation profiles are planned, to derive information on both the absorbing and backscattering properties of the aerosol layer, as are required in determining aerosol climatic effects. The lidar/tower study provides a good example of the increasingly useful role that lidar observations can play when combined with sparse but calibrated in situ measurements.

Stratospheric Aerosol Measurements. A layer of increased particulate concentration, typically several kilometers thick and centered near 20 km in altitude, is a quasi-permanent feature of the global stratosphere. The particles comprising this layer consist primarily of sulfate, and a significant fraction evidently forms locally from chemical reactions of sulfur-bearing gases. The concentration of these gases and of directly-injected particles is strongly increased by certain major volcanic eruptions, which thus can have a major effect on stratospheric aerosol behavior. During the past fifty years, the layer has been extensively studied by a number of remote-sensing and direct-sampling methods [4.153–7]. It is in this area that lidar observations have made one of their most extensive and best-documented contributions, beginning with the pioneering measurements of FIOCCO and GRAMS [4.158] in 1964, and continuing through a number of recent studies [4.159–164, 147] prompted by concern over possible stratospheric pollution by human activity [4.156, 157].

In applying the lidar technique to stratospheric studies, the vertical profile of received signal $P_r(z)$, resulting from elastic backscattering by stratospheric gases and particles, is compared to the signal profile that would result from the gas phase of the stratosphere alone. The fundamental result obtained from this comparison is a vertical profile of a quantity termed the "scattering ratio," defined as

$$\zeta(z) \equiv 1 + \beta_p(z)/\beta_g(z),$$

(4.56)

where $\beta_p(z)$ and $\beta_g(z)$ are, respectively, the elastic particulate and elastic gaseous (i.e., Rayleigh) volume backscattering coefficients, both at altitude z. As can be seen, scattering ratios in excess of unity provide a measure of particulate backscattering in the stratosphere, and hence

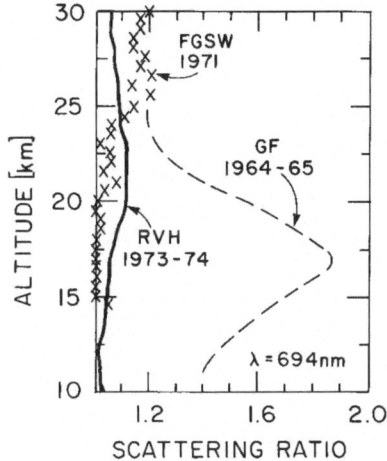

Fig. 4.17. Average vertical profiles of scattering ratio obtained in three lidar studies of the stratospheric aerosol. (Profile GF: Massachusetts, 1964–65 [4.111, 165, 166]; FGSW: Pacific Ocean, 1971 [4.180]; RVH: California, 1973–74 [4.164])

indicate the presence, vertical extent, and temporal and spatial variability of particulate scattering layers. In this manner, lidar observations provide a useful means of observing the variable structure and optical density of the stratospheric aerosol layer, with a coverage in space and time that would be extremely expensive to achieve with direct sensors.

Data from the 1964–1974 decade of stratospheric lidar observations, made by a number of groups in Massachussetts, Colorado, California, Alaska, Wyoming, Virginia, Jamaica, Japan, Brazil, Australia, Hawaii, the Soviet Union, and Israel [4.11, 165–179], as well as from aircraft flown over the Atlantic and Pacific oceans [4.163, 180], have been useful in documenting the general decline in stratospheric aerosol content during that period, and the dramatic increase that occurred late in 1974. The average profiles from three representative studies are reproduced in Fig. 4. 17. Profile GF is the average of 66 scattering ratio measurements made by GRAMS and FIOCCO [4.111, 158, 165, 166] during 1964–1965, using a ground-based ruby lidar in Lexington, Massachusetts. Profile FGSW is the average profile observed by FOX et al. [4.180] over the Pacific Ocean in August 1971, using an airborne dye lidar ($\lambda = 585$ nm; described in Subsection 4.2.2 and Table 4.1). (The profile has been converted to the ruby wavelength for this comparison, assuming $\beta_p \sim \lambda^{-1}$.) And finally, profile RVH is the average profile observed by RUSSELL et al. [4.164] on sixteen dates between June 1973 and March 1974, using a ground-based ruby lidar near San Francisco.

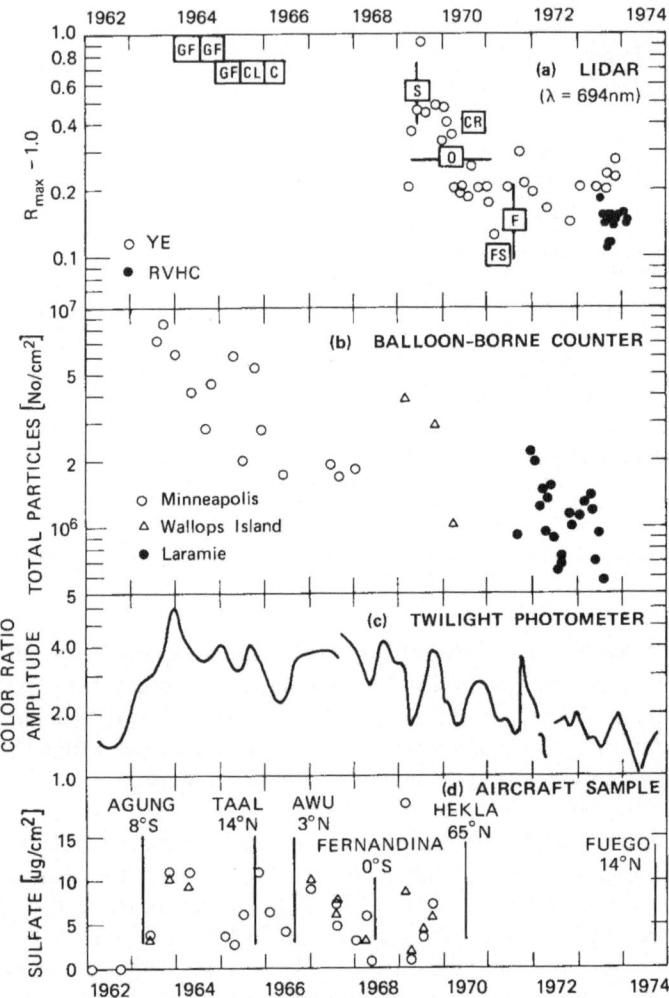

Fig. 4.18. Comparison of stratospheric aerosol measurements by four independent techniques, 1962–1974. a) Maximum ratio of particulate to gaseous backscattering, as observed by a number of lidar groups. GF: GRAMS and FIOCCO [4.111, 165, 166], Massachusetts; CL: COLLIS and LIGDA [4.167], California; C: CLEMESHA et al. [4.168], Jamaica; S: SCHUSTER [4.173], Colorado; CR: CLEMESHA and RODRIGUES [4.170], Brazil; O: OTTWAY [4.175], Jamaica, FS: FRUSH and SCHUSTER [unpublished], Colorado; F: FOX et al. [4.180], Hawaii and Bermuda; YE: YOUNG and ELFORD [4.179], Australia; RHV: RUSSELL et al. [4.164], California. b) Number of particles (radius ≳ 0.15 μm) above tropopause, as measured by photoelectric particle counter [4.182, 183]. c) Color ratio amplitude, as measured by twilight photometer at Weissenau, Germany, through October 1967; then at Bedford, Massachusetts [4.184–186]. d) Mass of sulfate per standard cubic meter (scm), as collected on aircraft-borne filters [4.187]

The most obvious difference between the 1964–1965 profile and the more recent ones is the size of the scattering ratios. GRAMS and FIOCCO observed an average peak value of β_p/β_g of nearly 90%, while the profiles of the early 1970's display peak values of only 10 to 20%. In addition, there is a significant difference in the shapes of the profiles. In 1964–1965 the average peak in scattering ratio was at 17 km, whereas in the early 1970's it was typically above 20 km. Both of these changes in scattering ratio profiles are probably attributable to the gradual removal of particles resulting from the major eruption of the volcano Agung on Bali in March 1963, and other major eruptions that occurred through May 1970 [4.181].

The gradual decline of scattering ratio between 1964 and mid-1974, as observed by a number of stratospheric lidar groups in both hemispheres, is illustrated in Fig. 4.18, which also includes some comparison data obtained by other observational methods. Considering that the four methods all observe different measures of stratospheric particulate content, and the variety of measurement locations, there is good mutual agreement in depicting the overall trend. Numerous short-term variations, superimposed in the general trend, are evident in the data of all four measurement techniques, especially the twilight photometry data, for which measurement frequency was the highest. Some of the major short-term variations are evidently associated with (and lag somewhat behind) the times of major volcanic eruptions, as indicated by the vertical lines in Figure 4.18d. However, there are doubtless other influences on the short-term variability of the stratospheric aerosol layer, and this variability is the subject of current studies using a number of techniques.

In late 1974 widespread observations indicated a dramatic increase in stratospheric particulate content, evidently in response to the October 1974 eruption of the volcano Fuego (14.5° N, 91° W) in Guatemala. At least four lidar groups [e.g., 3.161, 212–214] documented the intermittent appearance of multiple layers, with scattering ratios comparable to those observed in 1964 after the Agung eruption. Maximum scattering ratios were generally observed in and around February 1975, after which an overall decline was observed until near the end of 1975. Complete and systematic observations of the global spread and removal of the 1974 Fuego eruption cloud provide useful new data on stratospheric circulation and diffusion, as well as on particle formation and removal processes.

Upper Atmospheric Density. Above 50 km or so, the earth's atmosphere rarely contains particulate material and it is possible, with high performance lidar systems, to observe backscattering from the gaseous

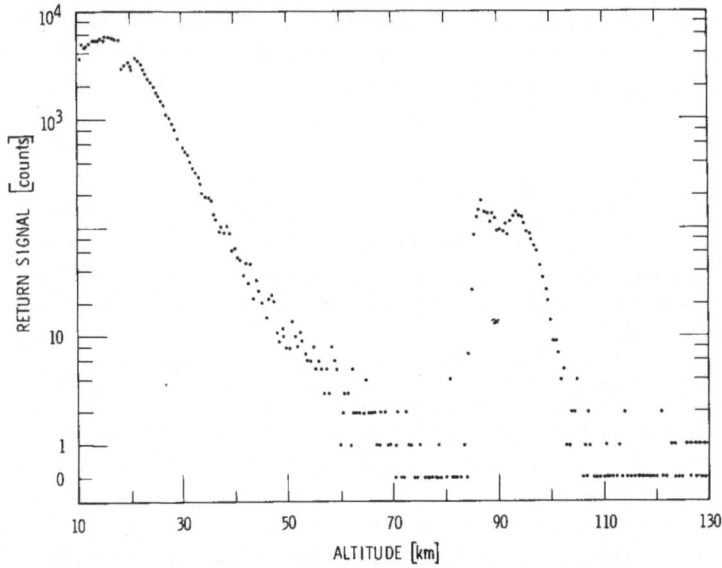

Fig. 4.19. Vertical profile of received resonantly scattered lidar signal, showing presence of atomic sodium layer between 85 and 100 km [4.191]

atmosphere with sufficient sensitivity to make density determination possible. With such a lidar (described in Subsection 4.2.2 and Table 4.1), KENT and his co-workers at the University of the West Indies have made extensive and systematic measurements for a number of years [4.188]. Under good atmospheric conditions their lidar is capable of providing relative density measurements whose accuracy is about 1% at 70 km and 10% at 90 km for 200 laser firings, which may be made in a period of about 30 min. Measurements, which are made in 4 km intervals, and are referred to returns from the height range 60–64 km, extend from 70 to 100 km.

Variations in relative density have been found to occur during the night, and these have been attributed to the action of atmospheric tides [4.189]. By averaging all readings obtained during a night (typically 2500 laser firings), a series of mean relative density profiles has been obtained [4.190]. Some 50 such profiles were obtained, fairly evenly spread over a two year period. By grouping these to form seasonal averages a number of interesting features have emerged. These are: i) that mean atmospheric density in the mesosphere and lower thermo-sphere (i.e., 70–100 km) over Kingston, Jamaica, showed a long term reduction during the period 1971–1973; and ii) that superimposed on this trend was a marked annual variation, with a maximum between

January and March in the two years. (The maximum to minimum swing in density was of the order of 10%, and the annual peak to peak reduction, indicating the general trend, was also about 10%).

Although other techniques of making measurements in the region are available (sounding rockets and radar meteor observations) these interesting results demonstrate the great value of lidars in making such observations on a continuous and consistent basis.

Mesospheric Constituents (by Resonance Scattering). Because it is strongly quenched by molecular collisions in the troposphere, resonance elastic scattering has been applied largely to observations of the upper atmosphere. The most common application has been to ground-based observations of the layer of free atomic sodium near 90 km. Numerous resonance scattering observations of the layer, under both nighttime and daytime conditions, have been reported in the literature [4.190–197]. Some of these are summarized by HAKE et al. [4.191], who made night-time measurements with a tunable dye-laser radar. A typical profile thus obtained is shown in Figure 4.19. An interesting result noted by HAKE et al. was the enhancement of resonantly scattered returns at the time of the Geminids meteor shower on the night of December 13–14, 1971. A similar result was noted in Japan just after the Jacobini meteor shower on October 8, 1972. Together, these observations give evidence for meteor production as one source of the 90-km sodium layer.

FELIX et al. [4.194] also report the detection of atomic potassium at a height of 80–100 km. See Chapter 5 for a more thorough discussion of their observations.

Observations of Atmospheric Motion

Air Displacement Determination. The possibility of remotely detecting atmospheric turbulence, particularly Clear Air Turbulence (CAT), was an early objective of some lidar researchers. It could readily be shown, however, that backscattering from dielectric inhomogeneities due to thermal gradients associated with turbulence was many orders of magnitude smaller than Rayleigh backscattering from the atmospheric gases [4.198]. A more realistic approach appeared to be that in which variation in particulate concentration might be used to reveal significant discontinuities in airflow [4.199]. LAWRENCE et al. [4.200] were, in fact, able to relate aircraft observation of turbulence with lidar-detected atmospheric layers at night. It seems probable that the returns in question were from particulate layers, marking areas of convective overturning.

The possibility of observing turbulence and measuring wind velocity by actual tracking of inhomogeneities in the atmospheric particulate content was also recognized at an early stage. COLLIS [4.133, 201] reported tracking smoke puffs generated by pyrotechnics to measure horizontal and vertical airflow, and also [4.123], by tracking clouds of pesticide spray, to measure air flow velocities in forested valleys. Such tracking methods depend upon higher data acquisition rates than have been available until recently. An interesting demonstration of how inhomogeneities in the apparently clear atmosphere may be used to measure air flow has been reported by DERR and LITTLE [4.202]. In their technique a cw laser beam was split to illuminate two divergent paths above the ground. These were intersected by two separate receiver beams in such a way that returns were received from two common volumes at the same altitude by the respective receivers. The signals in the receivers were shown to have a high correlation, with a time delay corresponding to the wind velocity along the path between the elevated common volumes. To accommodate winds from any direction, a conical scanning arrangement was proposed.

Other methods of deriving air motion from successive observations of fortuitous inhomogeneities in "clear air" particles or in mist or fog are also being investigated.

Doppler Techniques. The application of the Doppler principle to determine the relative radial velocity of atmospheric volumes from which laser energy is backscattered offers possibilities for remotely measuring wind velocity or for detecting turbulent motion in the atmosphere.

The Doppler frequency shift Δf is related to radial velocity v_r of the scatterers returning energy as follows

$$\Delta f = 2v_r/\lambda. \tag{4.57}$$

In the case of atmospheric backscattering, several types of motion are potentially detectable: random motion of gas molecules; Brownian motion of suspended particles; falling motion of larger particles and hydrometeors; turbulent motion of the air and suspended particles; and air flow or wind. In most cases the motions involved, especially when resolved into radial velocities, extend over a broad range of magnitudes, and thus produce an extended spectrum of Doppler-shifted frequencies. In the case of random motion, the spectrum of the shifted frequencies will be symmetrical about that of the transmitted energy. Motions due to air flow caused by wind (or in the very short term, by large eddies) will, however, show a shift corresponding to the group velocity of scatterers causing the return. In the case of falling particles or hydrometeors a similar shift will result, but because of the differing fall rates of different-sized scatterers the returns will again be spread over a significant spectrum (thus providing the possibility of assessing particle size distributions of falling precipitation).

The detection and interpretation of such Doppler returns presents considerable difficulty. Although most progress has been made in applying the concept at very short ranges, e.g., in wind tunnels, engine exhaust, test chambers, etc., in so-called *velocimetry* techniques, some encouraging experimental results have been made in atmospheric remote probing applications.

Various techniques have been utilized for detecting the frequency shift in the energy backscattering by the particles of the "clear air" or even by the gaseous atmosphere, using cw lasers (argon or CO_2) or pulsed coherent CO_2 lasers.

The approaches used differ in the manner in which the frequency shift is detected and measured. Three main alternatives are available:

i) direct detection of the frequency shift in the received signal compared with the laser output by high resolution interferometric techniques,

ii) coherent heterodyne or homodyne detection, where the backscattered signal is compared with the signal from a local oscillator, and

iii) the comparison of the signals received from a single volume, by two beams originating from the same transmitter, in an arrangement where the geometry of the paths gives rise to a differential in received frequencies due to the Doppler effect. (This geometry restricts its application to close range, however). HUFFAKER [4.203] discussed the application of the heterodyne and differential approaches to the remote measurement of wind velocity and turbulence from the ground or from aircraft, as well as for observing trailing vortices generated by large aircraft. LAWRENCE et al. [4.204] described a specific instrumental approach employing a CO_2 laser ($\lambda = 10.6\,\mu m$) in a homodyne arrangement for measuring wind velocity. BENEDETTI-MICHELANGELI et al. [4.205], using an argon lidar and the interferometric technique, described their success in measuring wind velocity with good accuracy in the lower troposphere in North Italy. (The same group has also addressed the problem of measuring atmospheric temperature remotely, by observing the spread of the Doppler-shifted frequencies backscattered by gaseous molecules [4.206].)

More recently, ABSHIRE et al. [4.207] have reported application of the homodyne concept to the observation of hydrometeors in the atmosphere. They show how spectra of fall velocities of rain and snow may be measured, but note that further development is necessary to extend the technique for analyzing drop and particle size distributions or precipitation rate.

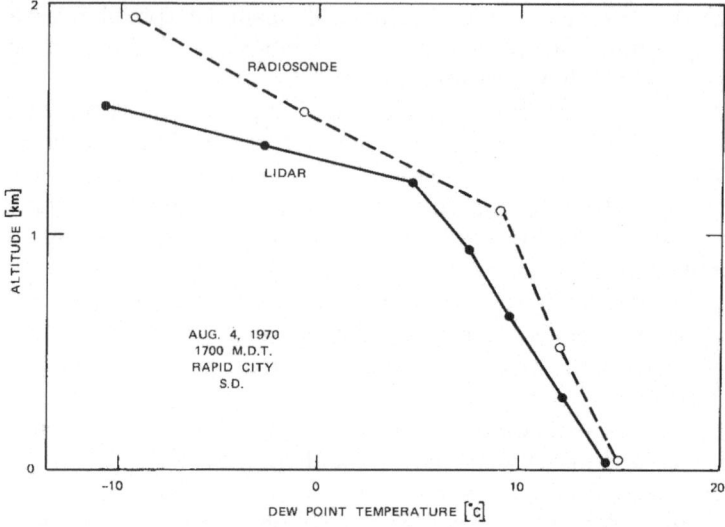

Fig. 4.20. Comparison of atmospheric water vapor vertical profiles (expressed as dew point temperature) measured by differential absorption lidar and radiosonde [4.82]

4.6.2 Differential Absorption Applications

Water Vapor

The relative abundance of water vapor in the lower atmosphere, together with the fact that it has isolated absorption lines in the visible spectrum at output frequencies of early lasers, made it an obvious candidate for remote measurement using the differential absorption lidar technique. As early as 1964, Schotland [4.80, 81] made measurements of water vapor vertical profiles, by thermally tuning a ruby laser on and off the water vapor absorption line at 694.38 nm. Figure 4.20 shows an example of a vertical profile (expressed as dew point temperature) obtained in this manner, together with a concurrent profile obtained by a radiosonde launched five miles away. Differences between the two profiles could possibly be ascribed to the differing locations, or, on the other hand, to difficulties in analysis of the differential lidar signals (see Subsection 4.4.3). Subsequent research has addressed this question, and also the problem of engineering a field-usable system of reasonable cost. Tests of a later research system, for example, have yielded water vapor densities accurate to $\pm 15\%$ over 100-meter path increments for typical New York summer moisture conditions [Schotland, private communication].

The possibility of using other tunable lasers to measure water vapor absorption in the vicinity of other spectral bands (e.g., near 940 nm) is also under consideration by several groups.

Pollutant Gases

The theoretical minimum detectable concentrations listed in Table 4.2 for a range resolution of 100 m make differential absorption the most promising of the remote spectroscopic techniques for determining range-resolved trace gas profiles at typical ambient concentrations. Moreover, if range resolution can be sacrificed, greatly reduced average concentrations can be measured over extended paths by using a topographical reflector (e.g., a hill) and quite modest laser pulse energy.

Actual application of the differential absorption technique to detection of trace gases does not yet match the predicted potential, but some successful atmospheric measurements have been made. ROTHE et al. [4.208] reported the first application of the technique to the measurement of ambient atmospheric NO_2. Using a tunable dye laser and backscattering from the atmospheric aerosol, NO_2 concentrations of 0.2 ppm at distances up to 4 km were measured in Köln, F. R. Germany. In a later measurement [4.209], the same investigators mapped the spatial distribution of NO_2 concentration around the smoke stack of a chemical factory by horizontally scanning the lidar. The contours of NO_2 concentration obtained in this manner are shown in Fig. 4.21. Using a pulse energy of 1 mJ and a repetition rate of 1 Hz, 40 on- and off-line shots were fired along a given direction before aiming along an adjacent direction and repeating the process. To obtain the results shown in Fig. 4.21, signals from a total of 8000 laser shots were averaged along each of the five indicated directions.

A calibrated differential absorption measurement of NO_2 was reported by GRANT et al. [4.94], who used elastic backscattering from the atmosphere on both sides of a chamber containing a controlled amount of NO_2. Measurements were made at two peak-and-valley wavelength pairs (441.8–444.8 nm and 446.5–448.1 nm) using a tunable dye laser. The NO_2 concentrations measured in this manner were compared to values measured by a conventional transmissometer, giving the results shown in Fig. 4.22. At a concentration of 20 ppm, the differential absorption measurements gave a signal-to-noise ratio of 1 in the chamber length of 2.45 m. This is equivalent to an uncertainty of 0.5 ppm in 100 m, or 0.05 km-ppm. This measurement uncertainty includes all significant sources of error, as discussed in Subsection 4.4.3. GRANT et al. showed how reasonable system improvements could reduce the

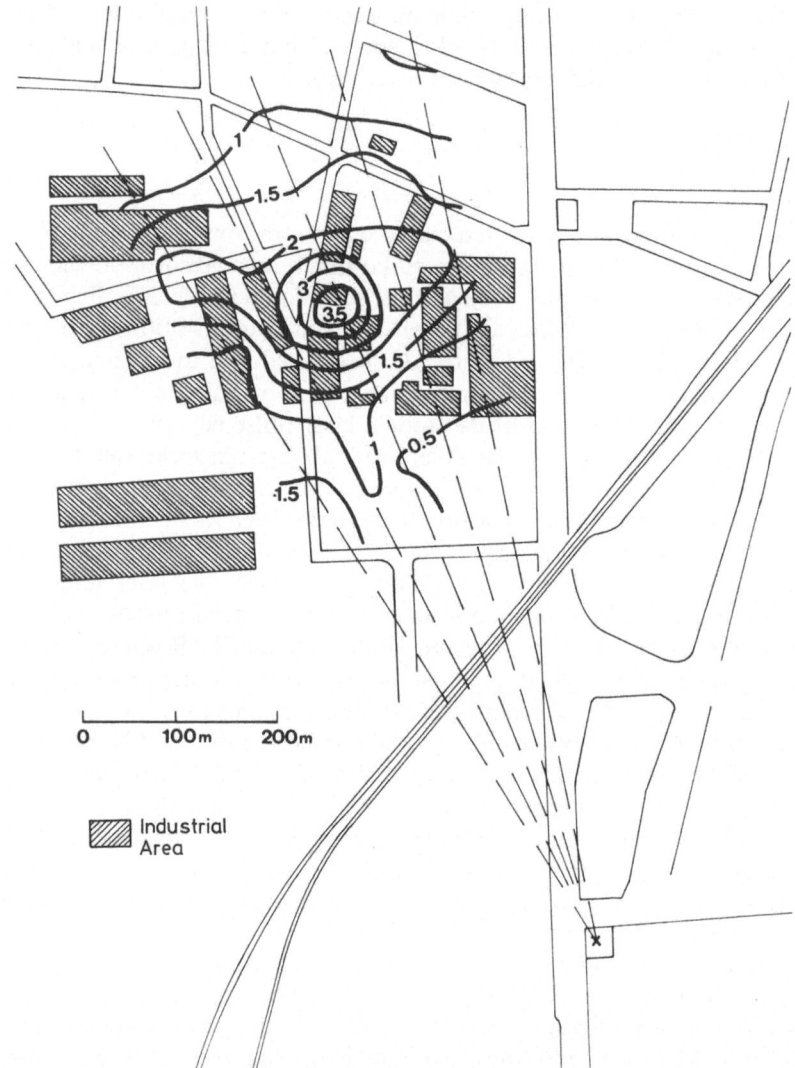

Fig. 4.21. Contours of NO_2 concentration [ppm] in the vicinity of a chemical plant, as measured by differential absorption lidar [4.209]

uncertainty to 0.005 km-ppm, and that this could be achieved in daytime measurements by use of narrow interference filters.

In a subsequent experiment [4.210] GRANT and HAKE made similar calibrated measurements of SO_2 and O_3 using peak-and-valley wavelength pairs at 292.3–293.3 nm and 292.3–294.0 nm, respectively. The laser actually used permitted only sequential measurements at peak and

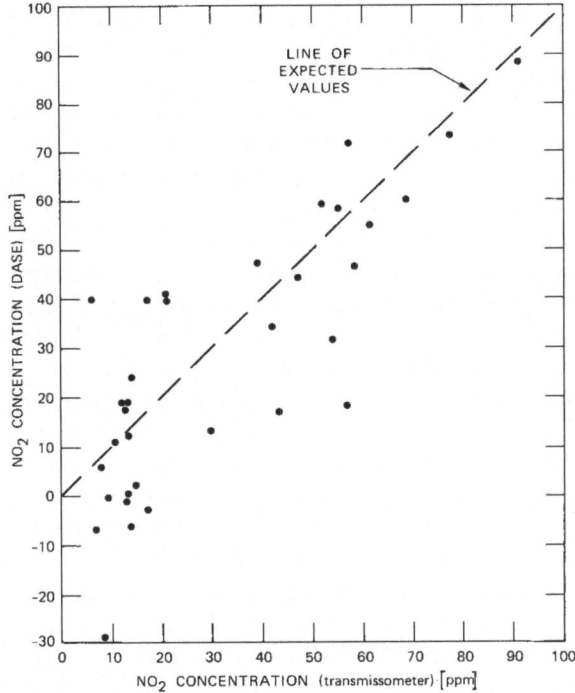

Fig. 4.22. Comparison of NO_2 concentration in a sample chamber, as measured by differential absorption lidar and local transmissometer [4.210]

valley wavelengths. However, GRANT and HAKE estimated that a lidar transmitting simultaneously at two wavelengths would provide a sensitivity of ± 0.01 km-ppm for SO_2 and ± 0.03 km-ppm for O_3.

An example of an IR measurement has been reported by HENNINGSEN et al. [4.92], who remotely measured the concentration of CO in plastic bags by monitoring the differential absorption of the signal backscattered from a topographical target at wavelengths near 2.3 μm. The range and sensitivity of the technique were in good agreement with the predictions noted above.

4.7 Conclusions

4.7.1 Present Capabilities and Limitations

The demonstrated capabilities of lidar and the problems limiting full realization of lidar's potential in atmospheric probing have been indicated in the preceding section (Section 4.6). In brief, lidar has much to offer in a wide range of applications concerned with meteorology, air pollution control, and atmospheric research.

With present technology, however, lidar techniques are almost exclusively confined to research applications. The reasons for this are primarily economic, partly concerned with eye-safety, and to a lesser (but significant) degree involve the fact that quantitative techniques for using lidar in certain applications have not yet been fully worked out.

For many applications where the lidar technique has obvious capability—e.g., ceilometry, transmissometry, atmospheric stratification— alternative methods are either adequate or considerably less costly. In other applications where lidar has potential that is yet to be fully demonstrated—e.g., remote smoke plume opacity measurement, the measurement of gaseous pollutant emissions, the determination of water vapor profiles, etc. (to name only a few)—the relatively high cost of the lidar approach has discouraged rapid development. There remain the various research applications where lidar can make unique contributions, and thus justify the cost and complexity of the equipment and procedures needed. Among these we note many applications in the investigation of the structure and dynamics of the natural atmosphere (especially measurements of mixing depth), the role of particles in studies of radiative energy transfer or in transport and diffusion mechanisms such as plume rise and dispersion, and in monitoring the presence and distribution of pollutant gases of various types.

4.7.2 Future Prospects

The future of lidar techniques clearly depends upon the development of simpler, eye-safe systems that can meet observational and measurement requirements both at an acceptable cost and without eye hazards. For the many tasks of a routine operational type that lidar could perform—ceilometry, visibility measurement, remote smoke stack monitoring, atmospheric stratification observation—only reliable, simple-to-use or automatic devices of relatively low cost have any expectation of acceptance. This is because a) the "value" of the task to the operational user is limited, and b) alternative devices or methods of at least minimum efficacy are already generally available at modest cost. In view of the rapid changes that are occurring in modern technology—not only of lasers but of digital computational and display devices, it is very conceivable, however, that appropriate operational devices for routine observations will be developed within the next decade. In the area of research—both in atmospheric studies in general and in air pollution investigations in particular—the already well-demonstrated capability of lidar can be expected to be expanded and extended, and to provide an ever increasing input to a wide range of such studies.

Acknowledgments. In this chapter we have drawn heavily on the work of others as indicated by the references cited. We are however, particularly indebted to our colleagues at SRI and wish to acknowledge especially the inputs of E. E. UTHE, W. B. GRANT, R. D. HAKE, JR., and M. L. WRIGHT.

References

4.1   E. O. HULBERT: J. Opt. Soc. Am. **27**, 344 (1937)
4.2   E. A. JOHNSON, R. C. MEYER, R. E. HOPKINS, W. H. MOCK: J. Opt. Soc. Am. **29**, 512 (1939)
4.3   F. E. JONES: Roy. Aero. Soc. J. **53**, 433 (1949)
4.4   S. S. FRIEDLAND, J. KATZENSTEIN, M. R. ZATZICK: J. Geophys. Res. **61**, 415 (1956)
4.5   R. BUREAU: La Météorologie **3**, 292 (1946)
4.6   M. G. H. LIGDA: Proc. lst Conf. Laser Technology, US Navy ONR (1963) pp. 63–72
4.7   W. E. K. MIDDLETON, A. F. SPILHAUS: *Meteorological Instruments* (Univ. Toronto Press, Toronto 1953) p. 208
4.8   C. A. NORTHEND, R. C. HONEY, W. E. EVANS: Rev. Sci. Instr. **37**, 393 (1966)
4.9   B. R. CLEMESHA, G. S. KENT, R. W. H. WRIGHT: J. Appl. Meteorol. **6**, 386 (1967)
4.10   G. S. KENT, P. SANDLAND, R. W. H. WRIGHT: J. Appl. Meteorol. **10**, 443 (1971)
4.11   R. J. ALLEN, W. E. EVANS: Rev. Sci. Instr. **43**, 1422 (1972)
4.12   P. W. WYMAN: Appl. Opt. **8**, 383 (1969)
4.13   G. W. GRAMS, C. WYMAN: J. Appl. Meteorol. **11**, 1108 (1972)
4.14   R. T. BROWN, Jr., J. APPL. Meteorol. **12**, 698 (1973)
4.15   M. L. WRIGHT, E. K. PROCTOR, L. S. GASIOREK, E. M. LISTON: "A Preliminary Study of Air Pollution Measurement by Active Remote Sensing Techniques", Final Report, SRI Project 1966, Contrast NAS 1-11657, prepared for NASA Langley Research Center, Hampton, Virginia (1975)
4.16   S. E. HARRIS: Proc. IEEE **57**, 1096 (1969)
4.17   R. L. BYER: In *Treatise on Quantum Electronics*, ed. by H. RABIN and C. L. TANG. (Academic Press, New York 1973)
4.18   R. A. FERGUSON: "Feasibility of a cw Lidar Technique for Measurement of Plume Opacity", Final Report SRI Project 1979; EPA No. 650/2-73-037 (Nov. 1973)
4.19   E. E. UTHE, R. J. ALLEN: J. Opt. Quant. Electron. **7**, 121 (1975)
4.20   R. J. FOX, G. W. GRAMS, B. G. SCHUSTER, J. A. WEINMAN: J. Geophys Res. **78**, 7789 (1973)
4.21   M. L. WOLBARSHT (editor): *Laser Applications in Medicine and Biology,* Vol. I (Plenum Press, New York 1971)
4.22   American National Standards Institute: American National Standard for the Safe Use of Lasers (1430 Broadway, New York, April 1973);
       · See also D. H. SLINEY, B. C. GREASIER: Appl. Opt. **12**, 1 (1973)
4.23   H. C. VAN DE HULST: *Light Scattering by Small Particles* (Wiley and Sons, New York 1957)
4.24   M. KERKER: *The Scattering of Light and Other Electromagnetic Radiation* (Academic Press, New York 1969)
4.25   R. C. FLEAGLE, J. A. BUSINGER: *An Introduction to Atmospheric Physics* (Academic Press, New York 1963)
4.26   L. ELTERMAN: "UV, Visible, and IR Attenuation for Altitudes to 50 km, 1968"; Environmental Research Papers, No. 285 (April 1968) Air Force Cambridge Research Laboratories, Bedford, Mass.
4.27   M. C. W. SANDFORD: J. Atmos. Terrest Phys. **29**, 1651 (1967)

4.28 W.C.Bain, M.C.W.Sandford: J. Atmos. Terrest. Phys. **28**, 543 (1966)
4.29 R.L.Schwiesow: In *Remote Sensing of the Troposphere*, ed. by V.E.Derr (U.S. Gov't Printing Office 1972) Chaps. 10 and 42
4.30 R.D.Cadle: In *Chemistry of the Lower Atmosphere*, ed. by S.I.Rasool (Plenum Press, New York, London 1973) Chap. 2
4.31 C.E.Junge: *Air Chemistry and Radioactivity* (Academic Press, New York 1963)
4.32 G.Mie: Ann. Physik **30**, 57 (1909)
4.33 D.Deirmendjian: *Electromagnetic Scattering on Spherical Polydispersions* (American Elsevier, New York 1969)
4.34 M.Born, M.E.Wolf: *Principles of Optics* (Pergamon Press, New York 1959)
4.35 J.V.Dave: "Subroutines for Computing the Parameters of Electromagnetic Radiation Scattered by a Sphere", IBM Report No. 320–3237 (IBM Corporation, PID, 40 Saw Mill River Rd., Hawthorne, New York)
4.36 J.J.DeLuisi, I.H.Blifford,Jr., J.A.Takamine: J. Geophys. Res. **27**, 4529 (1972)
4.37 I.H.Blifford, L.D.Ringer: J. Atmos. Sci. **26**, 716 (1969)
4.38 C.E.Junge: J. Geophys. Res. **77**, 5183 (1972)
4.39 K.Bullrich: In *Advances in Geophysics, Vol* 10, ed. by H.E.Landsberg and J.Van Mieghem, (Academic Press, New York 1964) Chap. 3
4.40 M.P.McCormick, J.D.Lawrence,Jr., F.R.Crownfield,Jr: Appl. Opt. **7**, 2424 (1968)
4.41 H.Harrison, J.Herbert, A.P.Waggoner: Appl. Opt. **11**, 2880 (1972)
4.42 B.J.Brinkworth: Atoms. Environ. **5**, 605 (1971)
4.43 H.C.Bryant, A.J.Cox: J. Opt. Soc. Am. **56**, 1529 (1966)
4.44 J.E.McDonald: Quart. J. Roy. Meteor. Soc. **88**, 183 (1962)
4.45 A.C.Holland, G.Gagne: Appl. Opt. **9**, 1113 (1970)
4.46 P.Huffman, W.R.Thursby, J. Atmos. Sci. **26**, 1073 (1969)
4.47 P.Huffman: J. Atmos. Sci. **27**, 1207 (1970)
4.48 K.N.Liou: Appl. Opt. **11**, 667 (1972)
4.49 J.R.Wait: Canad. J. Phys. **33**, 189 (1955)
4.50 K.N.Liou: J. Atmos. Sci. **29**, 524 (1972)
4.51 A.P.Waggoner, N.C.Ahlquist, R.J.Charlson: Appl. **12**, 2886 (1972)
4.52 J.A.Reagan, B.M.Herman: Joint Conf. on Sens. of Environ. Pollutants, Palo Alto, Calif. (8–10 November 1971) AIAA Paper No. 71–1057
4.53 F.G.Fernald, B.M.Herman, J.A.Reagan: J. Appl. Meteorol. **11**, 482 (1972)
4.54 P.M.Hamilton: Atmos. Environ. **3**, 221 (1969)
4.55 G.Ward, K.M.Cushing, R.D.McPeters, A.E.S.Green: Appl. Opt. **12**, 2585 (1973)
4.56 J.D.Lindberg, L.S.Laude: Appl. Opt. **13**, 1923 (1974)
4.57 L.S.Ivlev, S.I.Popova: Izv. Atmos. Ocean. Phys. **9**, 1034 (1973)
4.58 G.W.Grams, I.H.Blifford, D.E.Gillette, P.B.Russell: J. Appl. Meteorol. **13**, 459 (1974)
4.59 A.P.Waggoner, M.B.Baker, R.J.Charlson: Appl. Opt. **12**, 896 (1973)
4.60 K.Fischer: Beitr. Phys. Atmos. **46**, 89 (1973)
4.61 G.Hänel: Aerosol. Sci. **3**, 377 (1972)
4.62 P.B.Russell, G.W.Grams: J. Appl. Meteorol. **14**, 1037 (1975)
4.63 K.Sassen: J. Appl. Meteorol. **13**, 923 (1974)
4.64 R.M.Schotland, K.Sassen, R.Stone: J. Appl. Meteorol. **10**, 1011 (1971)
4.65 K.N.Liou, H.Lahore: J. Appl. Meteorol. **13**, 257 (1974)
4.66 R.W.Bergstrom: Beitr. Phys. Atmos. **46**, 198 (1973)
4.67 R.A.McClatchey, R.W.Fenn, J.E.A.Selby, F.E.Volz, J.S.Garing: Optical Properties of the Atmosphere (Revised); Environmental Research Papers, No. 354, May 1971, Air Force Cambridge Research

Laboratories, Bedford, Mass. R. A. McCLATCHEY, J. E. A. SELBY: Atmospheric Attenuation of Laser Radiation from 0.76 to 31.25 μm, NTIS No. AD-779 726 (January 1974) Air Force Cambridge Research Laboratories, Bedford, Mass

4.68   W. E. K. MIDDLETON: *Vision Through The Atmosphere* (Univ. of Toronto Press, Toronto 1952)

4.69   H. HORVATH: Atmos. Environ. **5**, 177 (1971)

4.70   P. W. KRUSE, L. D. McGLUCHLIN, R. B. McQUISTON: *Elements of Infrared Technology* (Wiley, New York 1962)

4.71   D. P. WOODMAN: Appl. Opt. **13**, 2193 (1974)

4.72   W. M. PORCH, D. S. ENSOR, R. J. CHARLSON, J. HEINTZENBERG; Appl. Opt. **12**, 34 (1973)

4.73   H. QUENZEL: J. Geophys. Res. **75**, 2915 (1970)

4.74   A. MANI, O. CHACKO, S. HARISHARAN: Tellus **21**, 829 (1969)

4.75   G. PLASS, G. KATTAWAR: Appl. Opt. **10**, 2304 (1972)

4.76   B. M. GOLUBITSKIY, T. ZHAD'KO, M. TANTASHEV: Izv. Atmos. Oceanic Phys. **8**, 1226 (1973)

4.77   K. N. LIOU, R. M. SCHOTLAND: J. Atmos. Sci. **28**, 772 (1971)

4.78   E. ELORANTA: Calculation of Doubly-Scattered Lidar Returns. Thesis, Dept. of Meteorology, University of Wisconsin, Madison, Wisconsin (1972)

4.80   R. M. SCHOTLAND: *Proc. Third Symp. on Remote Sensing of the Environment* (Univ. of Michigan, Ann Arbor, Michigan, 1964) pp. 215–224

4.81   R. M. SCHOTLAND, E. E. CHERMACK, D. J. CHANG: *Proc. First Inst. Symp. on Humidity and Moisture* (Reinhold Book Division, New York 1964) pp. 569–582

4.82   R. M. SCHOTLAND: *Proc. 14th Weather Radar Conference* (Univ. of Arizona, Tucson, Ariz. 1971)

4.83   R. M. SCHOTLAND: J. Appl. Meteor. **13**, 71 (1974)

4.84   M. L. WRIGHT, E. K. PROCTOR, L. S. GASIOREK, E. M. LISTON: "A Preliminary Study of Air Pollution Measurement by Active Remote Sensing Techniques," Final Report, SRI Project 1966, Contrast NAS 1-11657, prepared for NASA Langley Research Center, Hampton, Virginia (1975)

4.85   M. L. WRIGHT (Stanford Research Institute, Menlo Park, California, unpublished)

4.86   J. A. HODGESON, W. A. McCLENNY, P. L. HANST: Science **182**, 248 (1973)

4.87   R. M. MEASURES, G. PILON: Opto-Electron. **4**, 141 (1972)

4.88   R. L. BYER, M. GARBUNY: Appl. Opt. **12**, 1496 (1973)

4.89   S. A. AHMED: Appl. Opt. **12**, 901 (1973)

4.90   R. L. BYER: J. Opt. Quant. Electron. **7**, 147 (1975)

4.91   H. KILDAL, R. L. BYER: Proc IEEE **59**, 1644 (1971)

4.92   T. HENNINGSEN, M. GARBUNY, R. L. BYER: Appl. Phys. Lett. **24**, 242 (1974)

4.93   E. D. HINKLEY: Symp. Remote Sens. Env. Air Pollutants (1974), Pittsburgh Conf. Analytical Chemistry and Applied Spectroscopy (Cleveland, Ohio 1974)

4.94   W. B. GRANT, R. D. HAKE, JR., E. M. LISTON, R. C. ROBBINS, E. K. PROCTOR, JR.: Appl. Phys. Lett. **24**, 550 (1974)

4.95   P. WARNECK, F. F. MARMO, J. O. SULLIVAN: J. Chem. Phys. **40**, 1132 (1964)

4.96   P. L. HANST: In *Adv. in Environ. Sci. and Tech.* Vol. 2, ed. by J. N. PITTS and R. L. METCALF (Wiley-Interscience, New York 1971)

4.97   J. H. CALLOMAN, T. M. DUNN, I. M. MILLS: Phil. Trans. Roy. Soc. (London) 259 *A*, 499 (1966)

4.98   M. GRIGGS: J. Chem. Phys. **49**, 857 (1968)

4.99   B. A. THOMPSON, R. HARTECK, R. R. REEVES, JR.: J. Geophys. Res. **68**, 6431 (1963)

4.100 R. T. MENZIES: Opto-Electron. **4**, 179 (1972)

4.101 D. N. JAYNES, B. H. BEAM: Appl. Opt. **8**, 1741 (1969)

4.102 T. D. WILKERSON, B. ERCOLI, F. S. TOMKINS: Absorption Spectra of Atmospheric Gases, Technical Note BN-784, Institute of Fluid Dynamics and Applied Mathematics, University of Maryland, College Park (February 1974) **46**, 3040 (1975)

4.103 R. T. THOMPSON, JR., J. M. HOELL, JR., W. R. WADE: J. Appl. Phys.

4.104 E. W. BARRETT, O. BEN-DOV: J. Appl. Meteorol. **6**, 500 (1967)

4.105 F. G. FERNALD, B. M. HERMAN, J. A. REAGAN: J. Appl. Meteorol. **11**, 482 (1972)

4.106 W. B. JOHNSON, E. E. UTHE: Atmos. Environ. **5**, 703 (1971)

4.107 P. A. DAVIS: Appl. Opt. **8**, 2099 (1969)

4.109 W. VIEZEE, J. OBLANAS, R. T. H. COLLIS: Evaluation of the lidar technique of determining slant range visibility for aircraft landing operations, SRI Report AFCRL-TR-73-0708 (1973)

4.110 B. M. HERMAN, S. R. BROWNING, J. A. REAGAN: J. Atmos. Sci. **28**, 763 (1971)

4.111 G. W. GRAMS: Optical Radar Studies of Stratospheric Aerosols Ph. D. Thesis, MIT (1966)

4.112 P. B. RUSSELL, W. VIEZEE, R. D. HAKE, JR., R. T. H. COLLIS: Lidar Measurements of the Stratospheric Aerosol over Menlo Park, California, October 1972-March 1974. Stanford Research Institute, Menlo Park, Calif. Final Report Project 2217 (June 1973)

4.113 D. J. GAMBLING, K. BARTUSEK: Atmos. Environ. **6**, 181 and 869 (1972)

4.114. G. W. GRAMS, I. H. BLIFFORD, B. G. SCHUSTER, J. J. DELUISI: J. Atmos. Sci. **29**, 900 (1972)

4.115 P. A. DAVIS: Appl. Opt. **10**, 1314 (1971)

4.116 E. E. UTHE, C. E. LAPPLE: "Study of Laser Backscatter by Particulates in Stack Emissions"; SRI Report EPA Contract CPA-70-173 (1972)

4.117 R. T. H. COLLIS, E. E. UTHE: Opto-Electron. **4**, 87 (1972)

4.118 E. E. UTHE, W. B. JOHNSON: "Lidar Observations of Lower Tropospheric Aerosol Structure during Bomex"; SRI Report AEC Contract AT (04-3) 115 (1971)

4.119 See Ref. [4.160] below

4.120 Dynatrend Inc. (Eds.): Laramie Comparative Experiment, Data Report and Preliminary Report of Conclusions. Report CIAP DOT March (1973)

4.121 P. B. RUSSELL, W. VIEZEE, R. D. HAKE: "Lidar Measurements of Stratospheric Aerosols over Menlo Park, California: October 1972-March 1974". Final Report to NASA; SRI Project 2217; Stanford Research Institute

4.122 See Ref. [4.121]

4.123 R. T. H. COLLIS: Bull. Am. Meteorol. Soc. **49**, 918 (1968)

4.124 P. M. HAMILTON: Phil. Trans. Roy. Soc. (London) A**265**, 153 (1969)

4.125 C. M. R. PLATT: J. Atmos. Sci. **30**, 1191 (1973)

4.126 I. ANDERMO, F. FORNAEUS, I. SVENSSON, G. WIBORG: "Laser Ceilometer: Description and Measurements made until September 1964", FOA 2 Report A 2345-285, Forsvarets Forkningsanstalt, Stockholm (1965)

4.127 L. G. BIRD, N. E. RIDER: Meteorol. Mag. (London) **9.7**, 107 (1968)

4.128 R. T. H. COLLIS: Bull. Amer. Meteorol. Soc. **50**, 688 (1969)

4.129 R. T. H. COLLIS: Science **149**, 978 (1965)

4.130 W. VIEZEE, R. T. H. COLLIS, J. D. LAWRENCE: J. Appl. Meteor. **12**, 140 (1973)

4.131 R. T. H. COLLIS, W. VIEZEE, E. UTHE, J. OBLANAS, R. A. ROBERTS: J. Geophys. Res. **76**, 5194 (1971)

4.132 W. VIEZEE, J. OBLANAS, R. T. H. COLLIS: "Evaluation of the lidar technique of determining slant range visibility for aircraft landing operations", SRI Report AFCRL-TR-73-0708 (1973)

4.133 R. T. H. COLLIS: Q. J. Roy. Meteor. Soc. **92**, 220 (1966)

4.134 S. R. PAL, A. I. CARSWELL: Appl. Opt. **12**, 1530 (1973)

4.135 E. E. UTHE, R. J. ALLEN, P. B. RUSSELL: Stanford Research Institute Final Report, Project 2859 (December 1974)
4.136 R. T. H. COLLIS, M. G. H. LIGDA: Nature **203**, 508 (1964)
4.137 E. E. UTHE: Bull. Am. Meteorol. Soc. **53**, 358 (1972)
4.138 E. E. UTHE, P. B. RUSSELL: Bull. Am. Meteorol. Soc. **55**, 115 (1974)
4.139 V. R. NOONKESTER, D. R. JENSEN, J. H. RICHTER, W. VIEZEE, R. T. H. COLLIS: J. Appl. Meteorol. **13**, 249 (1974)
4.140 D. F. NATUSCH, J. R. WALLACE: Science **186**, 695 (1974)
4.141 T. F. HATCH, P. GRASS: *Pulmonary Deposition and Retention of Inhaled Aerosols* Academic Press, New York 1964)
4.142 P. V. HOBBS, H. HARRISON, E. ROBINSON: Science **183**, 909 (1974)
4.143 J. M. MITCHELL, JR.: J. Appl. Meteorol **10**, 703 (1971)
4.144 J. V. DAVE: J. Appl. Meteorol. **12**, 601 (1973)
4.145 P. CHÝLEK, J. A. COAKLEY, JR.: Science **183**, 75 (1974)
4.146 S. H. SCHNEIDER, R. E. DICKINSON: Rev. Geophys. Space Phys. **12**, 447 (1974)
4.147 See also P. B. RUSSELL, W. VIEZEE, R. D. HAKE, JR., R. T. H. COLLIS: Q. J. Roy. Meteor. Soc. (July 1976) in press
4.148 W. E. EVANS: Development of Lidar Stack Effluent Opacity Measurement Systems, Final Contract Report, Edison Electric Institute, SRI Menlo Park, Ca. (1967)
4.149 S. COOK, G. W. BETHKE, W. D. CONNER: Appl. Opt. **11**, 1972 (1972)
4.150 W. B. JOHNSON: J. Appl. Meteorol. **8**, 443 (1969)
4.151 R. T. H. COLLIS: In *Advances in Geophysics,* Vol. 13, ed. by H. LANDSBERG (Academic Press, New York 1969) p. 113
4.152 P. B. RUSSELL, E. E. UTHE: The Mt. Sutro Tower Aerosol and Radiation Study. Fall Annual Meeting, Amer. Geophys. Union, San Francisco, Calif. (1974)
4.153 J. M. ROSEN: Space Sci. Rev. **9**, 58 (1969)
4.154 C. E. JUNGE, C. W. CHAGNON, J. E. MANSON: J. Geophys. Res. **66**, 2163 (1961)
4.155 P. GRUNER, H. KLEINERT: Probleme der Kosmischen Physik **10**, 1 (1927)
4.156 A. J. BRODERICK (editor): *Proc. Second Conf. Climatic Impact Assessment Program,* 14–17 November 1972 (National Technical Information Service, Springfield, Virginia, No. DPT-TSC-OST-73-4)
4.157 A. J. BRODERICK, T. M. HARD (editors): *Proc. Third Conf. Climatic Impact Assessment Program,* 26 February-1 March 1974 (National Technical Information Service, Springfield, Virginia No. DPT-TSC-OST-74-15)
4.158 G. FIOCCO, G. GRAMS: J. Atmos. Sci. **21**, 323 (1964)
4.159 P. B. RUSSELL, W. VIEZEE, R. D. HAKE, JR., R. T. H. COLLIS: *Proc. Fourth Conf. Climatic Impact Assessment Program,* Cambridge, Mass. (February 1975)
4.160 G. B. NORTHAM, J. M. ROSEN, S. H. MELFI, T. J. PEPIN, M. P. MCCORMICK, D. J. HOFMANN, W. H. FULLER, JR.: Appl. Opt. **13**, 2416 (1974)
4.161 M. P. MCCORMICK, W. H. FULLER, JR.: Appl. Opt. **14**, 4 (1975)
4.162 R. FEGLEY: Sixth Conf. on Laser Radar Studies of Atmos; Sendai, Japan (1974)
4.163 F. G. FERNALD, B. G. SCHUSTER, C. FRUSH: Opt. Quant. Electron. **7**, 141 (1975)
4.164 P. B. RUSSELL, W. VIEZEE, R. D. HAKE: Lidar Measurements of Stratospheric Aerosols over Menlo Park, California; October 1972-March 1974. Final Report to NASA; SRI Project 2217; Stanford Research Institute, Menlo Park, California
4.165 G. FIOCCO, G. GRAMS: Tellus **18**, 34 (1966)
4.166 G. GRAMS, G. FIOCCO: J. Geophys. Res. **72**, 3523 (1967)
4.167 R. T. H. COLLIS, M. G. H. LIGDA: J. Atmos. Sci. **23**, 257 (1966)
4.168 B. R. CLEMESHA, G. S. KENT, R. W. H. WRIGHT: Nature **209**, 184 (1966)
4.169 B. R. CLEMESHA, J. NAKAMURA: Nature **237**, 328 (1972)
4.170 D. R. CLEMESHA, S. N. RODRIGUES: J. Atmos. Terrest. Phys. **33**, 1119 (1971)

4.171 G.S.Kent, B.R.Clemesha, R.W.Wright: J. Atmos. Terrest. Phys. **29**, 169 (1967)

4.172 S.Pilipowskyj, J.A.Weinman, B.R.Clemesha, G.S.Kent, R.W.Wright: J. Geophys. Rev. **73**, 7553 (1968)

4.173 B.G.Schuster: J. Geophys. Res. **75**, 2123 (1970)

4.174 K.Bartusek, D.J.Gambling, W.G.Elford: J. Atmos. Terrestr. Phys. **32**, 1535 (1970)

4.175 M.T.Ottway: 4th Conf. on Laser Radar Studies of Atmos.; Tucson, Ariz. (1972)

4.176 M.Hirono, M.Fujiwara, O.Uchino, T.Itabe: see Ref. 4.214: Rept. Ionosph. Space Res. Japan **26**, 237 (1972); see also Canad. J. Chem. **52**, 1560 (1974)

4.177 U.M.Zakharov, O.K.Kosto, U.S.Portasov: 5th Conf. on Laser Radar Studies of Atmos.; Williamsburg, Va (1973)

4.178 A.Cohen, M.Graber: 5th Conf. on Laser Radar Studies of Atmos.; Williamsburg, Va. (1973)

4.179 S.A.Young, W.G.Elford: Internal Rept. ADP 119, Dept. of Phys., Univ. of Adelaide, Adelaide, Australia (1975)

4.180 R.J.Fox, G.W.Grams, B.G.Schuster, J.A.Weinman: J. Geophys. Res. **78**, 7789 (1973)

4.181 J.F.Cronin: Science **172**, 847 (1971)

4.182 D.J.Hofmann, J.M.Rosen, T.J.Pepin, J.L.Kroening: *Proc. of 2nd Conf. on CIAP*, 14–17 November, 1972, ed. by A.J.Broderick, (available from NTIS, No. DPT-TSC-OST-74-4) pp. 23–33

4.183 D.J.Hofmann, J.M.Rosen, T.J.Pepin, J.L.Kroening: Global Monitoring of Stratospheric Aerosol, Ozone, and Water Vapor. Progress Report No. GM-10 and No. GM-17, Dept. of Physics and Astronomy, Univ. of Wyoming (June 1973) and February 1974)

4.184 F.E.Volz: J. Geophys. Res. **75**, 1641 (1970)

4.185 F.E.Volz: J. Geophys. Res. **78**, 2619 (1973)

4.186 F.E.Volz: Stratospheric Background Aerosol from Twilight Data (draft manuscript, 1974)

4.187 A.W.Castleman: Space Sci. Rev. **15**, 547 (1974)

4.188 G.S.Kent, P.Sandland, R.H.Wright: J. Appl. Meteorol. **10**, 443 (1971)

4.189 G.S.Kent, W.Keenliside, M.C.W.Sandford, R.H.Wright: J. Atmos. Terrestr. Phys. **34**, 373 (1972)

4.190 G.S.Kent, W.Keenliside: J. Atmos. Sci. **31**, 1409 (1974)

4.191 R.D.Hake, D.E.Arnold, D.W.Jackson, W.E.Evans, B.P.Ficklin, R.A.Long: J. Geophys. Res. **77**, 6839 (1972)

4.192 J.E.Blamont, M.L.Chanin, G.Megie: Ann. Geophys. **28**, 833 (1972)

4.193 M.R.Bowman, A.J.Gibson, M.C.W.Sandford: Nature **221**, 456 (1969)

4.194 F.Felix, W.Keenliside, G.Kent: Nature **246**, 345 (1973)

4.195 A.J.Gibson, M.C.W.Sandford: Nature **239**, 509 (1972)

4.196 C.J.Shuler, C.T.Pike, H.A.Miranda: Appl. Opt. **10**, 1689 (1971)

4.197 M.C.W.Sandford, A.J.Gibson: J. Atmos. Terrestr. Phys. **32**, 1423 (1970)

4.198 R.J.Munick: J. Opt. Soc. Am. **55**, 893 (1965)

4.199 R.T.H.Collis: Astron Aeron. **2**, 52 (1964)

4.200 J.D.Lawrence, M.P.McCormick, S.H.Melfi, D.P.Woodman: Appl. Phys. Lett. **12**, 72 (1968)

4.201 R.T.H.Collis: Bull. Am. Meteorol. Soc. **49**, 918 (1968)

4.202 V.E.Derr, C.G.Little: Appl. Opt. **9**, 1976 (1970)

4.203 R.M.Huffaker: Appl. Opt. **9**, 1026 (1970)

4.204 T.R.Lawrence, D.J.Wilson, C.E.Craven, I.P.Jones, R.M.Huffaker, J.A.L. Thomson: Rev. Sci. Instr. **43**, 512 (1972)

4.205 G. BENEDETTI-MICHELANGELI, F. CONGEDUTI, G. FIOCCO: J. Atmos. Sci. **29**, 906 (1973)
4.206 G. FIOCCO, G. BENEDETTI-MICHELANGELI, K. MAISCHBERGER, E. MADONNA: Nature **22**, 78 (1971)
4.207 N. L. ABSHIRE, R. L. SCHWIESOW, V. E. DERR: J. Appl. Meteorol. **13**, 951 (1974)
4.208 K. W. ROTHE, U. BRINKMAN, H. WALTHER: Appl. Phys. **3**, 115 (1974)
4.209 K. W. ROTHE, U. BRINKMAN, H. WALTHER: Appl. Phys. **4**, 181 (1974)
4.210 W. B. GRANT, R. D. HAKE, JR.: J. Appl. Phys. **46**, 3019 (1975)
4.211 P. B. RUSSELL, E. E. UTHE, F. L. LUDWIG, N. A. SHAW: J. Geophys. Res. **79**, 5555 (1974)
4.212 R. W. FEGLEY, H. T. ELLIS: Geophys. Res. Lett. **2**, 139 (1975)
4.213 M. FUJIWARA, T. ITABE, M. HIRONO: Rept. Ionosph. Space Res. **29**, 74 (1975)
4.214 P. B. RUSSELL, W. VIEZEE, R. D. HAKE, JR.: "Lidar Measurements of the Stratospheric Aerosol"; 3rd Quarterly Rept., Project 4019, Stanford Research Institute, Menlo Park, Calif. (November 1975)

5. Detection of Atoms and Molecules by Raman Scattering and Resonance Fluorescence

H. INABA

With 26 Figures

This chapter describes Raman scattering and resonance fluorescence techniques for the remote detection of atoms and molecules. Section 5.1 distinguishes between the fundamental processes involved in the several possible scattering mechanisms and fluorescence, and contains tabulations of theoretical and experimental values of the relevant parameters. The concepts and requirements for operational laser radar systems are described in Section 5.2, which also contains a discussion of propagation effects and several techniques for the remote monitoring of atmospheric temperature. Section 5.3 outlines in detail actual laser radar systems which have been used for remote sensing of atoms and molecules, not only as major components in the ordinary atmosphere, but as trace contaminants in polluted air. Some experimental work on remote temperature monitoring is also mentioned. Section 5.4 discusses the sensitivity of Raman scattering and resonance fluorescence systems, their limitations and extent of applicability, for a wide range of applications, from ground-based to satellite systems. Different signal processing schemes are considered, as well as eye-safety requirements.

5.1 Scattering and Fluorescence Mechanisms

As described in Chapter 2, there are various optical techniques employing lasers upon which spectroscopic methods for remotely detecting atoms and molecules present in the atmosphere can be based. The applicability of lasers for single-ended remote probing derives primarily from their capability to produce very short, but powerful, bursts of coherent radiation. This permits spatially-resolved measurements to be made in a manner termed laser radar or lidar—the optical counterpart of radar.

Raman scattering and resonance fluorescence techniques are the main topics of this chapter. In order to place them in proper perspective with respect to other optical interactions, Table 5.1 summarizes typical characteristics for all of the major interaction regimes of optical waves with atmospheric constituents. Optical scattering is usually divided into three kinds of processes, called Mie, Rayleigh, and Raman, depending

Table 5.1. Comparison of optical interaction processes applicable to laser remote sensing methods in the atmosphere

Interaction	Process	Frequency relation	Cross section $\frac{d\sigma}{d\Omega}$ [cm^2/sr]	Detectable constituents
Scattering	Mie	$v_r = v_0$	10^{-26}–10^{-8} (Aerosols)	Particulate matter
	Rayleigh	$v_r = v_0$	10^{-26} (Non-resonance) 10^{-23} (Resonance)	Atoms and molecules
	Raman	$v_r \neq v_0$	10^{-29} (Non-resonance) 10^{-26} (Resonance)	Molecules (atoms)
Emission	Fluorescence	$v_r = v_0$ $v_r \neq v_0$	10^{-26} (Quenched) 10^{-24} (Quenched)	Atoms and molecules
Absorption		$v_r = v_0$	10^{-20}	Atoms and molecules

v_r: Detection frequency,
v_0: Laser frequency

on the species and sizes of scatterers compared with the wavelength. Although the emission process includes both spontaneous (fluorescence) and stimulated emission, the latter process seems to contribute nothing to the observable scattered energy in the atmosphere. The fourth column of Table 5.1 shows typical values of interaction cross sections ($d\sigma/d\Omega$) for the SO$_2$ molecule (except for the case of Mie scattering, which pertains to aerosols only) for optical waves in the region from 300 to 400 nm.

Mie scattering, described in detail in Chapters 3 and 4, is classical, elastic scattering occuring at the transmitted wavelength, and takes place when the dimensions of the particle are close to or larger than the optical wavelength. The Mie-scattered light is concentrated in the forward direction, with a much smaller intensity backward. Although the cross section of this scattering is customarily very large, so that high sensitivity is achieved for detecting particulate matter such as dust, water droplets, and other aerosols [5.1–6], Mie scattering does not lead to a quantitative analysis of atomic or molecular constituents in the atmosphere.

The differential absorption process employing a remote reflector can be used for the detection of atoms and molecules in the atmosphere either by using coincidences with fixed-frequency laser beams or by tuning laser lines to absorption lines of the specific species of interest, as described in Chapter 6. Although the absorption cross sections for several molecular species are large enough to permit their detection at ambient levels, and they are not modified by quenching due to inelastic collisions in the air (as is the case for fluorescence), this technique does not have the

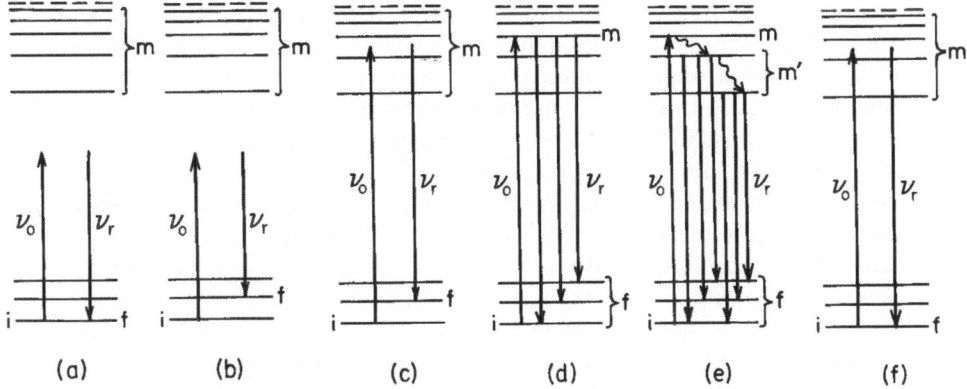

Fig. 5.1a–f. Schematic energy level diagrams of the atom and molecule, and various interaction processes associated with scattering and fluorescence. (a) Rayleigh scattering, (b) ordinary Raman scattering, (c) resonance Raman scattering, (d) resonance fluorescence, (e) broad fluorescence, and (f) resonance scattering

capability of range resolution, but allows only the measurement of *average* concentration along a laser beam path using well-known techniques [5.7–10]. This disadvantage can be overcome by measuring the difference of elastically backscattered intensities from distributed Mie and Rayleigh scatterers between two transmitted wavelengths *on* and *off* an absorption line of the specific atom or molecule [5.11–18]. This method, described in detail in Chapter 4, is called range-resolved differential absorption or differential absorption of scattered energy, and combines the single-ended technique with the high sensitivity of resonance absorption not only in the near ultraviolet and visible, but also in the infrared [5.19].

Rayleigh scattering is known to occur in phase with the incident radiation without any appreciable energy exchange with internal states of the atom or molecule, as depicted by Fig. 5.1a. In this figure, v_0 is the incident optical frequency (wavenumber) and v_r is the scattered frequency. The figure illustrates electronic ground and excited states, with the individual levels designated i for initial, m for intermediate, and f for final. These levels are supposed to be established for atoms by interactions producing fine and hyper-fine structures; and for molecules they correspond to vibrational-rotational levels. The Rayleigh-scattered energy is concentrated along the direction of the incident beam, with equal intensities forward and backward. Since the central wavelength of the Raleigh-scattered component is the same as that of the Mie-scattered component, and its intensity is a smooth function inversely proportional to the fourth power of the wavelength, the scattered light does not identify the scatterer; thus its usefulness in monitoring and analyzing the atmosphere is limited. Moreover, although the spectral distribution of Rayleigh scattering contains temperature information via the Doppler effect, it always requires rather high sensitivity measurements because only very small amounts of energy are exchanged by this effect.

Figure 5.1 also presents the possible types of transitions involved in the atomic and molecular processes by which a photon is emitted inelastically. Raman scattering is the process involving an exchange of a significant amount of energy between the scattered photon and the scattering species, as shown in Fig. 5.1b. Thus the Raman scattering component is shifted from the incident beam frequency v_0 by an amount corresponding to the internal energy of the species. The cross section for Raman scattering is usually smaller than that for Rayleigh scattering by about three orders of magnitude. However, the Raman scattering scheme is a useful laser radar technique [5.20] because it provides a means of identifying and monitoring atmospheric constituents from a single location, and because the effect is independent of the transmitted wavelength. The Raman-scattered intensity is proportional to the number of molecules in their initial states producing the spectral band.

When the exciting frequency v_0 is tuned near or close to a proper resonance of the atom or molecule, as shown in Fig. 5.1c, a great increase in the Raman scattering cross section is generally expected by resonance enhancement of the polarizability tensor. This process is conventionally called resonance Raman scattering [5.21, 22], and includes not only exactly *on*-resonance but also *near*-resonance interactions. Although this effect has been recognized and discussed for many years, it has been only recently that the availability of tunable laser sources has opened the way for quantitative measurements of this process in a variety of gases, liquids and solids. Three to six orders of magnitude enhancement over the ordinary Raman scattering cross section for the N_2 molecule was reported in going to the resonance Raman effect for gases such as I_2, NO_2, and O_3. Hence, resonance Raman scattering will be of great interest in significantly improving scattering efficiency, and will, therefore, be useful in allowing one to remotely detect lower concentrations of atmospheric constituents as long as the transmitted beam is not severely attenuated by absorption.

Fluorescence is the spontaneous emission of a photon following excitation into an excited state by absorption of incident radiation at a frequency v_0 within a specific absorption line or band of an atomic or molecular species. In Fig. 5.1d the excited level decays by re-emitting photons via transitions to the original and different lower levels. These emissions exhibit discrete peaks conventionally called resonance fluorescence. The excited atoms and molecules also suffer collisions which redistribute them into other excited levels through non-radiative transitions, as denoted by the wavy arrows in Fig. 5.1e. This process usually yields broad fluorescence as a near continuum. As a consequence, the excitation of fluorescence always requires the use of tunable coherent optical sources (tunable lasers and optical parametric devices) to match the proper resonance frequency of the species. The re-emitted radiation, measured at the appropriate frequency, is useful in identifying and monitoring the atomic or molecular species responsible for the fluorescence.

All kinds of fluorescence generally undergo quenching in the atmosphere because of collisions with air molecules, which lowers the intensity by several orders of magnitude compared with the low pressure case (either in the laboratory or in the upper atmosphere). Nevertheless, the cross section of quenched fluorescence is usually still larger than Rayleigh and ordinary Raman cross sections. It should also be noted that collision processes, in general, also give rise to other effects, such as spectral broadening of fluorescence, increasing its depolarization, and decreasing the anisotropy of its angular distribution, due to "loss of memory" of the direction and polarization of the exciting beam.

Fluorescence is customarily thought of as two single-photon processes; that is, a two-step interaction consisting of absorption of a single photon with frequency v_0 followed by spontaneous emission of a photon with frequency v_r. Hence it involves intrinsically a time uncertainty between the two processes, and its intensity displays an exponential decay at low gas pressures. On the other hand, scattering associated with an individual atom or molecule, like Rayleigh and Raman scattering, is generally considered to be a two-photon process described by a single-step interaction yielding effectively the simultaneous destruction of one photon with frequency v_0 and the creation of a different photon with frequency v_r. All experimental evidence so far indicates that the collisional quenching previously mentioned with respect to fluorescence has no meaning for two-photon scattering since this process is effectively instantaneous.

There is another situation, depicted in Fig. 5.1f, in which excitation in a region of strong absorption produces simultaneous emission of a photon equal to or in close proximity with the exciting frequency v_0. HEITLER [5.23] referred to this two-photon process as *resonance fluorescence* in his well-known book, and recently HUBER [5.24] called it *resonant scattering*. Strictly speaking, use of the term *resonance fluorescence* seems to be somewhat confusing because fluorescence should be regarded legitimately as two single-photon processes. Consequently, we will refer to it as *resonance scattering*.

Since the frequency of resonance scattering is coincident with or very close to that of Mie scattering, as well as resonance fluorescence, a background problem due to the former component usually restricts usefulness of this scheme in the lower atmosphere. However, for atomic vapors, which possess large cross sections for the resonant radiation, and at high altitudes where background interference is not serious, this resonance scattering scheme operated in the near ultraviolet and visible regions offers good sensitivity for laser radar probing.

5.2 Basic Properties of Raman Scattering and Resonance Fluorescence

5.2.1 Raman Scattering

Raman scattering of incident radiation from gases, liquids, and solids has been of interest for a long time as a source of information about various materials, especially their molecular structure and chemical composition [5.25]. A spectral analysis of a scattered wave reveals the existence of a series of sideband frequencies v_r, shifted down and up by amounts equal to the vibrational-rotational frequencies $\tilde{v}_r$ of the molecules irradiated

$$v_r = v_0 \pm \tilde{v}_r, \tag{5.1}$$

where v_0 is the frequency (wavenumber) of the incident wave. Sideband frequencies below v_0 are called Stokes lines, while those above v_0 are called anti-Stokes lines. Detailed explanations and listings of Raman lines are found in textbooks [5.26]. The current renaissance in Raman spectroscopy has been achieved to a large extent by the availability of the laser, and recently there has been considerable activity in both the basic and applied aspects of this field.

Figure 5.2 summarizes the frequency shifts of the vibrational-rotational Raman spectra of typical trace molecules present in the atmosphere, with respect to the transmitted laser frequency. (The values listed refer to the central frequency of the Q-branch of the particular Raman band.) Also included are molecules such as nitrogen, oxygen, and water vapor which are present in higher quantities. An estimate of the absolute concentration of each molecular species can be performed by comparing the Raman backscattered intensity with that of the Raman line from N_2 molecules which occupy the same volume [5.27–29].

The selection rules for Raman transitions for diatomic and linear molecules are known to be $\Delta v = 0, \pm 1$, and $\Delta J = 0, \pm 2$, where v and J indicate the vibrational and rotational quantum numbers, respectively. The transition specified by $\Delta v = 0$, $\Delta J = \pm 2$ corresponds to pure rotational Raman scattering; the transitions $\Delta v = \pm 1$, $\Delta J = 0, \pm 2$ to vibra-

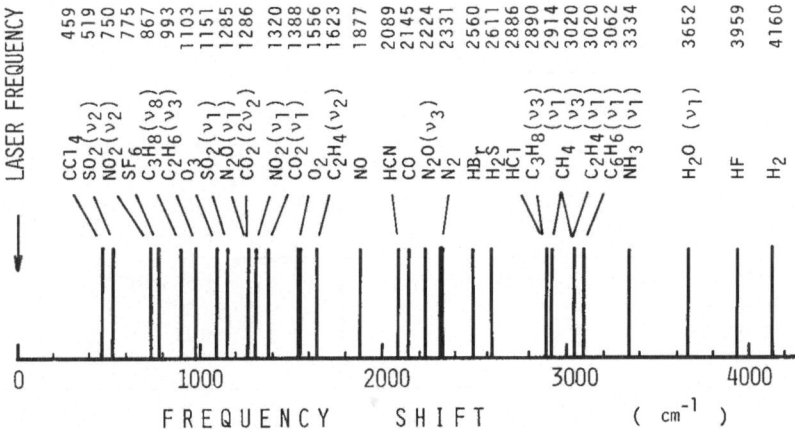

Fig. 5.2. Frequency shifts of Q-branch of vibrational-rotational Raman spectra of typical molecules present in ordinary and polluted atmosphere relative to the exciting laser frequency

tional-rotational Raman scattering; whereas the transition $\Delta v = 0$, $\Delta J = 0$ gives rise to Rayleigh scattering as a natural consequence.

In general, the scattering of radiation from atoms, molecules, and particles is not isotropic. The cross section is then specified by the differential cross section $\sigma(\theta, \phi) = d\sigma/d\Omega$ per unit solid angle, where θ is the scattering angle and ϕ the polarization angle. The total scattered power $P_{\Omega,j}$ per unit solid angle for a Raman band corresponding to the jth vibrational mode of the molecules, is related to the incident intensity I_0 [power per unit area] by the differential Raman scattering cross section: $P_{\Omega,j} = (d\sigma_j/d\Omega)I_0$. According to Placzek's polarizability theory [5.30], the differential backscattering cross sections for vibrational-rotational transitions are generally given by the following expressions when Raman scattering is observed in a direction perpendicular to the polarization direction of linearly-polarized incident light:

a) Vibrational Raman backscattering cross section for Q-branch ($\Delta v = +1$, $\Delta J = 0$)

$$\left(\frac{d\sigma_j}{d\Omega}\right)^Q_{\text{Ram}} = \frac{(2\pi)^4 b_j^2 (v_0 - \tilde{v}_j)^4}{[1 - \exp(-hc\tilde{v}_j/kT)]} g_j \left(\hat{i}_j^2 + \frac{7}{180} \hat{a}_j^2\right). \tag{5.2}$$

b) Vibrational Raman backscattering cross section for O- and S-branches ($\Delta v = +1$, $\Delta J = -2$; and $\Delta v = +1$, $\Delta J = +2$, respectively)

$$\left(\frac{d\sigma_j}{d\Omega}\right)^{O+S}_{\text{Ram}} = \frac{(2\pi)^4 b_j^2 (v_0 - \tilde{v}_j)^4}{[1 - \exp(-hc\tilde{v}_j/kT)]} g_j \frac{7}{60} \hat{a}_j^2. \tag{5.3}$$

Table 5.2. Theoretical values of differential Raman backscattering cross section for Q-branch, O- and S-branches, and total of these three branches for the vibrational transition $v=0\rightarrow1$ excited at the wavelength of 337.1 nm; together with related molecular constants for N_2, O_2, CO_2, and CH_4. (N_A: Avogadro's number)

Molecule	$\tilde{v}_j$ [cm^{-1}]	$g_j\hat{i}_j^2$ [cm^4/g]	$g_j\hat{a}_j^2$ [cm^4/g]	Q-branch $\left(\dfrac{d\sigma}{d\Omega}\right)^Q_R$ [cm^2/sr]	O- and S-branches $\left(\dfrac{d\sigma}{d\Omega}\right)^{O+S}_R$ [cm^2/sr]	Total $\left(\dfrac{d\sigma}{d\Omega}\right)^T_R$ [cm^2/sr]
N_2	2329.66	$0.45\times10^{-32}N_A$	$0.64\times10^{-32}N_A$	2.9×10^{-30}	0.55×10^{-30}	3.5×10^{-30}
O_2	1556.26	0.27	1.08	3.3	1.3	4.6
$CO_2(v_1)$	1388.15	0.26	0.52	3.4	0.73	4.2
$CH_4(v_1)$	2914.2	4.56	0	21.0	0	21.0

c) The total vibrational Raman backscattering cross section, obtained by adding (5.2) and (5.3) is

$$\left(\frac{d\sigma_j}{d\Omega}\right)^T_{Ram} = \frac{(2\pi)^4 b_j^2(v_0-\tilde{v}_j)^4}{[1-\exp(-hc\tilde{v}_j/kT)]}\, g_j\left(\hat{i}_j^2 + \frac{7}{45}\hat{a}_j^2\right). \tag{5.4}$$

Here $\tilde{v}_j$ is the frequency [cm^{-1}] of the j-th vibrational mode of the molecule, $b_j[=(h/8\pi^2 c\tilde{v}_j)^{1/2}]$ is the zero-point vibrational amplitude of this mode, g_j is the degeneracy of the j-th vibrational mode, $\hat{i}_j^2$ and $\hat{a}_j^2$ are the isotropic and anisotropic parts of the derived polarizability tensor associated with the normal coordinate Q_j, and T is the absolute temperature. The other symbols have their usual meaning.

It should be noted that the differential backscattering cross sections for *pure rotational* Raman scattering and Rayleigh scattering are also given by expressions (5.2) and (5.3) by letting $b_j^2 g_j/[1-\exp(-hc\tilde{v}_j/kT)]=1$, and interpreting $\hat{i}_j^2$ and $\hat{a}_j^2$ as the isotropic and anisotropic parts $\hat{i}^2$ and $\hat{a}^2$ of the polarizability tensor for these scatterings specified by $\Delta v=0$. The above equations are valid under the assumption that the exciting frequency v_0 is very much less than any transition frequency between electronic states of the molecule, i.e., the *off*-resonant condition.

In Table 5.2 are listed values of the three kinds of differential Raman scattering cross sections, calculated according to (5.2–4) for the N_2, O_2, CO_2, and CH_4 molecules. Also listed are values of the Raman frequency shift $\tilde{v}_j$ and components involved in the Raman polarizability tensor $g_j\hat{i}_j^2$ and $g_j\hat{a}_j^2$ [5.25, 31] in terms of Avogadro's number N_A. These results are referred to an incident wavelength of 337.1 nm where necessary, in proportion to $(1/\lambda_r)^4$, λ_r being the Raman-shifted wavelength. Given in Table 5.3 for these molecules are values for the differential cross section of Rayleigh backscattering, pure rotational Raman backscattering,

Table 5.3. Theoretical values of differential cross sections for Rayleigh, pure rotational Raman backscattering, and their total backscattering, excited at the wavelength of 337.1 nm; and the components of the polarizability tensor for N_2, O_2, CO_2, and CH_4 molecules

Molecule	$\hat{\imath}^2$ [cm^{-3}]	$\hat{a}^2$ [cm^{-3}]	Rayleigh $d\sigma/d\Omega$ [cm^2/sr]	Pure rotational $d\sigma/d\Omega$ [cm^2/sr]	Total $(d\sigma/d\Omega)^{\mathrm{T}}$ [cm^2/sr]
N_2	3.20×10^{-48}	0.90×10^{-48}	3.90×10^{-27}	1.14×10^{-28}	4.01×10^{-27}
O_2	2.66	1.40	3.28	1.96	3.47
CO_2	7.24	5.94	9.02	8.3	9.85
CH_4	7.13	0	8.60	0	8.60

total backscattering, and the components $\hat{\imath}^2$ and $\hat{a}^2$ [5.32–34] of the polarizability tensor relevent to these processes.

As an example, Fig. 5.3 represents the theoretical distribution of a typical *vibrational-rotational* Raman spectrum for the N_2 molecule at a temperature of 300 K as a function of wavenumber corresponding to the Stokes shift [5.20]. The ordinate gives the value of differential scattering cross section under the excitation of incident radiation at 337.1 nm. In this figure, which corresponds to the vibrational transition $v = 0 \rightarrow 1$, all lines in the Q-branch lie very close to each other and are not resolved except with extremely high resolution spectroscopy. The S- and O-branches, however, are well separated in energy and appear as sidebands on either side of the intense Q-branch. Figure 5.4 shows the calculated spectrum of typical *rotational* Raman scattering for the N_2 molecule at 300 K, excited by an incident beam at 337.1 nm. Two components, S- and O-branches, are also positioned symmetrically on either side of the central Rayleigh component. Both Figs. 5.3 and 5.4 explain quite well the measured results of N_2 vibrational-rotational and pure rotational Raman spectra obtained using high-resolution Raman spectrometers with Ar ion laser sources [5.35, 36]. Since the rotational and vibrational degrees of freedom of the molecular species are sensitive to the molecular temperature, techniques based on Raman signatures can be implemented for the remote measurement of gaseous temperature, both in thermal and non-thermal equilibrium conditions. This application will be described later.

For the case of atoms, in which scattering generally involves the electronic excited states, quantum-mechanical expressions for Rayleigh and Raman scattering cross sections were derived in terms of oscillator strengths and vector-coupling coefficients by PENNEY [5.37]. His results contain previously neglected contributions of the anti-symmetric parts of the polarizability tensor, which are significant over wide wavelength ranges for atoms with non-zero electronic angular momentum. One can

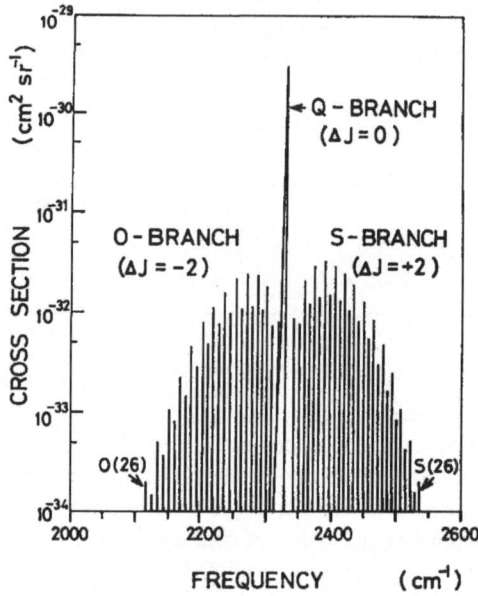

Fig. 5.3. Calculated distribution of vibrational-rotational Raman spectrum at 300 K showing the O-, Q-, and S-branch structures, and differential Raman scattering cross section for N_2 molecule excited at the wavelength of 337.1 nm

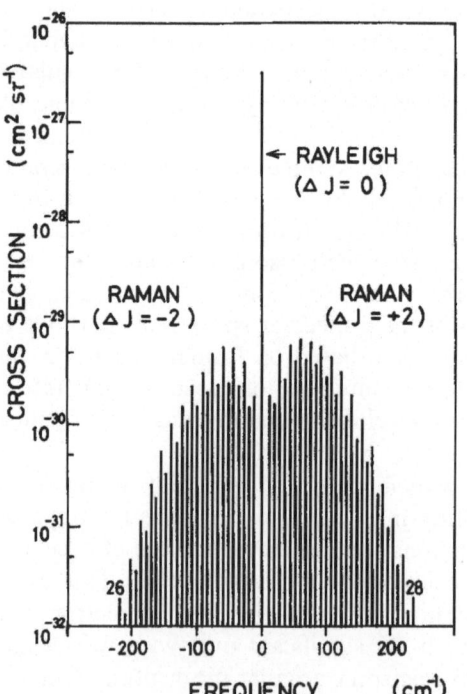

Fig. 5.4. Calculated distribution of pure rotational Raman and Rayleigh spectra at 300 K, showing the O- and S-branch structures and central Rayleigh component. and differential scattering cross section for N_2 molecule excited at the wavelength of 337.1 nm

Table 5.4. Summaries of the experimental values for absolute differential scattering cross section of the 2331 cm^{-1} Q-branch and total Raman transitions in the N_2 molecule. Each value measured at the incident wavelength shown in the last column has been converted, for comparison, to values for the excitation wavelengths of 514.5, 488.0, and 337.1 nm by assuming that the cross section is proportional to $1/\lambda_r^4$

Wavelength referred to excitation source	$d\sigma/d\Omega$ ($\times 10^{-31}$ cm^2/sr)				Excitation wavelength for the measurement
	514.5 nm	488.0 nm	337.1 nm		
Workers	Q-branch	Q-branch	Q-branch	Total	
MURPHY et al. [5.31]	4.4 ± 0.3	5.5 ± 0.4	28 ± 2	35	435.8 nm
STANSBURY et al. [5.32]	3.8	4.8	25	—	435.8 nm
PENNEY et al. [5.38]	4.4 ± 0.2	5.6 ± 0.3	29 ± 2	—	514.5 nm
FOUCHE et al. [5.39]	4.4 ± 1.7	5.6 ± 2.2	29 ± 11	—	514.5 nm
FENNER et al. [5.40]	2.6 ± 0.9	3.3 ± 1.1	17 ± 6	—	488.0 nm
HYATT et al. [5.41]	4.2 ± 0.2	5.4 ± 0.3	28 ± 2	—	514.5 and 488.0 nm

utilize his expressions for a quantitative evaluation of the differential cross section not only for ordinary Raman scattering, but also for near-resonance Raman scattering in the proximity of a strong, isolated atomic line, as will be shown below.

The practical measurement of species concentration in the atmosphere can be greatly facilitated if the Raman backscattering cross sections of the various gases relative to that of N_2 are accurately known. Namely, once the precise ratio is obtained, the concentration of a specific species is then equal to the N_2 concentration scaled by the ratio between the observed Raman-backscattered intensity of the species and that of N_2 gas. The absolute Raman scattering cross section of the N_2 molecule corresponding to the 2331 cm^{-1} Q-branch has been measured by several different spectroscopic methods, including comparison with the known cross section for the 992 cm^{-1} line in liquid benzene, comparisons with the absolute Rayleigh cross section and the absolute cross section for the rotational Raman transition ($J = 1 \rightarrow 3$) in the hydrogen molecule, and direct measurement with a calibrated spectrometer. Table 5.4 gives a summary of the experimental values for the absolute differential Raman scattering cross section for the Q-branch, along with published absolute values of the total differential cross section for the N_2 molecule determined by several workers [5.31, 32, 38–41]. The estimated values for excitation wavelengths of 514.5, 488.0 and 337.1 nm are tabulated by assuming the cross section to be proportional to $1/\lambda_r^4$, for comparison with the theoretical values shown in Table 5.2. The pre-laser values of MURPHY et al. [5.31], and STANSBURY et al. [5.32] have been converted to values for the three excitation wavelengths in the same manner.

It should be pointed out that most of the recent laser-Raman data for the absolute scattering cross section for N_2 which were measured independently give good agreement with the pre-laser N_2 value by MURPHY et al. and the theoretical values tabulated in Table 5.2, within experimental errors. They also reveal good coincidence with each other, using different techniques, except for the value reported by FENNER et al. [5.40]. Moreover, HYATT et al. [5.41] reported that their measurement of the frequency dependence of the scattering closely follows the fourth power of the Raman frequency for N_2, and no resonance effect was observed in the visible region.

Raman scattering cross sections of a variety of gaseous molecules have been measured using visible light sources such as mercury arc lamps and gas and solid-state lasers operated either cw or pulsed. An extensive list of pre-laser data of vibrational Raman scattering cross sections obtained with a mercury lamp source (435.8 nm) has been published by MURPHY et al. [5.31] for a large number of molecules. Recently, some groups, including LEONARD [5.42], FOUCHE and CHANG [5.39], CHEN and WU [5.43], PENNEY et al. [5.38], PORTO et al. [5.40, 41], SCHWIESOW [5.44], and LEVATTER et al. [5.45], have measured the relative values of Raman scattering cross section for various important molecules using visible and near-ultraviolet laser sources. We have also measured the Raman scattering cross sections relative to N_2 for a number of gases of interest in air pollution studies by means of a pulsed Raman spectrometer incorporating a pulsed N_2 laser (337.1 nm) and a pulse-gated photon counting method developed for extremely high-sensitivity detection [5.20, 46, 47].

Table 5.5 summarizes the measured values of differential Raman backscattering cross sections of various molecules present in polluted as well as ordinary atmospheres, together with the frequency shifts of the Raman bands for each molecule. All listed values have an accuracy of about 10 %, and are referred to the excitation wavelength of 337.1 nm for the comparison. They can be readily converted to any wavelengths according to (5.2) or (5.4) unless a resonance effect takes place at a particular wavelength region. The sign (Q) indicates the value of the Q-branch vibrational Raman backscattering cross section instead of that of the total cross section.

5.2.2 Difference Between Resonance Raman Scattering and Fluorescence

Resonance Raman scattering, that is, Raman scattering enhanced by proximity of the exciting line to resonance, has received considerable attention in the last two decades. Although resonance excitation raises the scattering power of many substances by several orders of magnitude, the resonance Raman effect cannot always be observed with ease because the scattered radiation may be substantially attenuated by the simultaneously-intensified absorption, and possibly concealed by superimposed fluorescence emission.

With regard to the distinction between resonance Raman scattering and fluorescence, there have been considerable discussions from various points of view in the last few years [5.48]. Several experimental and analytical features were mentioned previously with respect to this distinction; for instance, Raman scattering primarily takes place during an extremely short time which could be recognized as effectively instantaneous, and its frequency shift is independent of the incident frequency, while fluorescence shows an exponential decay, with its frequency characteristic of the specific transition of the species of interest. Resonance Raman scattering is usually strongly polarized and always insensitive to quenching due to inelastic collisions [5.22] (as observed for resonance scattering of molecular excitations above the dissociation limit), in contrast to fluorescence. Also, there is an argument that the re-emission dominated by transitions through a single intermediate state should be called fluorescence, whereas the scattering process significantly involves at least several intermediate states [5.49]. Since these definitions are apparently applicable to certain specific materials, it seems less meaningful to search for unified rules for general occasions covering not only gases, but solids and liquids as well.

Recent experimental studies performed by WILLIAMS et al. [5.50], in which the scattering lifetime for the transition below the $B(^3\Pi_{0u})$ dissociation limit of the I_2 molecule was measured as a function of the incident laser frequency, demonstrated that at low gas pressure (~ 0.03 Torr) the essential character of the re-emission accomplishes a continual transition from that of Raman scattering to resonance fluorescence when the frequency approaches on-resonance. Their result is substantially indicative of the following situation: For fluorescence under conditions where the incident laser frequency coincides with a transition between a ground and an excited state, the time response for the re-emission process is slow and governed by the natural lifetime of the excited state at low gas pressure, but the lifetime is quenched by collisions at high gas pressure. As the incident frequency is tuned away from resonance, beyond substantial overlap with the spectral absorption, the response appears to be instantaneous in that it follows the exciting laser pulse, and the re-emission displays the usual properties of Raman scattering by a gradual decrease in intensity.

When the incident frequency is only slightly off-resonance, the lifetime Δt for re-emission is expected to be limited by the frequency difference $\Delta\omega$ between the excited state and the incident radiation; so that it should

Table 5.5. List of the measured differential Raman backscattering cross sections for various gaseous molecules present in the atmosphere relative to that for the N_2 molecule, together with the Raman frequency shifts. All of these data have been referred to the excitation wavelength of 337.1 nm, where necessary, in proportion to $1/\lambda_r^4$. The wavelength indicated in the parentheses below the worker's name in each column is the excitation wavelength used for the measurement

Molecule	Frequency shift $\tilde{\nu}_j$ [cm^{-1}]	MURPHY et al. [5.31] (435.8 nm)	STANSBURY et al. [5.32] (404.7, 435.8 nm)	LEONARD [5.42] (337.1 nm) and CHEN and WU* [5.43] (694.3 nm)	FOUCHE et al. [5.39] (514.5 nm)	INABA et al. [5.20,46,47] (337.1 nm)	PENNEY et al. [5.38] (514.5 nm)	FENNER et al. [5.40,41] (488.0 nm)	SCHWIESOW [5.44] (488.0 nm)	LEVATTER et al. [5.45] (337.1 nm)
CCl_4	459	7.4				7.7				
$SO_2(\nu_2)$	519							0.11(Q)		
(ν_1)	1151			2.4(Q)	5.0(Q)	5.7(Q) 4.9	5.4(Q)	4.3(Q)	5.1(Q)	5.46(Q)
$NO_2(\nu_2)$	750					7.2				
(ν_1)	1320					15				
SF_6	775	3.3				3.4				
$C_3H_8(\nu_8)$	867	2.2			2.1(Q)					3.19(Q)
(ν_{11}, ν_5)	1451									
(ν_3)	2890				6.1(Q)					
$C_3H_6(\nu_{13})$	920									1.16(Q)
(ν_{10})	1297									2.63(Q)
$C_6H_6(\nu_2)$	992	12.2				11.8		7.4(Q)		
(ν_1)	3062	10.4				10.1(Q)		8.3(Q)		
$C_2H_6(\nu_3)$	993	2.7						1.5(Q)		
O_3	1103				3.6(Q)					
$N_2O(\nu_1)$	1285				2.5(Q)	1.1		2.0(Q)		
(ν_3)	2224				0.53(Q)			0.51(Q)		
$CO_2(2\nu_2)$	1286	0.86(Q) 0.9			1.4(Q)	0.84(Q) 1.1		0.83(Q)	0.93(Q)	
(ν_1)	1388	1.2(Q) 1.1			1.2(Q)	1.2(Q) 1.?	1.4(Q)	1.3(Q)		1.36(Q)

Molecule	$\tilde\nu$							
$C_2H_4(\nu_3)$ (ν_2) (ν_1)	1342 1623 3020	3.1(Q) 2.0(Q) 5.8(Q)			3.2(Q) 2.0(Q) 5.3(Q)			
N_2O_4	1360	0.051*						
O_2	1556	1.2(Q) 1.4	1.2(Q)		1.4(Q) 1.4	1.1(Q)	1.2(Q)	1.22(Q)
NO	1877	0.5(Q)	0.53(Q)	0.46(Q)	0.44(Q)	0.26(Q)	0.43(Q)	0.47(Q)
CO	2145	0.92(Q)		0.91(Q)	0.90(Q)	0.97(Q)	0.98(Q)	0.99(Q)
N_2	2331	1(Q) 1	1(Q) 1	1(Q)	1(Q) 1	1(Q)	1(Q) 1	1(Q)
$ND_3(\nu_1)$	2420						3.1(Q)	
HBr	2560	4.4	6.2(Q)					
H_2S	2611	5.6		7.2(Q)			6.5(Q)	
CH_3OH (ν_2) $(2\nu_6)$	2846 2955	4.0 2.2	3.9 2.1					
HCl	2886	2.7	3.4(Q)					
$CH_4(\nu_1)$ (ν_3)	2914 3020	7.3(Q) 1.1(Q)		8.4(Q) 0.84(Q)	7.5(Q) 1.2(Q)	7.7(Q)	6.3(Q)	
C_2H_5OH	2943	5.5		5.5				
D_2	2986	2.4(Q) 2.1	1.8(Q)					
$NH_3(\nu_1)$	3334	3.2					5.4(Q)	
$H_2O(\nu_1)$	3652	1.6			2.5(Q)		2.1(Q)	
H_2	4160	3.1	2.9(Q)	2.6(Q)	3.2(Q) 2.3		1.9(Q)	

be determined approximately by the uncertainty principle, $\Delta t \simeq 1/\Delta\omega = 1/2\pi c \Delta v$. In other words, if the energy is not conserved by an amount $\hbar\Delta\omega$ in the transition to the excited state, then the time the molecule or atom can spend in the state is limited to $1/\Delta\omega$. From these considerations we infer that exact-resonance Raman scattering, for which $\Delta\omega = 0$, and resonance fluorescence are the same long-lived process associated with the intrinsic lifetime of the excited state of interest, and that the decay time varies continuously from it to an instantaneous value to provide more Raman-like behavior at off-resonance. Furthermore, the difference between so-called resonance Raman scattering below [5.50–53] and above [5.22, 54, 55] the dissociation limit can now be properly understood. The former process is strongly dependent on the exciting frequency, whereas the latter process does not have a large exciting-frequency dependence because any initial state is in resonance with a transition to some continuum states.

Nevertheless, many experimental results have shown that the intensity of Raman scattering can be substantially enhanced in approaching a strong, isolated absorption line or band before the transition to fluorescence-like properties takes place. As this type of scattering, which has been called near-resonance or pre-resonance Raman scattering, should be attractive and useful in the remote detection of atomic and molecular species in the atmosphere by the laser radar technique, we will give a more detailed description of this process in the next section.

In order to avoid confusion, we shall adopt the following point of view throughout this chapter, similar to PENNEY et al. [5.48] and WILLIAMS et al. [5.50]: Scattering is effectively an instantaneous process, insensitive to effects of atomic and molecular interaction such as quenching, collisional depolarization, and relaxation depending upon species and pressure. On the other hand, fluorescence is associated with a process accompanying a measurable time decay at low gas pressure, and is quenched significantly through collision effects at high pressure. We also note that in the case of pulse excitation, the two properties involved in scattering would appear to be closely related if there is insufficient time for a species to experience a collision.

5.2.3 Near-Resonance Raman Scattering

Analytical discussions of enhanced Raman scattering due to near-resonance excitation have been performed by several workers [5.56–59] by introducing a damping term into the resonance denominator for the expression of the polarizability tensor or the scattering cross section. The basic expression for the differential Raman scattering cross section can generally be written in the form

$$
\left(\frac{d\sigma}{d\Omega}\right)_{RR} = \frac{\omega_r^4}{c^4\hbar^2}\frac{N_i}{N}\sum_f\left|\sum_m\left[\frac{\langle f|\mu_r|m\rangle\langle m|\mu_0|i\rangle}{(\omega_m-\omega_i)-\omega_0+i\Gamma_m/2}\right.\right.
$$
$$
\left.\left.+\frac{\langle f|\mu_0|m\rangle\langle m|\mu_r|i\rangle}{(\omega_m-\omega_f)+\omega_0-i\Gamma_m/2}\right]\right|^2. \tag{5.5}
$$

Here μ_0 and μ_r are components of the dipole moment vector in the directions of electric field polarization of the incident and scattered radiation with angular frequencies of ω_0 and ω_r, respectively; and $\langle m|\mu_0|i\rangle$, for instance, designates the matrix element of μ_0 between the initial state i and an intermediate state m, while $\langle f|\mu_r|m\rangle$ is the matrix element of μ_r between m and the final state f. $\hbar\omega_i$, $\hbar\omega_m$ and $\hbar\omega_f$ are the energies of the respective states i, m and f; and N_i/N represents the fraction of molecular or atomic species in the initial state. The quantity Γ_m denotes the width (FWHM) of state m in angular frequency units [rad/s]. This expression contains the sum over all final states, for each initial state, which contribute to the scattering within the observed spectral line. Moreover, to correspond to a practical situation, this cross section must also be summed over the distribution of initial states of the specific atom or molecule in question.

For an exciting radiation with sufficiently narrow spectral width close to an isolated absorption line, this equation can be simplified to permit quantitative estimates to be made, since one or a few terms in the sum over m in (5.5) will customarily predominate. In such cases, the second term in (5.5) is usually much smaller than the first, as a result of energy conservation, and one can write the cross section in the form

$$
\left(\frac{d\sigma}{d\Omega}\right)_{RR} = \frac{\omega_r^4}{c^4\hbar^2}\frac{N_i}{N}\frac{|\langle f|\mu_r|m\rangle|^2\,|\langle m|\mu_0|i\rangle|^2}{(\omega_L-\omega_0)^2+\Gamma_m^2/4}. \tag{5.6}
$$

Here we have assumed that only one intermediate state is involved in the transition associated with a single pair of initial and final states, and $\omega_L=\omega_m-\omega_i$ designates the central frequency of the absorption line.

We can rewrite (5.6) by expressing the absolute square of the matrix elements in terms of the oscillator strengths a_{mi} and a_{fm} and by introducing the degeneracies of relevant energy states g_m and g_f

$$
\left(\frac{d\sigma}{d\Omega}\right)_{RR} = 2\pi\,\frac{N_i}{N}\frac{g_f}{g_m}\frac{\omega_r^3}{\omega_0}\left(\frac{e^2}{m_ec^2}\right)^2\frac{a_{fm}a_{mi}}{(\omega_L-\omega_0)^2+\Gamma_m^2/4}, \tag{5.7}
$$

where m_e is the electron mass. Rosen et al. [5.60] derived a similar equation for the differential scattering cross section containing the

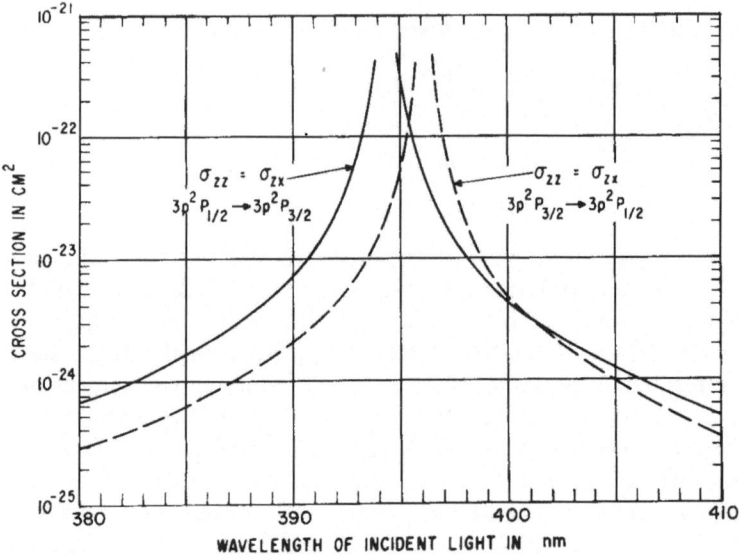

Fig. 5.5. Theoretical wavelength dependence of electronic Raman scattering for aluminum atoms for the transitions $3p^2 P_{1/2} \rightarrow 3p^2 P_{3/2}$ and $3p^2 P_{3/2} \rightarrow 3p^2 P_{1/2}$. The cross section σ_{zz} corresponds to scattering with polarization parallel to that of incident light, whereas σ_{zx} corresponds to scattering with polarization perpendicular to that of incident light. (From PENNEY [5.37])

oscillator strength and the Franck-Condon factor. These expressions are valid when Γ_m is dominated by inelastic collisions leading to a pressure-broadened linewidth which is larger than that due to the combination of natural, Doppler and hyperfine broadening. As a matter of fact, the linewidth is primarily due to pressure broadening for most molecules and atoms at atmospheric pressure.

For atomic species, considerable enhancement of the cross section is predicted by (5.7) when an incident laser line is tuned near a strong, isolated electronic resonance. As an example, Fig. 5.5 shows the wavelength dependence of the electronic Raman scattering cross section for aluminum atoms, as calculated by PENNEY [5.37] based on his formula expressed in terms of oscillator strengths and vector-coupling coefficients. In this analysis, only one intermediate higher level was assumed to contribute significantly to the resonant scattering process. The large increase of about 10^8 and 10^6 over the ordinary vibrational Raman scattering cross section for the N_2 molecule is expected at separations from resonance on the order of 1 nm and 10 nm, respectively, whereas the corresponding Raman shift near resonance is nearly 2 nm.

Similar behaviour could also be predicted for many atoms, such as B, C, Ga, Tl, Fe, Ni, Cu, In, and the rare earths, whose energy levels and

Table 5.6. Comparison of experimental and theoretical values for two kinds of differential cross section and the depolarization ratio, δ, for near-resonance electronic Raman scattering in the Na atom excited at the D_1 and D_2 lines, respectively

Excited line of Na I		$\left(\dfrac{d\sigma}{d\Omega}\right)_{RR_\parallel}$ [cm^2/sr]	$\left(\dfrac{d\sigma}{d\Omega}\right)_{RR_\perp}$ [cm^2/sr]	δ
D_1	Experimental	2.49×10^{-21}	2.21×10^{-21}	0.89
(589.6 nm)	Theoretical (I)	2.13×10^{-21}	2.13×10^{-21}	1.0
	(II)	2.01×10^{-25}	2.87×10^{-26}	0.14
D_2	Experimental	4.80×10^{-21}	4.32×10^{-21}	0.90
(589.0 nm)	Theoretical (I)	4.27×10^{-21}	4.27×10^{-21}	1.0
	(II)	2.72×10^{-24}	1.43×10^{-24}	0.53

(I) $3S_{1/2}$ ground state is taken as an intermediate state.
(II) $4D_{3/2}$ excited state is taken as an intermediate state.

selection rules make possible strong electronic Raman transitions. We have also evaluated theoretically the electronic Raman scattering cross sections, using PENNEY's formula, for some alkali metals, including Li, Na, K, Rb, Cs atoms, leading to resonance enhancement of several orders of magnitude [5.61].

For the Na atom, an experimental study of near-resonance Raman scattering excited at the D_1 and D_2 lines was performed by our group employing a tunable dye laser [5.62]. The differential scattering cross sections $(d\sigma/d\Omega)_{RR\parallel}$ and $(d\sigma/d\Omega)_{RR\perp}$, and the depolarization ratio δ at right angle to the exciting laser beam were measured, where $\parallel$ and $\perp$ denote the polarization directions of the scattered light parallel and perpendicular to that of the exciting laser beam, respectively. Table 5.6 shows these experimental results together with the theoretical values. Theoretical calculations of these quantities were based on two different processes. One is the transition involving the ground state $3S_{1/2}$ as an intermediate state by taking into account the degeneracy of magnetic sublevels in the atomic states concerned. The other is the ordinary process via an intermediate state corresponding to a higher excited state $4D_{3/2}$. It was found that the observed result is well explained by the former process, for which the scattering cross sections are more than three orders of magnitude larger, as shown in Table 5.6.

Experimental verifications of the characteristics of this type of near-resonance scattering now seem feasible because of the availability of tunable lasers in the near ultraviolet and visible regions. However, there has been little study, to our knowledge, of the strong re-emission excited near resonance with an isolated atomic line.

Compared to atoms, which generally have quite large oscillator strengths and distinct, well-separated absorption lines, the situation for molecules is usually less favorable for accomplishing near-resonance Raman scattering. The oscillator strength for an individual molecular transition is typically much smaller (diluted by vibrational-rotational structure), and the transitions are closely spaced so that re-emission is divided among them. Furthermore, only a small fraction of molecules is effectively in the initial levels which contribute to a particular absorption line. These factors readily reduce the enhancement of the cross section by several orders of magnitude in typical cases. As a consequence, it is necessary to be much closer to a particular molecular line than in the case of a strong atomic transition in order to achieve significant enhancement in near resonance.

For molecular species, three kinds of near- and on-resonance interactions should generally be considered, depending upon the characteristics of their spectral structure:

i) individual vibrational-rotational transitions which are rather well separated, having accessible strong absorption at low gas pressure.

ii) the spectral structure of isolated bands, customarily displayed by a number of molecules, rather than individual lines.

iii) dissociation continuum caused by excitation beyond the dissociation limit of the molecule.

Most measurements of the so-called resonance Raman scattering have been made in I_2 vapor excited very near resonance (within 0.01 nm) with individual molecular lines. FOUCHE and CHANG [5.51], and ST. PETERS et al. [5.52] have reported the re-emitted intensity from I_2 molecules in air about six orders of magnitude stronger than the vibrational Raman scattering from N_2 molecules, although there have been discussions about the interpretation of the process involved in the observed phenomena. ROSEN et al. [5.60] recently measured a resonance Raman scattering cross section of 1.7×10^{-24} cm^2/sr in a vibrational spectrum of the $B^3\Pi_{0u}^+ - X^1\Sigma_{0g}^+$ electronic transition in I_2 vapor with an excitation wavelength of 546.636 nm.

There are some other molecules exhibiting strong absorption lines, such as Br_2 and radicals like OH found in flames and gaseous discharges. The large resonance enhancement of the scattering cross section for the molecular line structure should offer the capability for detecting and analyzing remotely these molecules in the atmosphere; for example, the I_2 molecule may serve as an indicator of marine biological activity [5.63].

There is a complication in evaluating the cross section for near-resonance Raman scattering from a known band structure arising from the interference between contributions to the absolute square in (5.5), where the sum over intermediate states involves terms with different signs. However, at a separation from resonance large compared to the width of the band, these interference effects tend to be so small that the band contributes to the scattering approximately proportional to its overall strength. Using this approximation, together with the formula derived from (5.6), PENNEY [5.48] evaluated the scattering cross section for the NO molecule. The predicted value of 2×10^{-28} cm^2/sr, corresponding to the contribution of the $\gamma(0, 0)$ band centered at 226.5 nm of the $A^2\Sigma^+ - X^2\Pi$ electronic transition, to the scattering excited at 230 nm, is about 40 times larger than the ordinary cross section, extrapolated from the measurements at 514.5 nm using the usual $(1/\lambda_r^4)$ dependence. Although this example for the NO molecule leads to only a modest increase in cross section, larger enhancement could be expected if the electronic transition were sufficiently strong.

Besides the near-resonance Raman scattering resulting from excitation below the dissociation limit, there is also a resonance type of Raman scattering associated with excitation into a dissociative continuum, as mentioned previously. In this case the re-emission takes place instantaneously, before the molecule dissociates, and thus retains the characteristics of scattering, independent of collisions and associated quenching. This process has been identified and studied experimentally by HOLZER et al. [5.22] in some heavier halogen molecules (I_2, Br_2, Cl_2, BrCl, IBr, and ICl) excited by an Ar ion laser. This scattering is characterized by strong overtones out to many harmonics, with complex structures. They reported that the cross section for the fundamental vibrational Raman spectrum of the I_2 molecule, excited at 488 nm, is about 4.4×10^{-28} cm^2/sr. This value is about 100 times larger than that for the ν_1 band of methane, or nearly 800 times stronger than that for the N_2 molecule at the same wavelength. WILLIAMS and ROUSSEAU [5.55] also found very marked systematic spectral band-shape variations in the resonance Raman spectra of the I_2 molecule as a function of the incident laser frequency above the $B^3\Pi_{0u}^+$-state dissociation limit. This frequency dependence and the presence of multiple overtones were explained by the dependence of the polarizability on Franck-Condon overlap integrals, which demonstrated semi-quantitative agreement between the observed and calculated spectra.

Other molecules which are conveniently excited above their lowest dissociation limits are also expected to accomplish resonance Raman scattering with substantially enhanced intensity. For example, the differential cross section for the O_3 molecule excited near its very strong Hartley band, centered at 250 nm with a width of about 50 nm, was estimated to be approximately 10^{-24} cm^2/sr by PENNEY [5.48, 64]. This value is about five orders of magnitude larger than that for the ordinary

Table 5.7. List of the calculated and experimental values of the differential cross sections for resonance Raman scattering, and relevant transitions, for various small molecules

Molecule	Transition	Excitation wavelength [nm]	Calculated value of $(d\sigma/d\Omega)_{RR}$ [cm^2/sr]	Experimental value of $(d\sigma/d\Omega)_{RR}$ [cm^2/sr]	References
I_2	Above $B\,^3\Pi_{0u}^+$ dissociation limit	488.0		4.4×10^{-28}	[5.22]
	$B\,^3\Pi_{0u}^+ - X\,^1\Sigma_g^+$	546.636	1.2×10^{-24}	1.7×10^{-24}	[5.60]
NO_2	$\tilde{A}\,^2B_1 - \tilde{X}\,^2A_1$	454.74		5.6×10^{-27}	[5.60]
NO	$A\,^2\Sigma^+ - X\,^2\Pi$	230	2×10^{-28}		[5.48]
O_3	$\tilde{D} - \tilde{X}$	255.0	$\sim 10^{-24}$		[5.48]
		286.6		1.1×10^{-26}	[5.149]
CO	$a\,^3\Pi - X\,^1\Sigma^+$	206.3	4.5×10^{-30}		[5.60]
Cl_2	$A\,^3\Pi_{0u}^+ - X\,^1\Sigma_g^+$	493.3	7×10^{-31}		[5.60]

N_2 vibrational Raman scattering predicted at the same wavelength. Such a large cross section would encourage possible laser radar detection of ambient O_3 levels (~ 10 ppb) in the lower atmosphere and similar observations of the stratospheric O_3 layer from a high altitude observatory.

As a summary, Table 5.7 lists experimental and theoretical values of the differential cross sections for resonance Raman scattering reported recently for various molecules.

5.2.4 Resonance Scattering

With tunable laser sources it is also possible to excite selectively various atoms and molecules through the resonant interaction shown in Fig. 5.1f. As this type of scattering is thought to involve, in principle, simultaneous two-photon exchange, the process may be considered in some aspects as resonantly-enhanced Rayleigh scattering near or inside an absorption, where the imaginary part of the polarizability also plays an important role [5.65–67].

The frequency dependence of the differential cross section characterizing this kind of resonance scattering by a species in a dilute gas, where interference between different scatterers is negligible, was studied analytically by HUBER [5.24]. He derived simple analytical expressions for the differential cross section in the following three cases, assuming Γ_N as the natural width (FWHM), Γ_E the width due to the elastic collision, and Γ_D the Doppler width:

(i) The case of $\Gamma_N \gg \Gamma_E$, Γ_D

$$\left(\frac{d\sigma}{d\Omega}\right)_{RS}^N = \frac{\Phi_{0r}\omega_0^4}{(\omega_0 - \omega_L)^2 + \Gamma_N^2/4};$$

(5.8)

(ii) The case of $\Gamma_D \gg \Gamma_N, \Gamma_E$

$$\left(\frac{d\sigma}{d\Omega}\right)^D_{RS} = \frac{4(\ln 2)^{1/2}\pi^{3/2}\Phi_{0r}\omega_0^3\omega_L}{\Gamma_D\Gamma_N} \exp[-(\ln 2)(\omega_0-\omega_L)^2/(\Gamma_D^2/4)];$$

$$(5.9)$$

iii) The case of $\Gamma_E \gg \Gamma_N, \Gamma_D$

$$\left(\frac{d\sigma}{d\Omega}\right)^E_{RS} = \frac{2\pi\Phi_{0r}\omega_0^4}{(\omega_0-\overline{\omega}_L)^2+\Gamma_E^2/4}\frac{\Gamma_E}{\Gamma_N},$$

$$(5.10)$$

where the Doppler width is given by $\Gamma_D = 2\omega_L[(2\ln 2)kT/m_a c^2]^{1/2}$ and $\overline{\omega}_L = \omega_L + \Delta\omega_L^E$, with $\Delta\omega_L^E$ being the frequency shift caused by the collisions. In deriving (5.10), only elastic collisions with the atoms of a buffer gas were assumed, neglecting any inelastic effects. In these expressions, Φ_{0r}, given by

$$\Phi_{0r} = |\langle f|\mu_r|m\rangle\langle m|\mu_0|i\rangle|^2/\hbar^2 c^4,$$

$$(5.11)$$

is independent of frequency, T is the temperature, and m_a is the mass of the atom.

The differential cross section per unit bandwidth of the scattering frequency ω_r, $d^2\sigma(\omega_0)/d\Omega d\omega_r$, for the case of the collision-broadened limit (iii) consists of two terms: The first term contains the delta function $\delta(\omega_0 - \omega_r)$ and characterizes the coherent elastic scattering of a photon, while the second term, describing the incoherent re-emission, corresponds to the resonance fluorescence. Consequently, the integral intensity of the fluorescence over ω_r is Γ_E/Γ_N times as large as that for coherent scattering, which is ignored in (5.10) because $\Gamma_E \gg \Gamma_N$.

The frequency dependence of (5.8), (5.9) and (5.10) is found to be characteristic of the absorption cross section. Hence integrating over the frequency and direction of the outgoing photon yields the total absorption cross section unless there were non-radiative processes, i.e., inelastic collisions. For the case of low gas pressure, where the non-radiative process causing quenching of the re-emission is negligible, the differential cross section of the resonance scattering is also calculated from the absorption cross section [5.68]. For instance, the differential cross section at line center $\omega_0 = \omega_L$ for the Doppler broadening limit (ii) is expressed by

$$\left(\frac{d\sigma}{d\Omega}\right)^D_{RS} = \frac{(\pi\ln 2)^{1/2}c^2 A_{mi}^2}{2\omega_L^2\Gamma_D\Gamma_N}\frac{g_m}{g_i}P(\theta),$$

$$(5.12)$$

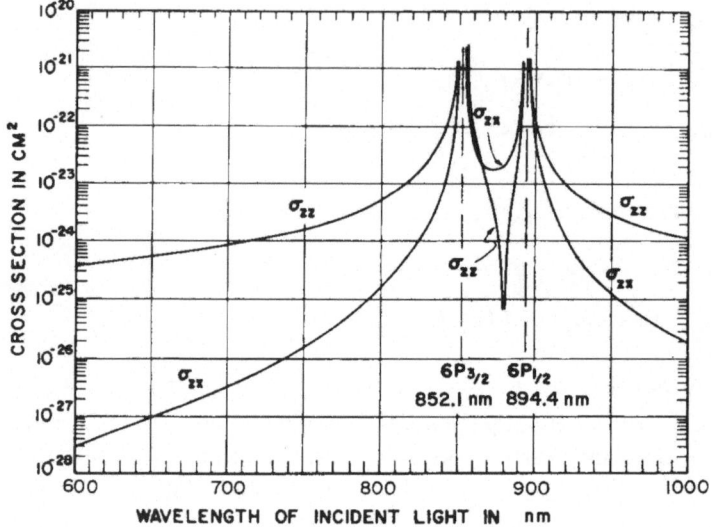

Fig. 5.6. Theoretical wavelength dependence of Rayleigh scattering across the electronic resonances for cesium atoms in the ground states $6P_{3/2}$ and $6P_{1/2}$. (From PENNEY [5.37])

where A_{mi} is the probability of spontaneous emission from the state m to the state i, g_m and g_i denote the statistical weight indicating degeneracies of both states, and $P(\theta)$ is the phase function describing the angular dependence of the scattered light as a function of the scattered angle θ [5.69, 70].

The cross sections of Rayleigh scattering for atomic species across absorption lines were also analyzed by PENNEY [5.37, 67]. Figure 5.6 illustrates the wavelength dependence of cross section for cesium atoms in the ground states $6P_{3/2}$ and $6P_{1/2}$. It is readily seen that there is considerable enhancement of the cross sections when the exciting frequency approaches the electronic resonance of the atom.

The evaluation of differential cross section of resonance scattering for some atomic vapors was performed using (5.8–12) [5.12, 13]. Table 5.8 summarizes typical calculated values for both Doppler- and collision-broadened limits (ii) and (iii), together with related parameters such as the radiative lifetime of the upper level τ_N, Γ_N/π, Γ_D/π, and Γ_E/π. The value of Γ_D corresponds to a temperature of 300 K, and that of Γ_E is evaluated at atmospheric pressure, assuming Lorentz broadening resulting from elastic collisions with air molecules. As collisions broaden the linewidth wider than the Doppler effect, the cross section eventually (at atmospheric pressure) diminishes to about one tenth its value in the Doppler-broadening limit. We note that the former case ($\Gamma_E \gg \Gamma_D$) will apply in the lower

Table 5.8. Calculated differential cross sections of some atomic vapors, and related physical parameters, for resonance scattering and resonance fluorescence

Atom	Wavelength [nm]	τ_N [s]	Γ_N/π [Hz]	Γ_D/π [Hz]	$\left(\dfrac{d\sigma}{d\Omega}\right)^{\mathrm{D}}_{\mathrm{RS}}$ [cm²/sr]	Γ_E/π [Hz]	$\left(\dfrac{d\sigma}{d\Omega}\right)^{\mathrm{E}}_{\mathrm{RS}}$ [cm²/sr]	$\left(\dfrac{d\sigma}{d\Omega}\right)_{\mathrm{RF}}$ [cm²/sr]
Na	589.0	1.6×10^{-8}	2.0×10^{7}	1.3×10^{9}	9.8×10^{-13}	1.2×10^{9}	1.5×10^{-13}	4.5×10^{-16}
Hg	253.7	1.1×10^{-7}	2.9×10^{7}	1.0×10^{9}	5.1×10^{-14}	8.7×10^{9}	8.1×10^{-15}	1.9×10^{-16}
K	766.5	2.7×10^{-8}	1.2×10^{7}	7.8×10^{8}	1.7×10^{-12}	1×10^{9}	1.8×10^{-13}	1.8×10^{-16}
Cd	326.1	2.5×10^{-6}	1.3×10^{7}	1.1×10^{9}	3.5×10^{-15}	$\sim 10^{10}$	$\sim 2 \times 10^{-17}$	
Cd	228.8	2.0×10^{-9}	1.6×10^{7}	1.5×10^{9}	1.5×10^{-13}	$\sim 10^{10}$	$\sim 6 \times 10^{-15}$	
Pb	283.3	1×10^{-8}	3.2×10^{7}	1.3×10^{9}	5.7×10^{-13}	$\sim 10^{10}$	$\sim 3 \times 10^{-15}$	

atmosphere, whereas the latter case ($\Gamma_D \gg \Gamma_E$) usually corresponds to the upper atmosphere and also to low pressure laboratory experiments where the collision frequency is quite small. For comparison, Table 5.8 also includes estimated values of cross section for resonance fluorescence, which is generally quenched by inelastic collisions at atmospheric pressure, as will be explained in the next section.

For further quantitative discussions, the resonance scattering cross sections of sodium D_1 and D_2 lines were measured carefully using a dye laser beam with a spectral width of 2 pm (2×10^{-12} m) tuned to either of these two lines of sodium vapor in a cell heated to 400 K [5.71]. The theoretical values were calculated on the basis of two different treatments using (5.9), which was extended to include the fine and hyperfine structures, and (5.12) [5.71]. Table 5.9 presents experimental and theoretical values of the total cross section over the whole spectrum of each D line. As the accuracy of the measurements is estimated to be approximately 10%, we conclude that the measured values compare quite well with the theoretical ones. For the resonance line of the potassium atom with a wavelength of 769.9 nm, the effective total cross section for resonance scattering is reported to be 1.6×10^{-20} cm²/sr for a laser linewidth of 10 pm [5.72].

The resonance scattering cross section for molecules is not usually expected to be as large as it is for atoms. This is primarily due to the fact that closely-spaced molecular transitions reduce appreciably the individual oscillator strengths and also distribute the re-emission among them. Moreover, only a small fraction of the molecules is in a particular initial level, similar to the case of near-resonance molecular Raman scattering. The reduced cross section for this type of scattering may be several orders of magnitude smaller than for atomic vapors, and thus limits the practical applicability of laser radar based on molecular resonance scattering to pollution monitoring [5.12, 13].

Table 5.9. Comparison of experimental and theoretical values of differential cross section for resonance scattering over the whole spectrum of D_1 and D_2 lines for the Na atom

Excited and scattered lines of Na I		$\left(\dfrac{d\sigma}{d\Omega}\right)^{D}_{RS}$ ($\times 10^{-13}$ cm^2/sr)	
		Scattered direction	
		$0, \pi$	$\pi/2, 3\pi/2$
D_1	Experimental	—	3.3
(589.6 nm)	Theoretical (I), (II), (III)	3.1	3.1
D_2	Experimental	—	6.4
(589.0 nm)	Theoretical (I)	6.5	5.9
	(II)	7.7	5.4
	(III)	6.9	5.7

(I)   Based on the resonance scattering process including the hyperfine structure.
(II) Based on the resonance scattering process including the fine structure.
(III) Based on the absorption cross section.

5.2.5 Fluorescence and Its Quenching in the Atmosphere

Fluorescence generally occurs only when the frequency of the exciting radiation lies within a particular absorption band of the material, as shown in Figs. 5.1d and e as a two-step photon interaction. At very low pressures, the probability Γ_e of re-emission from one of the excited states is usually constant, so its intensity decays exponentially with time constant Γ_e^{-1}. In more general cases, however, the excited molecules or atoms undergo collisions with each other and with other species before de-excitation occurs by re-emission of a photon. These collision processes usually result in changes of the fluorescence intensity as well as its spectral distribution and polarization, depending upon the constituents of the gas mixture and its total pressure. Quenching caused by collisions is known to reduce the fluorescence intensity and shorten the decay time of re-emission from the molecule or atom [5.73, 74].

The differential cross section per molecule for fluorescence excited by a monochromatic incident beam with frequency ω_0 is generally given by

$$\left(\frac{d\sigma}{d\Omega}\right)_{F} = \frac{c^2 \omega_r}{8\omega_0^3} \frac{\Gamma}{[(\omega_m - \omega_i) - \omega_0]^2 + \Gamma^2/4} \frac{A_{mi} A_{mf}}{\Gamma - \Gamma_E} \tag{5.13}$$

in the presence of Lorentz broadening [5.48, 75]. Here A_{mi} and A_{mf} are the transition probabilities for the radiative process from excited state m to initial state i and to final state f, respectively. The total bandwidth Γ of the upper state m is expressed by

$$\Gamma = \Gamma_N + \Gamma_E + \Gamma_Q, \tag{5.14}$$

where Γ_Q is the width for quenching by collisions, and both Γ_E and Γ_Q are usually proportional to the pressure.

The expression (5.13) can be rewritten in the following approximate forms

$$\left(\frac{d\sigma}{d\Omega}\right)_F \simeq \frac{\sigma_a}{4\pi} \frac{A_{mf}}{\Gamma - \Gamma_E} = \frac{\sigma_a}{4\pi} Q\eta_F,$$

and $\tag{5.15}$

$$\left(\frac{d\sigma}{d\Omega}\right)_{FT} \simeq \frac{(\sigma_a)_T}{4\pi} Q\eta_F,$$

using the absorption cross section given by

$$\sigma_a = \frac{\pi c^2}{2\omega_0^2} \frac{\Gamma}{[(\omega_m - \omega_i) - \omega_0]^2 + \Gamma^2/4} A_{mi},$$

and $\tag{5.16}$

$$(\sigma_a)_T = \frac{\pi e^2}{m_e c} a_{im}.$$

Here Q is the quenching factor given by the Stern-Volmer relation [5.76, 77]

$$Q = 1/(1 + \Gamma_Q/\Gamma_N) = 1/(1 + \sum_j q_j p_j), \tag{5.17}$$

where q_j denotes the quenching coefficient for species j associated with the collision, and p_j is the partial pressure of the species j. η_F is the fluorescence efficiency in the observed transition from the state m to f, determined by

$$\eta_F = A_{mf}/\Gamma_N = A_{mf}/\sum_n A_{mn}, \tag{5.18}$$

where the summation of the probability of spontaneous emission A_{mn} is taken over all possible transitions from the excited state m, because Γ_N characterizes the width of radiative decay for the state m.

For the molecule producing resonance fluorescence by re-emission from the state m directly excited by absorption of the incident radiation, the differential cross section is written in terms of the oscillator strength and the Franck-Condon factor

$$\left(\frac{d\sigma}{d\Omega}\right)_{RF} = \frac{c\omega_r}{4\omega_0}\left(\frac{e^2}{m_e c^2}\right)a_{im}|\langle i|m\rangle|^2\left(\frac{N_i}{N}\right)$$

$$\cdot\frac{\Gamma}{[(\omega_m-\omega_i)-\omega_0]^2+\Gamma^2/4}\frac{A_{mf}}{\Gamma-\Gamma_E}. \qquad (5.19)$$

Here a_{im} is the oscillator strength for the electronic transition, N_i/N denotes the fractional population of the initial vibrational-rotational state i, and $|\langle i|m\rangle|^2$ is the Franck-Condon factor described by the overlap integral between vibrational states m and i. The Franck-Condon factor is normalized so that

$$\sum_i|\langle i|m\rangle|^2 = \sum_m|\langle i|m\rangle|^2 = 1 \qquad (5.20)$$

for completeness of the vibrational states. A similar expression was also obtained by FOUCHE et al. [5.75] based on the Breit-Wigner formula.

For the process shown in Fig. 5.1e, which usually involves broad fluorescence due to many transitions from lower excited levels m' in the excited electronic state followed by non-radiative decays from the upper excited level m, the differential cross section is given by, after summing over all final states f and relevant excited states m',

$$\left(\frac{d\sigma}{d\Omega}\right)_{BF} = \left(\frac{c\omega_r}{4\omega_0}\right)\left(\frac{e^2}{m_e c^2}\right)a_{im}|\langle i|m\rangle|^2\left(\frac{N_i}{N}\right)\Gamma_F$$

$$\cdot\frac{\Gamma}{[(\omega_m-\omega_i)-\omega_0]^2+\Gamma^2/4}\cdot\frac{\Gamma_B}{(\Gamma_F+\Gamma_Q)(\Gamma_F+\Gamma_Q+\Gamma_B)}, \qquad (5.21)$$

where Γ_B is the width for collisions producing small changes in the excitation energy within the electronic state (which typically yields a broad band of fluorescence for the molecule), and Γ_F is the radiative width of the observed fluorescence. In this case the total bandwidth Γ is given by the sum of Γ_F, Γ_E, Γ_Q, and Γ_B instead of by (5.14). Although the pressure dependence of (5.19) follows the Stern-Volmer relation, that for (5.21) is more complicated in the presence of the factor including some of these widths.

Table 5.10 summarizes the differential cross sections and associated parameters of fluorescence for some typical molecules at atmospheric pressure. Indicated also are estimated values of lifetime for the radiative decay $\tau_N = \Gamma_N^{-1}$, lifetime $\tau = \Gamma^{-1}$ of the upper level, and quenching factor Q, along with the transition responsible for the fluorescence.

Table 5.10. List of calculated values of total differential cross section, total absorption cross section, quenching factor, and associated fluorescence parameters for some typical molecules under atmospheric pressure

Molecule	Transition	Excitation wavelength [nm]	τ_N [s]	τ [s]	Q	$(\sigma_{abs})_T$ [cm²]	$\left(\dfrac{d\sigma}{d\Omega}\right)_{FT}$ [cm²/sr]
SO_2	$\tilde{A}^1B_1 - \tilde{X}^1A_1$	290	4.2×10^{-5}	1.4×10^{-10}	1.2×10^{-5}	3.4×10^{-19}	3.2×10^{-25}
		300.1	4.2×10^{-5}	2.1×10^{-9}	4.9×10^{-5}	5.0×10^{-19}	2.0×10^{-24}
NO_2	$\tilde{A}^2B_1 - \tilde{X}^2A_1$	400	4.4×10^{-5}	1.4×10^{-10}	2.5×10^{-5}	2.8×10^{-19}	5.6×10^{-25}
		435.8	4.4×10^{-5}	1.3×10^{-9}	3.0×10^{-5}	3.0×10^{-19}	7.2×10^{-25}
I_2	$B^3\Pi_{0u}^+ - X^1\Sigma_g^+$	589.5	1.7×10^{-7}	2.6×10^{-9}	1.5×10^{-3}	4.6×10^{-18}	6.1×10^{-22}
NO	$A^2\Sigma^+ - X^2\Pi$	226.5	—	—	$\sim 3 \times 10^{-3}$	1.3×10^{-18}	3×10^{-22}
OH	$A^2\Sigma^+ - X^2\Pi$	282.6	8×10^{-7}	$(\sim 10^{-9})$	$\sim 10^{-2} \sim 10^{-3}$	1.2×10^{-17}	$\sim 10^{-20} \sim 10^{-21}$

The cross sections for these molecular fluorescences are the integrated values over the fluorescence band; that is, they correspond to the total cross sections when the fluorescence efficiency, η_F, equals unity. The values of $(d\sigma/d\Omega)_{FT}$ for SO_2, NO_2, I_2, and NO are estimated from (5.15) by utilizing the absorption cross sections $(\sigma_a)_T$ measured for SO_2, NO_2, and I_2 [5.78], and for NO molecules [5.79]. For the case of the OH molecule, the listed value of the fluorescence cross section was calculated by ROSEN et al. [5.60], while the absorption cross section was obtained by WANG and DAVIS [5.80]. In the evaluation of the quenching factor, we referred to some published data for the quenching coefficient q_{N_2} by assuming that the quenching of fluorescence in the atmosphere resulted mostly from collisions with N_2 molecules [5.12, 14, 29, 77]. FOUCHE et al. [5.75] measured the quenching coefficient for the NO_2 continuum by air to be 67 Torr^{-1}. This result gives good agreement with the quenching factor $Q \simeq 2 \times 10^{-5}$ reported by GELBWACHS et al. [5.81], and also with that listed in Table 5.10. For v_1-line fluorescence at 311 nm for SO_2, a value of 10^{-25} cm^2/sr was recently derived by PENNEY [5.48] for the differential cross section in N_2 at 700 Torr. The differential cross section for resonance fluorescence with the re-emitted wavelength the same as the exciting one (as is the case for resonance scattering) was estimated to be 5.6×10^{-23} cm^2/sr for the SO_2 molecule excited by the second harmonic beam of a dye laser at 300 nm [5.82].

There are many atoms which exhibit strong resonance fluorescence under excitation of resonance lines having large oscillator strengths. In Table 5.8 the evaluated values of fluorescence cross section for several atoms at the center of a Doppler-broadened line at a temperature of 300 K are listed, for which the quenching effect has been taken into account. The quenching factor in air was estimated to be approximately 1×10^{-3} and 2×10^{-3} for Na and Hg atoms [5.12], respectively, by reference to MITCHELL and ZEMANSKY [5.74]. The cross section for fluorescence is generally smaller by a few orders of magnitude than that for resonance scattering owing to quenching by air molecules.

From Tables 5.10 and 5.8 it is evident that the decay time of molecular and atomic fluorescence is sharply reduced, to a few nanoseconds or less, by quenching in air. This situation offers no serious problem in range resolution for single-ended measurements of these species based on the fluorescence scheme. Although a shortened fluorescence lifetime accompanied with a decrease of intensity by several orders of magnitude (compared to the low-pressure case) reduces the cross section, the cross section of quenched fluorescence is still larger, in general, than Rayleigh and ordinary Raman scattering cross sections, depending on the nature of the transition.

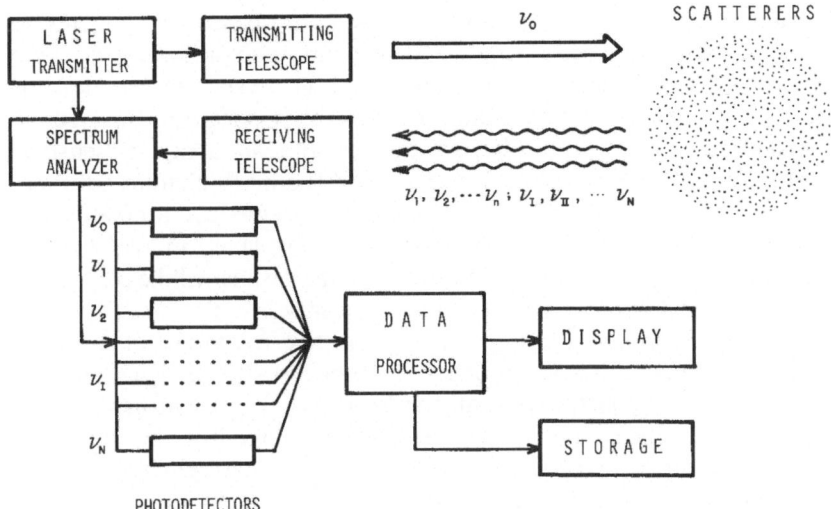

Fig. 5.7. Schematic diagram showing the basic concept of the laser radar system incorporated with Raman scattering and resonance fluorescence schemes for the range-resolved, remote detection of the atomic and molecular constituents in ordinary and polluted atmosphere in real time

5.3 Operational Principles and Characteristics of the Laser Radar Method Using Raman Scattering and Resonance Fluorescence

5.3.1 Basic Concept and Laser Radar Equation

The laser radar technique has made it possible to obtain single-ended, spatially-resolved measurements of optical signals backscattered from various materials existing in the atmosphere. The spectroscopic method incorporating these techniques, including spectral intensity and/or polarization analysis of the returned signal, provides information about atomic and molecular constituents and their concentrations.

In Fig. 5.7 the schematic diagram of a laser radar system based on such a spectroscopic method is shown in order to give a basic concept of the Raman scattering and fluorescence schemes for remote detection of atoms and molecules in the atmosphere [5.83, 27]. The transmitted optical beam from a laser through a collimating telescope is scattered from mixtures of particulate scatterers and gaseous constituents dispersed in the air. The spectrum of the backscattered energy is composed of Rayleigh- and Mie-scattered components of the frequency, centered at v_0, identical with the transmitted laser frequency as well as the Raman-shifted frequencies at v_1, v_2, ..., v_n, and various fluorescence components at v_I, v_{II}, ..., v_N superposed on the band or continuum structures due to a variety of gaseous species in the atmosphere. These spectral components are analyzed and detected simultaneously by a spectrum analyzer in conjunction with optical filtering devices and an array of sensitive photodetectors. Then, via a data processor, the multi-channel information (such as location and concentration of gaseous contaminants and their correlation with the particles) will be displayed in real time. If these data are for air pollution monitoring, the results will be sent to the pollution alarm/control system, when necessary.

For the Raman scattering scheme, selective excitation at any particular frequency is not demanded, but the backscattered signal includes Raman-scattered spectra from all the Raman-active constituents in the air, with intensities proportional to concentrations. For the resonance fluorescence and near-resonance Raman scattering schemes, on the other hand, tuning of the transmitted laser frequency onto or close to resonance with a specific transition of the species is always required. At present, tunable coherent optical sources applicable to laser radar schemes include dye lasers and their second harmonics for the wavelength region covering the near ultraviolet to the near infrared [5.84], and non-linear optical devices (such as optical parametric oscillators and sum- and difference-frequency generators) from the ultraviolet to the middle infrared near 30 μm [5.85, 86].

In the laser radar system depicted schematically in Fig. 5.7, the received power P_r in a spectral component of the frequency v_r back-scattered from the scatterer at range R is described by the generalized laser radar equation [5.83, 20]:

$$P_0(v_r, R) = P_0(v_0)K'T(v_0)T(v_r)Y(R)N(R)\Delta R(d\sigma/d\Omega)A_r/R^2 . \qquad (5.22)$$

This equation takes into account the different frequencies v_0 and v_r of the transmitted and received power. Here $P_0(v_0)$ is the transmitted power of the laser pulse, ΔR is the effective length of range resolution, K' is the efficiency of the total optical system, $T(v_0)$ and $T(v_r)$ are the one-way atmospheric transmittances at the transmitted and received frequencies, respectively, A_r is the effective receiver aperture, $Y(R)$ is the geometrical factor that accounts for overlap of the transmitted and received beam paths, $N(R)$ is the concentration, and $d\sigma/d\Omega$ is the backscattering differential cross section of the species concerned. From this equation, the received signal power is expected to decrease as the inverse square of the range in the region of interest characterized by $T(v_0)T(v_r)Y(R) \simeq 1$.

The range or distance of the target matter is determined by the time difference between the transmitted and received optical pulses. The effective length of range resolution is principally given by

$$\Delta R = c\tau/2 = c(\tau_p + \tau_s + \tau_c)/2, \tag{5.23}$$

where τ_p is the laser pulse length, τ_s is the response time of the optical interaction with the species, and τ_c is the time resolution of the detecting system. Since Raman scattering takes place effectively instantaneously, only the first two terms limit the range resolution; and then highly precise rangefinding can be achieved by shortening τ_p and decreasing τ_c. In the resonance fluorescence scheme, the range resolution is customarily restricted by τ_s. However, as described in Subsection 5.2.5, the fluorescence decay time is, in general, sharply reduced to the order of a few nanoseconds or less by quenching at atmospheric pressure, so the contribution of τ_s to ΔR is negligible for most cases.

Even at the low gas pressures encountered at high altitudes, where the probability for collisional quenching is quite small, the response time for many species (which may not be far from the natural decay time of the fluorescence) is not so long as to preclude useful range resolution for laser radar operation. For instance, a decay time of 500 ns corresponds to an effective length of range resolution of about 75 m.

A laser radar system using a cw laser is also available and useful, where a bistatic arrangement may be employed to yield range information by means of geometrical optics. In this case, the laser radar equation is described by inserting the range resolution ΔR determined by the geometry. Alternatively, the monostatic arrangement shown in Fig. 5.7 could also be used with a frequency-modulated cw laser, since the modulation frequency detected in the scattered wave carries range information.

It is to be noted here that the conventional laser radar based on Mie and Rayleigh scattering cannot resolve the values of N and $d\sigma/d\Omega$ appearing in (5.22), but can only obtain a value of $\bar{\beta}$ called the average volume backscattering coefficient. $\bar{\beta}$ is customarily defined by the product $\bar{\beta} = \bar{N}(d\bar{\sigma}/d\Omega)$, where $\bar{N}$ is the average total density of particles and atomic and molecular constituents responsible for Mie and Rayleigh scattering, respectively, and $d\bar{\sigma}/d\Omega$ is the average differential scattering cross section for the total number of particles and constituents in unit volume. In contrast, a laser radar system based on the spectroscopic method can characteristically determine the concentration of each constituent because the value of $d\sigma/d\Omega$ is obtained experimentally or theoretically according to the kinds of atomic and molecular species.

In the cases of resonance scattering and resonance fluorescence (characterized by $v_0 \simeq v_r$) where the value of $d\sigma/d\Omega$ is known for the specific transition of an individual species, elastic Mie and Rayleigh backscattering yield a background signal which interferes with discrimination. Unless the differential cross section is sufficiently large compared to $d\bar{\sigma}/d\Omega$ for these latter scatterings, as is the case for atomic vapors in the lower atmosphere, or the latter contributions are small owing to a decrease in $\bar{N}$, as in the upper atmosphere, there is a severe limitation in applying these schemes to the remote probing of atomic and molecular species in the atmosphere, as will be discussed further in Section 5.5.

5.3.2 Atmospheric Transmittance and Extinction

Atmospheric transmittance $T(v)$ over the range R is usually expressed by

$$T(v) = \exp\left[-\int_0^R \alpha(v, r)dr\right] = \exp\left[-\alpha(v)R\right]. \tag{5.24}$$

Here $\alpha(v, r)$ and $\alpha(v)$ are the extinction coefficient at a particular location and overall, respectively; and $\alpha(v)$ is composed of a sum of terms

$$\alpha(v) = \alpha_M(v) + \alpha_R(v) + \alpha_a(v), \tag{5.25}$$

where α_M, α_R, and α_a are the volume extinction coefficients due to Mie scattering from particles and aerosols, Rayleigh scattering, and absorption of atomic and molecular constituents in the atmosphere. Other scattering contributions, such as Raman scattering, are customarily much weaker and are negligible when considering the transmission loss.

These extinction processes usually limit the detectable range of a laser radar system by attenuation of the transmitted optical beam, whereas the two kinds of elastic scattering also increase the background level at the detector. The extinction due to Rayleigh scattering is largest in the ultraviolet region. Its extinction coefficient can be calculated approximately as a function of altitude either for a standard atmosphere or if the pressure profile is known. Mie scattering is rather wavelength independent, and is customarily much larger than Rayleigh scattering in the visible and infrared regions. Its intensity depends on the density and size distribution of aerosols, which vary widely with conditions in the lower atmosphere, in contrast to molecular Rayleigh scattering which is fairly constant. This means that great care must be taken in using an extinction coefficient based on Mie scattering.

The complications associated with Mie-scattering extinction are conventionally avoided by the use of a simple empirical relation between the Mie scattering coefficient $\alpha_M[\text{km}^{-1}]$ and visibility $V_M[\text{km}]$:

$$\alpha_M(\lambda) = (3.91/V_M)(0.55/\lambda)^{0.585 V_M^{1/3}}, \tag{5.26}$$

where the wavelength λ is expressed in μm [5.87, 88]. This approximate relation explains rather well the experimental results in the wavelength range from 0.35 μm to 1.54 μm, although there is some question about its validity for a coherent laser beam compared to a white light source [5.89, 90].

As for atmospheric extinction in the visible and ultraviolet by absorption, no serious problem is ordinarily encountered for wavelengths longer than about 250 nm, whereas below this wavelength O_2 absorption becomes important for path lengths longer than 100 meters. At wavelengths shorter than 185 nm the atmosphere becomes totally opaque even for very short paths, as described in Subsection 3.1.3. In the infrared, absorption is the major contribution to the atmospheric volume extinction coefficient $\alpha(v)$. As a consequence, the useful spectral region for laser radar sensing of the atmosphere is usually limited to the atmospheric transparency windows discussed in detail in Chapter 3 and in [5.87], which avoid several strong absorption bands due to H_2O and CO_2. Within these "windows", however, the absorption loss over path lengths larger than about 1 km is significant even under relatively clear conditions.

In summary, the atmospheric volume extinction coefficient is typically $1-2\,\text{km}^{-1}$ for the ultraviolet and visible regions, while it is approximately $0.5\,\text{km}^{-1}$ for the infrared window regions [5.91]. Thus, the transmission loss determined by the extinction coefficient restricts the maximum range for laser spectroscopic sensing in a horizontal path to less than 10 km unless the attenuation does not significantly reduce the detection sensitivity.

5.3.3 Basic Characteristics and Comparison of Raman Scattering and Resonance Fluorescence Schemes

In this subsection we compare some of the advantages and disadvantages of the Raman scattering and resonance fluorescence schemes for detecting atoms and molecules in the atmosphere.

For normal Raman scattering, the following characteristics are of prime importance: 1) A particular frequency or tuned-frequency laser beam is not required, unlike the resonance fluorescence and resonance Raman scattering methods. Hence, one can choose the laser frequency in a spectral region free from atmospheric absorption, although a shorter wavelength is generally preferable for higher scattering efficiency. 2) A spatially-resolved measurement of preselected atmospheric constituents is performed from a single remote location. The wavelength can be tuned in order to observe distinctly-isolated Raman spectra produced by each molecular species, to realize interference-free or reduced-interference detection. 3) Ambiguity in the backscatter at the laser frequency due to Rayleigh and Mie scattering components can be avoided. This assures measurement of the density profile of any molecular species in the atmosphere, independent of the aerosol distribution. 4) Since the backscattered signals always include Raman echoes from major components of the atmosphere (such as N_2 and O_2) at exactly the same location, the absolute concentration of each minor species is easily obtainable by referring to this echo intensity and thus eliminating complex atmospheric and instrumental parameters given in the laser radar equation. 5) The lack of detection sensitivity over long ranges owing to small scattering cross sections makes the normal Raman scattering scheme difficult to use for detecting minor constituents, such as pollutants dispersed in the ambient air. This technique can be used, however, to detect concentrated effluents from stationary sources such as smoke stacks. 6) The requirement for high-power lasers in the visible and ultraviolet wavelength regions will lead to serious eye-safety problems in practical applications.

In the case of resonance fluorescence, we note the following: 1) Its attractive feature lies primarily in its cross section which, even when the fluorescence is quenched at atmospheric pressure, is generally a few orders of magnitude larger than that for ordinary Raman scattering, as discussed in Subsection 5.2.5. 2) Absolute concentrations are not so readily determined from intensity measurements of the returned fluorescence signal, as in the Raman scattering scheme, because of resonance absorption of the transmitted laser beam. Also, the cross section is not precisely known due to locally variable quenching factors along the optical path. This is especially true for infrared transitions, since the returned intensity cannot be compared with the infrared-inactive N_2 or O_2 signatures. 3) In order to derive quantitative information from the measurements and achieve efficient use of the transmitted laser beam, precise control of the wavelength and high spectral stability are always required, compared to the Raman technique.

Nevertheless, there should be circumstances where the above factors inherent in the fluorescence scheme are not serious barriers to laser radar probing of the atmosphere. Specifically, fulfillment of the following conditions will produce a favorable situation for wide applicability of this scheme:

i) The spectral distribution of quenched fluorescence in the atmosphere still retains characteristic features to allow identification of a species, with intensity much stronger than for ordinary Raman scattering.

ii) The species concentration is low relative to the major constituents, so that the spectral shape and quenching factor for fluorescence can be accurately predicted from the partial pressures of the major gases, which are primarily responsible for collisions.

In these situations the received fluorescence signal will be approximately proportional to the species concentration, and single-ended measurements can be performed with the effective range resolution determined by (5.23). Hence, the absolute concentration of each species can be determined by referring the received signal intensity to the Raman backscattered intensity of the N_2 molecule. With respect to condition ii), for quantitative measurements in the troposphere using resonance fluorescence, the spectral broadening and dominant quenching factors due to air molecules must be precisely determined as a function of pressure in the laboratory [5.48]. Furthermore, if excitation of the fluorescence takes place over fairly wide bands, then the requirement on spectral width and stability of the laser source is not critical to practical feasibility of this scheme.

5.3.4 Principle of Atmospheric Temperature Measurements by the Raman Scattering Method

In addition to remote sensing of atmospheric constituents, there is another remarkable field of application of the Raman scattering scheme—that is, it offers a single-ended method for measuring, instantaneously, temperature profiles of the atmosphere. The method is based upon the fact that Raman scattering signatures are direct measures of relative populations among the internal molecular modes, and these relative distributions in thermal equilibrium correspond to the fundamental definition of temperature. Thus, it is anticipated that this technique for remote temperature diagnostics has the potential for becoming the most basically precise means for non-perturbing, three-dimensional measurements and, moreover, can be applied to non-thermal equilibrium systems like flames and engine exhaust emissions as well.

In general, there are several techniques for remote temperature measurements based on rotational and vibrational Raman scattering, depending upon the manner of analyzing the spectra [5.92]. The rotational Raman scattering method includes: (a) Analysis of line-by-line intensity profiles or comparison of the envelope shape of all the lines in an observed band; (b) examination of the intensity ratio of selected spectral regions of the band, using a monochromator or filters; and (c) comparison of frequency shifts of the band peak intensity. For the vibrational Raman scattering method, the following two techniques can also be used, in addition to those corresponding to the above items (a), (b), and (c): (d) Measurement of the intensity ratio between Stokes and anti-Stokes components; and (e) examination of the width of a specific Q-branch band profile, such as the ground state band, which provides a convenient estimate of rotational excitation temperature, when available.

One clear advantage of the rotational scattering scheme over vibrational scattering lies in the relative strength of the total rotational spectrum, as is evident from the differential scattering cross sections shown in Tables 5.2 and 5.3. However, special caution should be taken in order to reject the overlapping interference because the pure rotational spectra distribute generally much closer to the excitation frequency than the vibrational spectra (c.f. Figs. 5.3 and 5.4 for N_2). The rotational Raman spectrum is particularly useful for remote measurements of temperatures to several hundred degrees Kelvin, since in this region it has the strongest scattering intensity and its relative line intensities vary most acutely with temperature. On the other hand, the vibrational Raman scattering method is, in general, suited for higher-temperature diagnostics of a hot, multi-component gas, such as in flames, whose composition is variable as a function of temperature. Furthermore, its relative freedom from spectral interferences due to the exciting line and different molecular species makes it more suitable for flame diagnostics.

Among three possible techniques (a), (b), and (c) involved in the rotational Raman scattering scheme, method (b), which was propounded by Cooney [5.93], is particularly attractive when applied to a remote, single-ended measurement system. It is known that the intensity of the spectral envelope of either the O- or S-branch of a pure rotational Raman spectrum is expressed by

$$I_{J'J''} = Ac_j w_j (2J+1) \exp[-(Bhc/kT)J(J+1)], \qquad (5.27)$$

where A is a normalizing parameter to establish the absolute value ($A \propto 1/T$), c_j is the relative line strength, w_j is the nuclear spin weight, B is the molecular rotational constant, and J is the rotational quantum number designating a mean value for the upper (J') and lower (J'') states. The frequency separation of the exciting and scattering line is $4B(J'' + 3/2)$, where $B = 1.83$ cm^{-1} for the N_2 molecule. Equation (5.27) represents the envelope of the rotational Raman spectrum as a function of temperature T which, in reality, is a series of discrete lines a few tenths of one nm apart.

Figure 5.8 illustrates qualitatively the shape dependence of the envelope of the spectrum for three different temperatures. Hence, comparison between the intensities of appropriately-chosen portions of the rotational Raman spectrum, as shown by the disposition of the two optical interference filters F_1 and F_2, backscattered from a given volume of the atmosphere provides, in effect, a measure of the temperature of that volume. This difference signal should be much more sensitive to

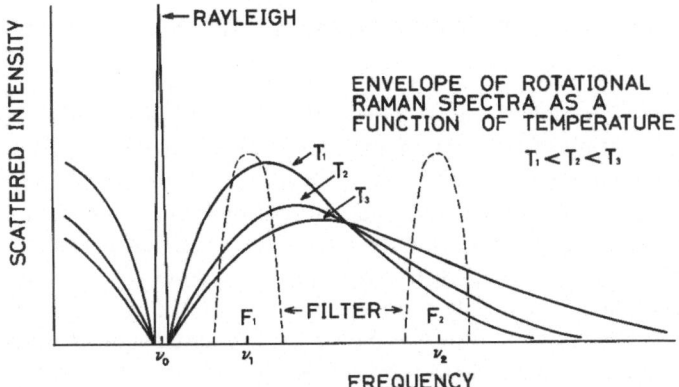

Fig. 5.8. Schematic representation of the shape changes of the envelope for pure rotational Raman spectra with the temperature around the exciting frequency v_0. Disposition of the two optical interference filters F_1 and F_2 and hence the two signals centered at the frequencies v_1 and v_2, respectively, is shown to be used in a remote measurement of the temperature profile in the atmosphere

changes in the shape of the spectral envelope than corresponding signals for techniques (a) and (c), and thus be more temperature-sensitive [5.94]. Furthermore, high temporal and spatial resolution can be obtained, and real-time measurements are possible. Also, because rotational Raman spectra of major molecular species in the atmosphere, such as N_2 and O_2, are customarily contained in a narrow wavelength region, they are, to a good approximation, affected equally by most extraneous and uncalibrated effects (such as the atmospheric transmittance and response characteristics of the detection system). Accordingly, when an intensity ratio is taken along with the spectra, these unknown factors are cancelled, in principle, and the result depends only on the rotational temperature to be measured.

In addition to the above rotational-Raman technique for absolute determination of temperature, STRAUCH et al. [5.95] found that the backscattered *vibrational* Raman signal from atmospheric N_2 correlated closely with temperature fluctuations. This result originates from a simple relation given by $[\Delta P_r(R)/P_r(R)]_{N_2} = [\Delta N(R)/N(R)]_{N_2} = -\Delta T(R)/T(R)$, based on the ideal gas law and the assumption that the partial pressure of N_2 is approximately constant at any altitude for the duration of a measurement, without requiring analysis of the spectral shape of the received signal $P_r(R)$. Moreover, by measuring the atmospheric pressure at the laser radar site, and assuming that the hydrostatic equation accurately predicts pressure at greater heights, they have shown analytically that this technique can be used to measure temperature profiles in the atmosphere.

5.4 Experimental Laser Radar Systems and Measurements

Practical operations and measurements using laser radar systems have been under way for several years. This followed feasibility studies to determine the capability for detecting and analyzing atomic and molecular constituents existing not only as major components in the ordinary atmosphere, but as trace contaminants in polluted air. This section describes typical arrangements and results of these experimental approaches based upon Raman scattering and resonance fluorescence schemes for detection of a variety of molecular and atomic species in the troposphere, stratosphere, and above. Some experimental work toward the development and implementation of equipment for the remote measurement of the atmospheric temperature distribution will also be presented. This latter application promises to be one of the most important in spectroscopic laser radar technology.

Laser radar systems based upon Raman scattering and resonance fluorescence operating in the visible and ultraviolet regions consist of:

(i) Laser transmitter and collimating optics, which transmit the laser pulse into the atmosphere along a highly-directional path.

(ii) Receiving optics consisting of a reflecting or refracting telescope to collect backscattered components from various substances dispersed in the atmosphere.

(iii) A spectral analyzer (either a mono- or polychromator), to discriminate Raman and/or resonance fluorescence components from unshifted Mie and Rayleigh components and a broad background signal. A set of narrow passband interference filters with high rejection performance is also useful for fixed frequencies.

(iv) A photodetector, together with data processing and display systems to convert the signal into a usable format.

(v) For the resonance fluorescence and resonance Raman scattering schemes, along with the resonance scattering scheme, a narrow-linewidth, stable, tunable laser is required.

5.4.1 Raman Scattering Scheme

Detection of Major Molecular Components in the Atmosphere

Raman backscattering of laser radiation from the major components of the atmosphere was first observed in 1967. In his pioneering work, LEONARD [5.96] detected the Raman scattering signal from atmospheric O_2 and N_2 molecules using a pulsed N_2 gas laser with peak power of 50–100 kW, oscillating at 337.1 nm. COONEY [5.97] measured the Raman intensity from N_2 molecules as a function of range up to 3 km, employing a Q-switched ruby laser with 250 MW peak output at 694.3 nm. Spectral measurements of Raman echoes from atmospheric N_2 and O_2 molecules were performed by INABA and KOBAYASI in 1970, using a Q-switched ruby laser beam with 5–10 kW peak power at 694.3 nm [5.27, 98, 99]. They later utilized a pulsed N_2 laser delivering 20 kW peak power at 337.1 nm in 10 ns pulses with 50 pps repetition for spectroscopic analysis of major molecules in an ordinary atmosphere as well as trace gases in polluted air [5.12, 20]. In order to reject fluorescence originating from the N_2 laser tube, a filtering device utilizing a diffraction grating was placed between the laser and collimating optics. The optical receiver was a 30-cm diameter, 150-cm focal-length Newtonian reflection telescope with a rotatable mount. The laser beam was transmitted into the atmosphere co-linearly with the axis of the receiving system. An $f/8.5$, 0.5-m single-grating monochromator was used as a spectral analyzer along with a rejection filter having 1×10^{-3} transmission at 337.1 nm. A photo-multiplier, HTV R 374, with S-20 spectral response served as the detector. The electronics were built to form a boxcar integrator consisting of sampling or pulse-gate circuitry having a gate time variable from 0.1 to 10 ns, time or range delay control, and an RC integrator. Introduction of this sampling technique made it possible to detect weak Raman signals with improved signal-to-noise ratio, but without sacrificing time or range resolution. The results were monitored on a 100-MHz oscilloscope and displayed on an XY recorder.

Figure 5.9 represents a typical result of spectral detection of Raman backscatter from an ordinary atmosphere at a range of about 30 m and a range depth of 3 m [5.20]. The Rayleigh and Mie components centered at the transmitted laser wavelength (337.1 nm) are also included in the figure. The vertical arrow indicates the expected central wavelength of the Q-branch of the vibrational-rotational Raman lines for each molecular species. Raman spectra from the major components (N_2, O_2, H_2O) are well identified, and CO_2 molecules are also weakly detected. The observed spectral width of the Q-branch for N_2 and O_2 is instrumental, and the O- and S-branches are distributed on both sides of each Q-branch.

Raman signals from N_2 molecules have been used to probe the atmosphere vertically through the troposphere into the stratosphere. Recently, a laser radar system using a pulsed ruby laser and a large optical collecting area of 16 m^2, constructed in Kingston, Jamaica, by KENT et al. [5.100], has extended the detection of its Raman backscatter up to 40 km height [5.101]. Comparisons were made between the observed density profiles and data from local radiosonde measurements and the U.S. Standard Atmosphere Supplement. The close agreement between them appears to indicate that the accuracy of remote measurements of atmospheric density by this Raman scattering scheme is comparable to

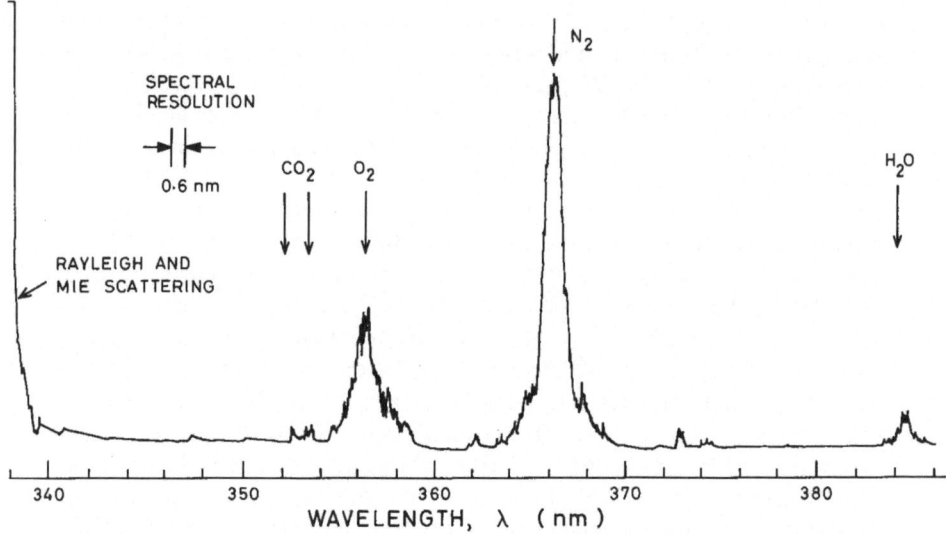

Fig. 5.9. Measured result of spectral analysis of Raman-shifted and unshifted backscatters from the ordinary atmosphere. (From INABA and KOBAYASI [5.20])

that obtained from meteorological soundings using balloons and radiosondes.

The significance of measuring the frequency-shifted Raman back-scatter from N_2 and O_2 molecules in the atmosphere can also be appreciated as follows. Unknown contributions to the backscattered signal at the laser frequency from both gaseous and aerosol constituents can be separated by simultaneous detection of returns at unshifted and shifted frequencies. Hence, a comparison permits the quantitative identification of the fractional aerosol contribution, and can be used to avoid the ambiguity inherent in the frequency-unshifted backscatter [5.102]. In conditions of poor visibility, such ambiguity is, in general, relatively small due to the preponderance of the aerosol return, but during good visibility the ambiguity is large. Moreover, it should be mentioned that the monitoring of Raman scattering from N_2 provides a measure of atmospheric transmission [5.27, 103], and also offers a useful means for studying aerosol scattering ratios and molecular mixing ratios in the earth's boundary layer, even under conditions of varying turbidity [5.104].

Measurement of H_2O Vapor Profile in the Atmosphere

As is seen in Fig. 5.9, the remote detection of water vapor in the atmosphere can also be performed by the laser-Raman radar technique. MELFI et al. [5.104, 105] observed Raman backscatter from atmospheric H_2O vapor

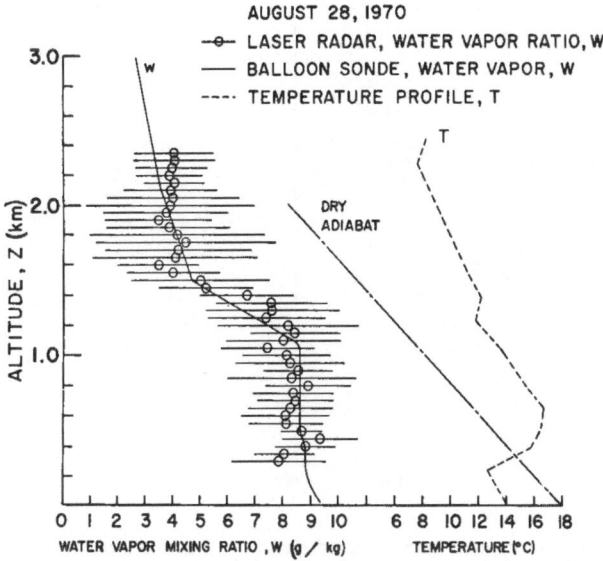

Fig. 5.10. Measured result of altitude profile of water vapor mixing ratio using Raman backscattered signals from H_2O and N_2 molecules in the atmosphere and its comparison with standard balloon-sonde data. (From MELFI [5.105])

up to 2–3 km, employing the frequency-doubled beam of a Q-switched ruby laser with an output power of about 0.04 J at 347.2 nm, and a Newtonian receiving telescope of 40 cm diameter. They also derived the water vapor mixing ratio (defined as the ratio of the mass of water vapor to the mass of dry air in a sample of the atmosphere) as a function of altitude using the Raman-backscattered intensities from H_2O and N_2 molecules. Figure 5.10 shows a typical example of their results, in which the ground-based laser radar ratio data were compared with a balloon-sonde measurement. Good agreement between the experimental ratio and independent meteorological measurements was generally found.

COONEY [5.106] also measured vertical profiles of atmospheric absolute humidity by the remote spectroscopic detection of Raman backscatter from H_2O vapor using a frequency-doubled, Q-switched ruby laser and 75-cm-diameter receiving optics. The comparison with data obtained by radio-sondes borne by balloons and helicopters, taken at essentially the same time and place, showed strong qualitative similarities and revealed a 10% relative accuracy and a 13% absolute accuracy in this measurement. It was also reported that the boundary layer height (or the top of the moist layer) can be tracked by this laser-Raman radar technique as the significant "folding over" of the water vapor profile.

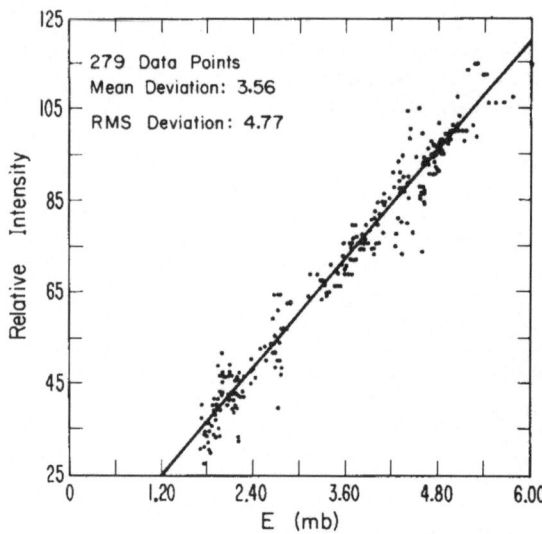

Fig. 5.11. Comparison of relative Raman backscattered intensity from atmospheric H_2O molecules with their partial pressure measured by a standard humidity meter using composite data from 10 different nights. (From Strauch et al. [5.107])

Direct comparison of Raman backscattering intensity from atmospheric H_2O vapor, using a pulsed N_2 laser (with 100 kW peak power and 100 pps repetition rate at 337.1 nm), with a standard humidity meter (the microwave refractometer) was performed by Strauch et al. [5.107] in order to indicate the usefulness and the potentiality of remote, range-resolved hygrometry by the laser-Raman radar scheme. The refractometer was located at the 30.5 m level of a meteorological tower, and the laser beam passed approximately 1 m from the microwave cavity so that the H_2O Raman signals were backscattered from a region about 20 cm in diameter and 5 m in length around it. In Fig. 5.11, their comparison of the relative Raman intensity with the observed partial pressure of water vapor (which is approximately proportional to the absolute water content) is shown, based on composite data from 10 different nights in November, 1970. The excellent agreement demonstrates feasibility of remote measurement not only of humidity as a function of position, over large volumes in almost real time, but also of the fluctuations, hence, the structure functions, of atmospheric H_2O vapor in the lower atmosphere.

Recently, the laser radar group of Tohoku University [5.108] has developed a system for simultaneous measurement and processing of multiple atmospheric parameters employing the second harmonic of a Q-switched Nd-YAG laser at 532 nm, and a 50-cm-diameter Fresnel collecting lens along with a transient recorder and on-line computer. This system, which is capable of detecting and analyzing the Raman signals of atmospheric water vapor and droplets, and the Mie backscattered signals, can yield information such as atmospheric extinction coefficient, Mie volume backscattering coefficient, and Mie total scatter-to-backscatter ratio. Such simultaneous measurements are important not only for remote hygrometry, but also in the identification and study of the physical behavior of Mie scatterers in the atmosphere.

Monitoring Molecular Pollutants in the Atmosphere

The Raman scattering scheme for range-resolved, remote analysis of molecular pollutants in the air has originally been proposed and discussed by INABA and KOBAYASI [5.83] in 1969 and others [5.109, 110]. The first experimental work was presented by INABA and KOBAYASI [5.27, 28, 99] in 1970, in which they employed a Q-switched ruby laser to perform spectroscopic remote detection of SO_2 and CO_2 molecules in the effluent from a small smoke stack. Furthermore, they detected Raman backscattered signals from an oil smoke plume and automobile exhaust gas at about 30 m range [5.12, 20]. Figure 5.12a shows typical results of their spectroscopic analysis of Raman returns from an oil smoke plume. Fuel oil was burned in a burner. The oil smoke plume from the chimney was so tenuous that it was barely visible to the eye. The intensity peaks were observed at Raman-shifted wavelengths corresponding to SO_2, C_2H_4, H_2CO, NO, CO, H_2S, CH_4, and also the major constituents, CO_2, O_2, N_2, and H_2O. The rather broad-band spectrum around 380 nm can be ascribed to the Raman band of liquid H_2O molecules, as confirmed by a laboratory measurement. Figure 5.12b shows the spectrum of Raman backscatter from automobile exhaust. The exhaust gas of the automobile was led through a chimney into the atmosphere. Raman bands due to the presence of C_2H_4, NO, CO and liquid H_2O were identified. Although the presence of H_2CO, H_2S, and CH_4 components were apparently observed, they were not definitely proved due to overlapping of the spectra. It should be noted that the integration time of the detection system was set to be 5 s in these experiments; and 10 s was enough to measure the central peak of the Raman band of a specific molecule.

As already pointed out in Subsection 5.3.3, one of the unique advantages of the Raman scattering scheme is that the absolute concentration of each constituent can be determined simply by comparing the backscattered intensity with that for N_2 at the same location, inasmuch as the atmospheric N_2 has a fixed, known concentration. This normalizing technique provides an efficient saving in equipment design and engineering, and large cancellation of instrumental and atmospheric vari-

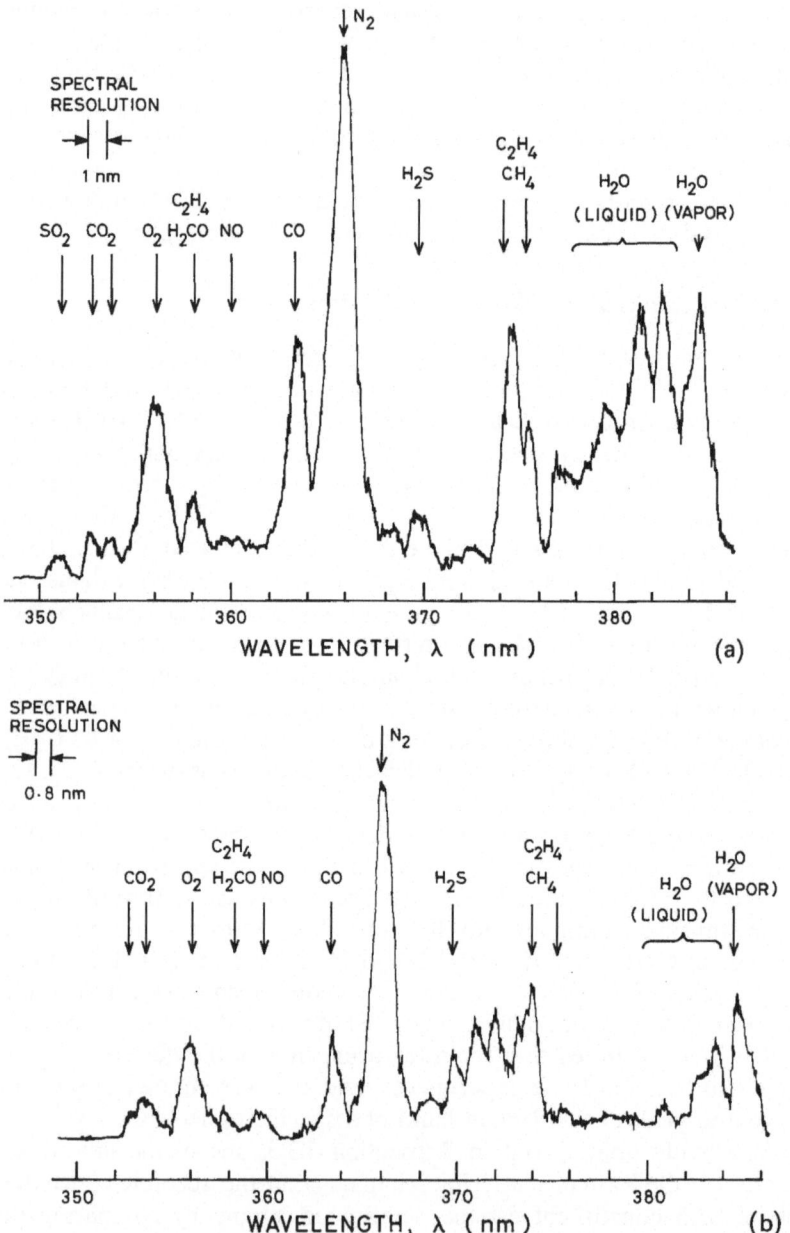

Fig. 5.12a and b. Measured spectral distributions of Raman-shifted components from a variety of molecular species in (a) an oil smoke plume, and (b) an automobile exhaust gas remotely analyzed by the laser-Raman radar scheme. (From INABA and KOBAYASI [5.20, 12])

Table 5.11. Experimental evaluation of molecular concentrations relative to that of atmospheric N_2 from Raman backscattered spectra remotely measured by the laser radar system shown in Figs. 5.9, and 5.12a and b

	Molecular species										
Observed spectrum	SO_2	CO_2	O_2	H_2CO + C_2H_4	NO	CO	N_2	H_2S	CH_4 + C_2H_4	H_2O (liquid)	H_2O (vapor)
Ordinary atmosphere (Fig. 5.9)	—	0.02	0.26	—	—	—	1	—	—	—	0.016
Oil smoke plume (Fig. 5.12a)	0.01	0.06	0.24	(0.12)	(0.03)	0.43	1	0.015	(0.069)	(0.11)	0.10
Automobile exhaust gas (Fig. 5.12b)	—	0.11	0.21	(0.04)	(0.09)	0.28	1	0.024	(0.057)	(0.05)	0.071

ables. By taking into account the laser radar equation (5.22), the spectral calibration of the receiving optics, and the response of the photodetector, the molecular concentration relative to that of N_2 has been evaluated from the observed Raman spectra shown in Figs. 5.9, and 12a, b [5.20].

Table 5.11 summarizes the results on the basis of the Raman scattering cross sections listed in Table 5.5. Since the Raman scattering cross sections for H_2CO, C_2H_4, and liquid H_2O molecules are not yet known, however, the assumption was made provisionally that the cross sections of the former two molecules are equal to that of N_2, and the latter equal to that of H_2O vapor. In this table, the relative concentrations of the O_2 and H_2O molecules in the ordinary atmosphere agree rather well with the values accepted for the standard atmosphere; however, the CO_2 concentration appears about ten times larger owing mainly to fluctuations in the measured spectra with small signal-to-noise ratios, as seen in Fig. 5.9. In the case of air pollutants, on the other hand, no reliable method exists at the present time to check the accuracy of the laser-Raman radar method. The values indicated inside the parentheses are less accurate, due to either large experimental errors or uncertainty of Raman scattering cross sections.

Based on this feasibility study of the Raman scattering scheme in identifying and analyzing a number of molecular species in the mixed gases from a single location, a practical, mobile laser-Raman radar system for monitoring stack effluent pollutants was constructed in Japan in 1971 [5.111, 112]. The system was installed in a small coach, and consisted basically of a frequency-doubled, Q-switched Nd-YAG

Fig. 5.13. Photograph of the mobile laser-Raman radar system, installed on a small coach, under operation for the routine monitoring of stack effluent pollutants and diffusion from an electric power plant

laser of 14 mJ output with 40 pps repetition rate at 532 nm, and a 50-cm-diameter Cassegrainian telescope along with a combination of narrow-passband interference filters and synchronous single-photoelectron counting equipment for high detection sensitivity. Interference filters having peak transmission at 566.7 and 607.3 nm were prepared for the SO_2 and N_2 channels, and the rejection ratio from 532.0 nm to 566.7 nm was over 10^3. The filtered photons at both channels were detected alternately by an RCA 4526 photomultiplier, followed by the synchronous single-photoelectron counting system which delivered net signal counts by subtracting background noise counts from the signal-plus-noise counts involved in the time gate for the required range resolution.

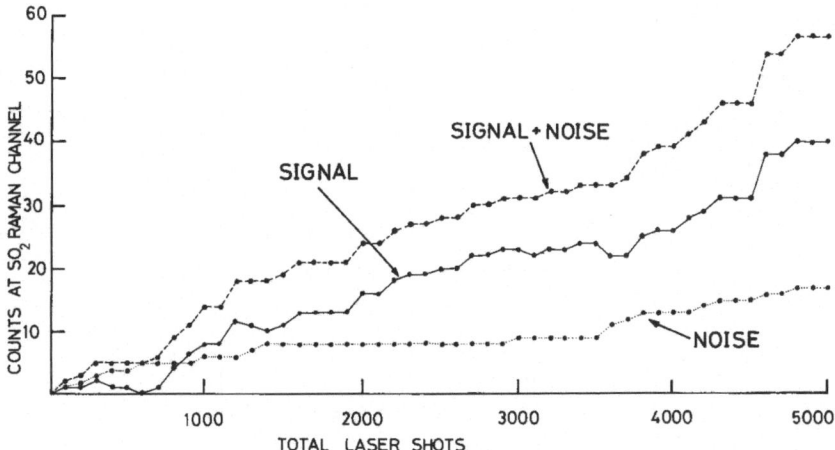

Fig. 5.14. Observed result of the remote detection of SO_2 molecules in the stack plume from an electric power plant by the Raman scattering scheme, comparing the measured counts at the SO_2 Raman channel as a function of accumulated laser shots. (From NAKAHARA et al. [5.112])

Field tests of this first practical system, performed by NAKAHARA et al. [5.112], demonstrated successfully the remote measurement of trace amounts of SO_2 in a smoke plume from a 150 m high stack of an operating power plant from a slant range of 228 m in less than a few minutes of observation time. Figure 5.13 shows a photograph of the exterior of this mobile laser-Raman radar system in operation for routine monitoring of stack effluent pollution and diffusion. In Fig. 5.14, a typical example of the experimental results for remote monitoring of SO_2 is shown by comparing the signal and noise counts at the SO_2 Raman channel. From this set of measurements, and the calibrated values of parameters involved in the laser radar equation, the SO_2 concentration was estimated to be 1850 ppm, which is rather favorably in the range of a nominal value of roughly 1000 ppm for this power plant. It should be mentioned that the SO_2 Raman signal was well discriminated from the nearby Stokes spectrum of CO_2 by taking special care with the interference filter for rejection, in spite of the fact that CO_2 is about 100 times more abundant than SO_2 in the smoke plume.

A similar observation of Raman scattering by SO_2 molecules in the stack plume of a coal-burning electric power generating plant was carried out by MELFI et al. [5.113]. They used a laser radar system consisting of a pulsed ruby laser of 1–1.5 J output power, with a maximum repetition of 1 pps at 694.3 nm, and a Newtonian receiving telescope of 61 cm diameter. The backscattered signals were analyzed by an interference filter ar-

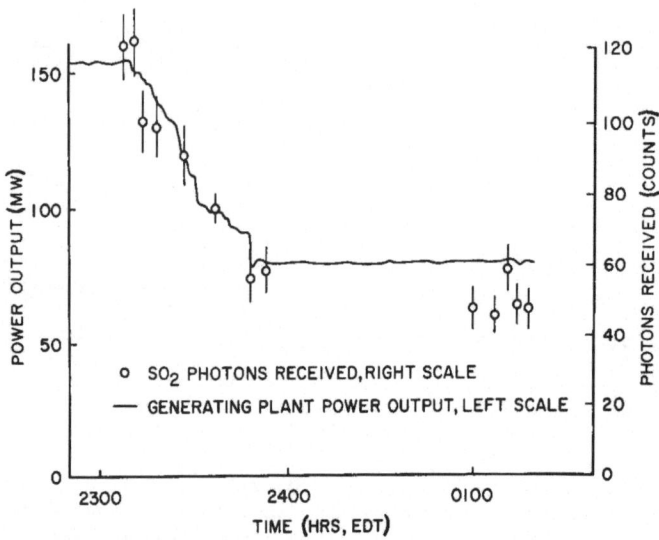

Fig. 5.15. Comparison of electrical generating plant power output with the apparent signal counts of SO_2 Raman backscatter from the stack plume. (From MELFI et al. [5.113])

rangement which blocks the intense Mie component from the fly ash in the plume, and then detected by an extended red-sensitive RCA 8852 photomultiplier. The output of the photomultiplier was processed by a wide-band amplifier and fed sequentially to a 15-channel photon counter to provide range resolution. The laser radar system was located approximately 210 m in slant range from the plume emitted from a stack 119 m high. On the basis of 100 laser firings, the SO_2 concentration was calculated to be about 800 ppm, which compared favorably with the value estimated from independent instrumentation mounted in the stack. Figure 5.15 shows the observed relation between the apparent SO_2 signal counts and the generating plant power output. This result is consistent with expectations, provided the generating efficiency of the plant and the sulfur content of the coal remained constant over the observation time.

Very careful design of the optical system, and sophisticated progress relevant to laser technology, have recently led to improvements in laser-Raman-radar system capability in achieving high signal levels and minimizing background effects. HIRSCHFELD et al. [5.114] reported a laser radar system, mounted on a truck trailer, which employs a frequency-doubled, pulsed ruby laser of 2 J output with 2 pps repetition rate at 347.2 nm, a 90-cm receiving aperture telescope together with a photomultiplier having a very high quantum efficiency (32%) photocathode, and a range-gated digitizer for photon counting. This system was tested for remote spectroscopic analysis of atmospheric H_2O and CO_2 molecules as well as SO_2 and kerosene vapor in controlled sources, with a range resolution of 10 m in daytime. Their estimated ability to detect 300 ppm-m of SO_2 and 17 ppm-m of kerosene seems to prove the adequacy of the laser-Raman radar technique for remote monitoring of air pollutant sources at appreciable ranges, even in full daytime. LEONARD [5.110] has also performed remote monitoring and analysis of industrial plant effluents from smoke stacks by the laser-Raman radar method. He has recently been involved in the use of Raman backscattering for remote detection of the composition and temperature of combustion emissions from a gas turbine aircraft engine under field conditions, employing a pulsed N_2 laser with 500 pps repetition rate at 337.1 nm [5.115]. GROSSMAN et al. [5.116] also built a mobile remote-sensing laboratory for identifying various molecular species in the atmosphere by their returned Raman signals. It consisted of a Q-switched ruby laser of 2 J output power and its second harmonic of 0.15 J output with 2 pps repetition rate, and a 30-cm-diameter Cassegrainian telescope. Practical designs and studies of the laser-Raman radar system using a Q-switched ruby laser were performed by KUPER et al. [5.117] by combining a high-gain image intensifier with a spectrometer for the simultaneous detection of different gaseous pollutants. Moreover, in recognition of the usefulness of the Raman scattering scheme for remote monitoring of concentrated pollutant sources such as smoke stacks and combustion engines, practical systems for field applications are being developed in various countries.

Remote Measurements of Atmospheric Temperature

As described in Subsection 5.3.4, Raman scattering signatures are inherently functions of the temperature of the gas responsible for the scattering, and are, therefore, of potential use as remote temperature probes. A ground-based method of measuring atmospheric temperature fluctuations using Raman backscatter from N_2 molecules was devised by STRAUCH et al. [5.95] in 1971, employing a pulsed N_2 laser with 100 pps repetition rate at 337.1 nm. The range was reduced to tower height of 30.5 m so that in situ comparative measurements of temperature could be made by thermistors along the length of the backscatter volume of 5 m. A pressure indicator located at ground level continuously monitored total pressure. Good correlation between the Raman-backscattered intensity of atmospheric N_2 molecules and the thermistor temperature changes was obtained, even when rapid fluctuations in temperature were experienced. They also showed that, with the existing equipment, the technique could possibly be extended to ranges of 5 or even 10 km if the observation time is increased to several minutes to obtain a sufficient number of backscattered photons [5.5].

Based on the concept using the intensity ratio of two selected portions of the rotational Raman spectrum, COONEY [5.93] estimated in his original paper that the temperature profile up to 2 km, with 100 m range resolution, could be observed by the ruby laser radar system of 20 J pulse output at 694.3 nm and 1 m^2 receiving area, utilizing atmospheric N_2 molecules. Following this idea, a field test for remotely measuring the atmospheric temperature was performed by SALZMAN and CONEY [5.118] using a laser radar system consisting of a ruby laser of 4 J output power with 10 ppm repetition, and a Schmidt-Cassegrain telescope of 25 cm aperture. The detection system processed three return signal wavelength intervals resolved by respective interference filters: two intervals along the rotational Raman spectrum, and one interval centered at the Mie/Rayleigh scattered wavelength. The measurement was made along a horizontal path at temperatures between $-20°$ C and $30°$ C and at ranges of about 100 m. The temperature of a controlled environment test zone was continuously monitored with a thermocouple rake assembly. From 10 laser shots (one minute of data acquisition), an accuracy of the absolute temperature measurement of $\pm 3°$ C was obtained with a range resolution of about 5 m. Because this accuracy compared well with that predicted for this particular laser radar unit, they revised it for vertical tests [5.119], and also suggested that a field-application system could be built with significant improvements in both absolute accuracy and range.

A laboratory experiment for measuring rotational temperatures of pure nitrogen near atmospheric pressure was performed using an Ar ion laser at 488 nm by HICKMAN and LIANG [5.120]. Reasonable agreement with temperatures monitored by a thermocouple placed inside an optical cell was found over a range of about 50 K.

Remote measurements of atmospheric temperature in laboratory and field testings were also reported by our group [5.94] using a laser radar system which consisted of a frequency-doubled, Q-switched Nd-YAG laser delivering 10 mJ, 20 pps output pulses at 532.0 nm, a 50-cm-diameter Fresnel-lens telescope, and a unit of optical detection and signal processing electronics. The following specific and important techniques were carefully devised to achieve very high signal-to-noise ratios for accurate measurements: (i) An I_2 molecular filter to block out adjacent intense Mie/Rayleigh scattering by more than a factor of 10^5; (ii) tuning and stabilization of the Nd-YAG laser oscillation to bring its second harmonic wavelength into coincidence with that of the absorption line of I_2; (iii) an angle-tuned interference filter to resolve two signal channels in the pure rotational Raman spectrum of air molecules; and

(iv) a detection and display system to allow real-time, analogue and digital processing of the intensity ratio of the two selected portions in the backscattered Raman signal, by eliminating pulse-to-pulse variation in the laser output and signal returns. The rotational temperature in a test cell, from 20 to 80° C, was measured and agreed with that obtained by thermistors within a few °C. Field experiments also indicated that about 10 s of averaging time were needed to attain a temperature accuracy of ~1% at a range of 1 km with the range resolution of 100 m, using these laser radar system parameters.

Thus, the potential of utilizing rotational Raman scattering as a real time, remote temperature sensing scheme for the atmosphere, in conjuction with the spectroscopic laser radar technique has been, in principle, established; but further development and testing are still required to exploit its usage on a large-scale, routine basis.

5.4.2 Resonance Fluorescence Scheme for Detection of Atoms in the Upper Atmosphere

With the availability of frequency-tunable coherent sources, e.g. dye lasers and optical parametric oscillators, it became much easier to take advantage of the resonance fluorescence and resonance scattering schemes for range-resolved, remote detection of atomic and molecular constituents in the atmosphere. However, experimental probing for these species in the troposphere proved to be much more difficult. This appears primarily due to the following situations: There is only a limited number of minor molecular constituents in the lower atmosphere, such as SO_2, NO_2, I_2, NO, and OH, in spectral regions accessible to the fluorescence scheme (which requires adequate transmitting powers and sufficient atmospheric transparency). Furthermore, practical difficulty or uncertainty occurs in evaluating the species concentration owing to unknown, locally-changeable quenching factors and attenuation of the transmitted laser beam by resonance absorption, together with interfering broad fluorescence from ambient atmospheric aerosols.

In contrast with the preceding case, remote monitoring by resonance fluorescence has a significant advantage in the upper atmosphere, which does not suffer from these problems, especially for the detection of atomic species possessing significantly larger cross sections than molecules.

In 1969, BOWMAN et al. [5.121] were able to detect atomic sodium as a minor constituent of the upper atmosphere by means of a laser radar system based on the resonance fluorescence scheme. With a flashlamp-pumped dye laser using Rhodamine 6G of a few mJ output with 0.2 pps repetition rate, matched to the D_1 line [5.122] and tuned by a tilted

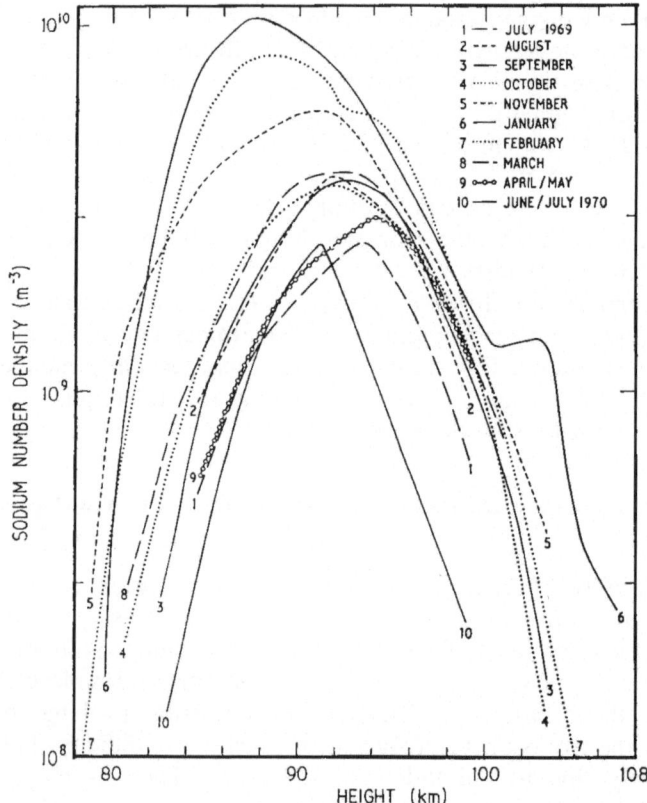

Fig. 5.16. Monthly averages of the Na number density as a function of height in the upper atmosphere observed by the laser radar system using a flashlamp-pumped dye laser. (From GIBSON and SANDFORD [5.123])

Fabry-Perot interferometer, and a 97-cm-aperture collecting mirror, the sodium layer existing near 90–100 km was found with good signal-to-noise ratio. The laser radar system was designed so that no overlap of the transmitter and receiver beams occured below 15 km; and because of the short fluorescence decay time for the dye laser, no mechanical schutter was needed to eliminate spurious signals, allowing for a relatively simple system. From an average of five night's probings, the total number of sodium atoms in a vertical column in the height range of 80–100 km was found to be 6×10^{13} m^{-2}, which coincides with typical column number densities (number density times column height) measured by an airglow photometer.

Since that time, regular observations by this laser radar technique have been performed of the height distribution of atomic sodium be-

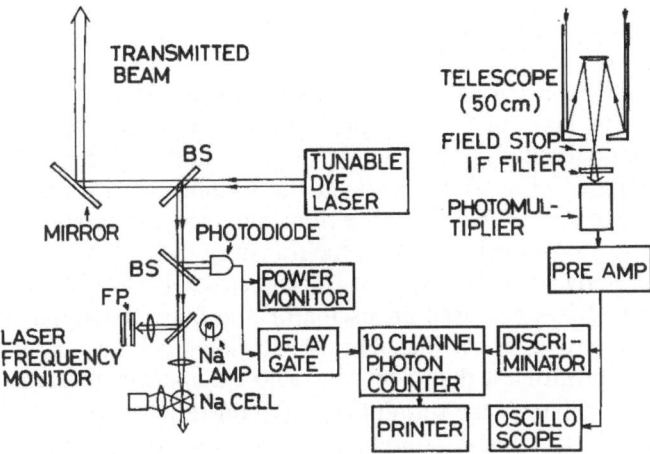

Fig. 5.17. Schematic diagram of the laser radar system employing a frequency-tunable dye laser for measuring the height distribution of Na atoms in the upper atmosphere based on the resonance fluorescence scheme. (From Jуuмonji et al. [5.71, 124])

tween 80 and 110 km altitudes by the group in Slough, England [5.123]. They reported a seasonal variation in abundance of the night-time sodium layer, exhibiting a maximum during winter and a minimum in summer, which agrees with the observed twilight glow from the layer. Figure 5.16 represents the monthly averages of the sodium number density as a function of height observed for a whole year, from July 1969 to July 1970, where a smoothed curve has been drawn through the scattered points for each month. A small secondary maximum at 103 km altitude was occasionally detected, which may not be typical of January, and may be associated with a layer of ionized sodium.

Following this dye-laser radar detection, several research groups distributed in England, USA, France, Brazil, and Japan have carried out observations of the upper atmospheric sodium layer. Figure 5.17 shows, as an example, the basic arrangement of the laser radar system employing a Rhodamine 6G dye laser pumped by a linear flashlamp, operated by a group at Tohoku University [5.71, 124]. Spectral narrowing of the oscillation width to about 2 pm was accomplished by the combination of diffraction grating and Fabry-Perot filter inserted in the laser cavity [5.125]. Frequency tuning was performed by changing the tilting angle of the filter and adjusted to the sodium D lines through the frequency monitor. With output power of about 2 mJ, the backscattered photons were collected by a 50-cm-aperture Cassegrainian telescope and detected by a high-gain photomultiplier, to be processed by the photon-counting mode to average the counts over many laser shots. A 10-channel range-gated photon counter with variable gate width was utilized, and the accumulated counts were displayed.

SCHULER et al. [5.126] used a similar arrangement, consisting of a Rhodamine 6G dye laser of 0.2 J output power with 12 pm spectral width and 45-cm-diameter Cassegrainian optics, and reported the temporary appearance of a double peak in the sodium density profile between 90 and 105 km. Furthermore, a fourfold increase in sodium layer content during a 4 h period around the transit of the radiant of the Geminids meteor shower on the night of December, 1971 was observed by HAKE et al. [5.127]. They used an oscillator and amplifier chain of dye lasers, each pumped by six flashlamps, yielding 0.5 J output per pulse with a spectral width less than 5 pm, and a 40-cm-aperture receiving telescope. Altitude profiles of the sodium layer obtained by the groups at Stanford Research Institute and Tohoku University are consistent with those reported by other groups. Figure 5.18 illustrates typical results of the sodium layer profile observed by the Tohoku University group surrounding transit of the Giacobini-Zinner meteor shower in October, 1972. The horizontal broken line shows the noise level resulting from dark current in the photomultiplier and also expected airglow. Figure 5.18a was obtained at about 3 h before the anticipated maximum of the meteor shower, and Fig. 5.18b corresponds to data on the following night, characterized by an enhancement of the sodium abundance in the normal layer at 90–95 km, and the appearance of the second layer above 100 km. This kind of observation may provide more evident correlation between the sodium layer structure and meteor activity, and also information concerning chemical reactions in the atmospheric sodium layer.

BLAMONT et al. [5.128] also measured the seasonal variation of sodium abundance, which showed evidence of stratification of the sodium layer; and atmospheric temperature in the range of 80–100 km deduced from the measured shape of the D_2 line. Their laser radar system involved a dye laser amplifier to deliver 0.3 J output power with 14 pm spectral width and 80-cm-diameter receiving optics. Moreover, nighttime laser radar observations of the sodium layer were made in the southern hemisphere by KIRCHHOFF and CLEMESHA [5.129] by collecting scattered photons with a 70-cm-aperture receiving telescope in combination with a dye laser of 5 mJ output power with 5 pm transmitted bandwidth. They reported that the peak sodium density was usually found at 95 km, the occasional appearance of a second peak between 100 and 105 km, and atmospheric tidal oscillations inferred from the observed structure of the sodium layer. Laser radar soundings of the sodium layer, which had been carried out only in nighttime, were extended to daytime by GIBSON and SANDFORD [5.130] by reducing the background by a factor of 200, and thus improving the signal-to-background ratio by a

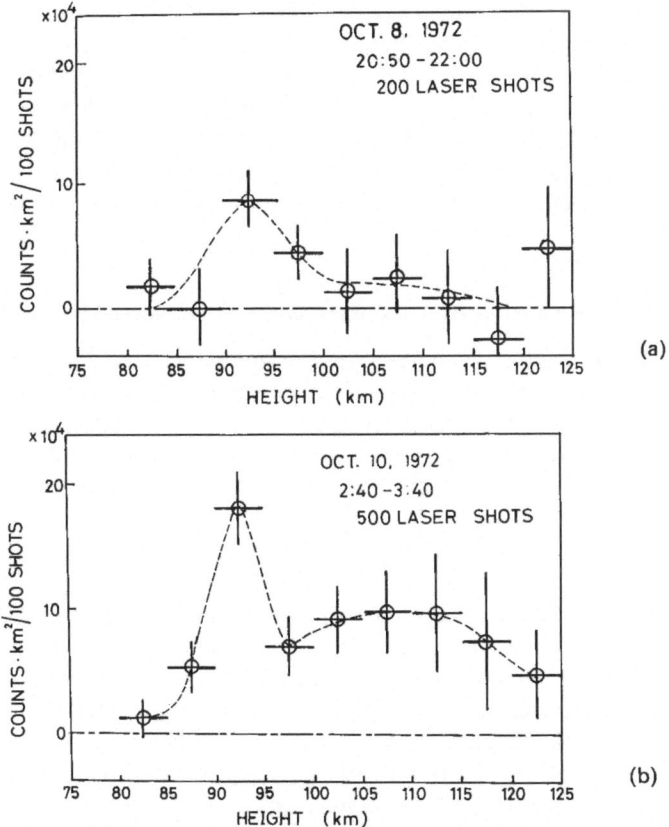

Fig. 5.18a and b. Observed results of the Na layer profile in the upper atmosphere (a) before, and (b) after the expected maximum of the Giacobini-Zinner meteor shower at about 0ʰ 40ᵐ J.S.T. on October 9, 1972. (From JYUMONJI et al. [5.71, 124])

factor of 100, using a newly designed flashlamp-pumped dye laser [5.131]. They noted that no evidence was found of a daytime enhancement of the sodium abundance in the layer about 90 km altitude.

Following these spectroscopic observations of the atmospheric sodium layer by the laser radar method, which has become mostly an established technique, a measurement of the density and distribution of atomic potassium in the region of 75–100 km was reported by FELIX et al. [5.72] in 1973. A dye laser, pumped by a Q-switched ruby laser, capable of transmitting a beam tuned to a resonance transition of potassium at 769.9 nm was employed, with an output of about 40 mJ and a bandwidth of the order of 10 pm. Since the returned signal due to resonance fluorescence was expected to be comparatively weak, the use of a large receiving area of 16 m² (consisting of the Mark II laser radar system at

Kingston, Jamaica [5.100]) was required. The results indicated the existence of a large fraction of K atoms below 84 km, which differs from measurements obtained so far for sodium. FELIX et al. also concluded that there was a broad potassium layer over the whole of the 24 km region studied, and estimated the column number density to be about 9×10^{11} m^{-2}, which is consistent with twilight photometric data. For recent simultaneous measurements of Na and K atoms, see "Additional References with Titles" at back of book.

5.4.3 Laboratory Experiments with the Resonance Fluorescence Method

Although the Raman scattering scheme for remote detection of atomic and molecular species in the lower atmosphere now has practical feasibility, the resonance fluorescence method has, except for laser radar sounding of atomic vapors in the upper atmosphere, so far been tested and evaluated only in the laboratory (as far as concentration measurements of molecular species are concerned).

In situ monitoring and analysis of sampled, atmospheric NO_2 and NO molecules in real time, based on resonance fluorescence excited by an Ar ion laser, were demonstrated by GELBWACHS et al. [5.81] and TUCKER et al. [5.132]. They were able to achieve a detection sensitivity of a few ppb with laser excitation of a few hundred mW output power at 488 nm. The experimental arrangement employed for effectively simultaneous measurements of both molecular concentrations is shown in Fig. 5.19. Air samples were drawn through a glass frit filter before entering the chamber in order to remove aerosols.

Measurement of the NO concentration in the air was accomplished by oxidation of the NO to NO_2, which was then monitored by its fluorescence. The NO_2 fluorescence in the filter passband of 0.70–0.83 μm was collected by a lens and focused onto the EMI 9558B photomultiplier, cooled to reduce the average dark count to one per second. The output of the photomultiplier was detected in the photon-counting mode and displayed on a strip chart recorder. An absolute calibration of the fluorescence count vs. the NO_2 concentration was obtained by comparing it with the standard Saltzman reagent test. It was confirmed that interfering signals were not generated by the presence of SO_2, NO, O_3, and water vapor in the air stream. However, aerosols, which are usually filtered out, result in interfering fluorescence generally several times greater than the fluorescence signal of the ambient NO_2 levels.

The NO_2 and NO concentrations in the air in the vicinity of Los Angeles were measured by the above technique by drawing air samples from outside the laboratory; and results of one such measurement are shown in Fig. 5.20. These data exhibit features typical of a smoggy day

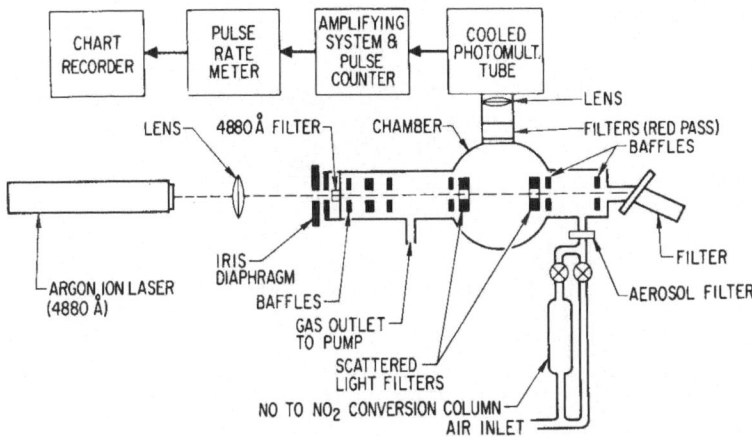

Fig. 5.19. Experimental arrangement for in situ measurement of sampled, atmospheric NO_2 and NO molecules using the fluorescence technique, with excitation by an Ar ion laser. (Fom TUCKER et al. [5.132])

in the Los Angeles basin, characterized by a strong temperature inversion. It is seen that the NO concentration rose to a high level with the onset of automotive traffic, and that conversion of NO to NO_2 in the air led to the subsequent increase of the NO_2 concentration.

The ambient concentration of OH molecules in the atmosphere is not accurately known due to the lack of any established measurement technique. This very reactive radical is considered to play a central role in smog chemistry and in controlling the global concentration of CO and CH_4. Recently, WANG and DAVIS [5.80] utilized resonance fluorescence of the OH molecule, excited by a second harmonic beam of a dye laser system (consisting of one oscillator and two amplifier stages using Rhodamine S), for in situ measurements of its concentration in air, following a preliminary experiment by BAARDSEN and TERHUNE [5.133]. The exciting beam, near 282.58 nm with 6 mJ output energy at 0.1 pps repetition rate, was focused in air through a lens, and the collected fluorescence around 309.0 nm was processed by the photon-counting apparatus after passing through tandem spectrometers. By performing the measurements in the laboratory, they observed that the OH concentration in air varied during the course of the day. On a particular day, it ranged from a value of 1.5×10^8 cm^{-3} in the early afternoon to 1.6×10^7 cm^{-3} around evening; and furthermore, to a level below their detection limit of 5×10^6 cm^{-3} (0.2 part per 10^{12} in air) at night. The OH concentration also dropped to a level below the detection limit when the outside weather was rainy or overcast, or when the ventilation system was shut off to prevent fresh outside air from entering the laboratory.

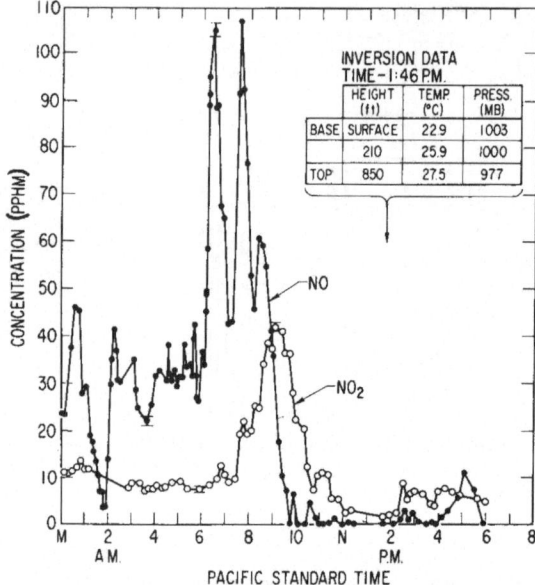

Fig. 5.20. Observed changes of sampled, atmospheric concentrations of NO_2 and NO molecules with time by the laser-excited fluorescence scheme, at site of laboratory building in El Segundo, California, USA, on October 6, 1972. (From Tucker et al. [5.132])

The relative accuracy of the measurement is presently estimated to be about 30% at high OH concentrations, and less accurate at lower concentrations. This experiment has, nevertheless, verified the advantages of real time capability and high sensitivity of the laser-excited fluorescence method for monitoring concentrations of certain atmospheric molecules.

Finally, we note that the technique of laser-induced fluorescence is being used for studies of various molecules in the laboratory, such as for the detection of reaction products and their internal state distributions, as well as in flame and combustion spectroscopy [5.134].

5.5 Detection Sensitivity and System Performance of Laser Radar Monitoring

For evaluation of their extent of applicability and their limitations, the laser radar atmospheric monitoring techniques based on Raman scattering and resonance fluorescence should be discussed by taking into account the laser characteristics and system performance presently available or anticipated in the near future. We will consider the minimum detectable concentrations of various atmospheric atoms and molecules on the basis of parameters for each system.

5.5.1 Detection Methods and Signal-to-Noise Ratio of the Receiving System

In laser radar operation, optical signals backscattered from atmospheric species are very weak over the entire spectrum. For the region encompassing the ultraviolet to the near infrared, which is accessible to photomultipliers, two types of detection methods are customarily being employed: digital photon (or photoelectron) counting, and analog boxcar integration. Both of these have sufficient response speed to attain good range resolution [5.20], and they permit increasing the signal-to-noise ratio by signal averaging.

Figure 5.21 depicts schematically a block diagram of the basic setup for both analog (a) and digital (b) detection methods. The returned signal received by the photomultiplier (PM) and processed by a pulse amplifier is sampled or range-gated at time $2R/c$ in a time interval τ_g which corresponds to the effective range resolution $\Delta R = c(\tau + \tau_g)/2$, where τ is usually defined by (5.23). The gating pulse is controlled by the pulse generator, triggered by the transmitted laser pulse, but shifted by a delay circuit in order to coincide with arrival of the scattered signals at $2R/c$. Thus it allows one to detect only the input photons within this gated interval and effectively screen out external noises outside τ_g. In the boxcar integration scheme of Fig. 5.21a, the signal current sampled by the gating pulse is integrated and stored, to be displayed by the recorder. With the photon counting scheme of Fig. 5.21b the number of photoelectrons converted from received photons is counted over the sampled time after passing through the discriminator, which rejects noise electron pulses on the basis of their height distribution.

The number of photoelectrons converted from the returned signal generated by a single laser pulse and detected in the total observation time τ_d is given by

$$\eta_r(\nu_r) = P_r(\nu_r)\tau_d\eta(\nu_r)/hc\nu_r , \qquad (5.28)$$

where $P_r(\nu_r)$ is the received power determined by the laser radar equation (5.22), and $\eta(\nu_r)$ is the quantum efficiency of the photodetector at frequency ν_r.

The signal-to-noise ratio for the pulse-gated photon counting scheme is expressed by

$$(S/N)_{PC} = n_r(f_p\tau_d\xi_S)^{1/2}/[n_r + 2(n_b + n_d\xi_N/\xi_S)]^{1/2} , \qquad (5.29)$$

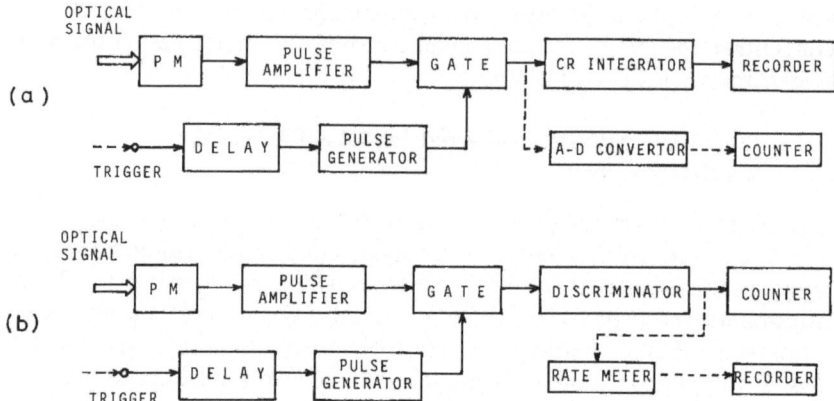

Fig. 5.21a and b. Block diagrams of basic arrangement for (a) an analog boxcar integration method, and (b) a digital photon-counting method for the detection of weak optical signals in conjunction with the photomultiplier (PM)

where f_p is the pulse repetition frequency of the transmitting laser, and n_b and n_d are the number of noise electron pulses due to background and dark current, respectively. ξ_S and ξ_N denote the counting efficiency for photoelectron and dark current pulses, which usually have values around 0.8–0.9 [5.135], determined by the pulse height distribution of each photomultiplier and the discriminator level of the receiving system. We should note that in this scheme, called synchronous pulse-gated photon counting [5.46, 47], the signal component is derived by subtraction of the total noise contribution from the total count rate, both of which are detected alternately in the same time duration of $f_p \tau_d$. Also, n_b is usually determined by

$$n_b(v_r) = B(v_r)\tau_d \eta(v_r) K' A_r \Omega_r \Delta\lambda, \tag{5.30}$$

where $B(v_r)$ is the spectral radiance of the background source, Ω_r is the solid angle of the receiver field of view, and $\Delta\lambda$ is the spectral bandwidth of the receiving system.

On the other hand, the signal-to-noise ratio for the boxcar integration scheme is written by

$$(S/N)_{BI} = n_r (f_p \tau_d/\mu_S)^{1/2}/[n_r + 2(n_b + n_d \mu_N/\mu_S)]^{1/2}, \tag{5.31}$$

where μ_S and μ_N are the photomultiplier noise factors for photoelectron and dark current pulses due to statistical fluctuation of the gain in the electron multiplication process, which generally ranges from about 1 to 5 [5.136]. Equation (5.31) is related to (5.29), under the condition $\xi_S \simeq \xi_N$ and $\mu_S \simeq \mu_N$ (which is usually encountered), by

$$(S/N)_{BI} \simeq (S/N)_{PC}/(\xi_S \mu_S)^{1/2} .\tag{5.32}$$

The above equation proves that the signal-to-noise ratio for the range-gated photon counting scheme is improved by a factor of $(\xi_S \mu_S)^{1/2}$ (>1) compared with that for the boxcar integration scheme. However, the dynamic range of the photon counting method is limited by the electronic bandwidth of the detection system when the incident light intensity approaches the region specified by the multi-photoelectron event [5.136, 47]. Thus, as far as dynamic range is concerned, the boxcar integration technique is preferable for detecting very wide ranges of returned signals, whereas the photon counting scheme improves detection sensitivity for extremely weak intensities characterized by a single photoelectron event [5.137].

The minimum detectable signal power for both schemes can be derived from (5.29) and (5.31) by letting $S/N = 1$

$$(P_r^{min})_{PC} = \{1 + [1 + 8 f_p \tau_d \xi_S (n_b + n_d \xi_N / \xi_S)]^{1/2}\} hcv_r / 2\eta f_p \tau_d \xi_S ,\tag{5.33}$$

and

$$(P_r^{min})_{BI} = \{1 + [1 + 8 f_p \tau_d (n_b + n_d \mu_N / \mu_S) / \mu_S]^{1/2}\} hcv_r \mu_S / 2\eta f_p \tau_d .\tag{5.34}$$

From these expressions the minimum detectable concentration for a specified species can be calculated through the laser radar equation (5.22) as a function of various detection and atmospheric parameters, as well as the laser characteristics.

As is seen from the above equations, atmospheric background radiation is of concern primarily with evaluation of signal-to-noise ratios and minimum detectable power. In addition to background radiation from natural sources, such as the sky and terrain, background signals also result from the elastic and inelastic backscattered radiation. As to the broad background spectrum, like natural light, (5.30) indicates that its intensity can be reduced by decreasing the field of view, the receiving spectral bandwidth, and the effective receiving area. The spectral radiance of the sky under clear daytime conditions [5.138] has a maximum in the visible region due to scattered solar radiation of about 10^{-5} W nm^{-1} cm^{-2} sr^{-1}. Below about 300 nm the O_3 absorption layer in the upper atmosphere screens virtually all the solar radiation (cf. Fig. 3.1), so that operation in an effective night condition (that is, without background interference from the sun) is possible in this "solar blind" region. In the infrared, the spectral radiance again peaks, owing to thermal radiation, at nearly 10^{-6} W nm^{-1} cm^{-2} sr^{-1} around 13 μm.

Background noise due to elastic Mie and Rayleigh backscattering is important for both Raman scattering and resonance fluorescence schemes in the visible and ultraviolet regions, while Mie scattering contributes mostly in the infrared. Filtering or rejection of this unwanted background radiation, which is clearly an important factor in remote atmospheric sounding by these techniques, can mostly be solved by the present state-of-the-art in optical technology and electronics.

It was recently reported by GELBWACHS and BIRNBAUM [5.139], and KERKER [5.140] that broadband fluorescence of ambient atmospheric aerosols falls in the detection band for Raman scattering and fluorescence of gaseous molecules and atoms in the atmosphere, and that aerosols thereby produce a background source of interference for laser radar methods based upon these interactions in the visible and near-infrared regions. For example, under severe aerosol loading in the Los Angeles basin, the fluorescence of aerosols contributed signals estimated to be equivalent to 600 ppm of SO_2 and 3000 ppm of CO, as measured by Raman scattering, and 1 ppm of NO_2 as determined by fluorescence, when excited at 488 nm in the laboratory. In order to overcome this detectivity limitation in laser radar probing of atmospheric molecular constituents, a two-wavelength excitation method was proposed to differentiate signals due to Raman scattering or resonance fluorescence from the continuum fluorescence caused by aerosols (because the latter excited at two wavelengths is approximately constant while the former signals differ). The effectiveness of this method was demonstrated in the laboratory by in situ measurements of atmospheric NO_2 in the presence of ambient aerosols by GELBWACHS and BIRNBAUM [5.139], as described in Subsection 5.4.3. This technique is expected to be useful in the field operation of spectroscopic laser radar systems employing Raman scattering and resonance fluorescence.

5.5.2 Detection Capability of the Raman Scattering Scheme for Ground-Based Monitoring of the Lower and Upper Atmosphere

To aid in the discussion of detection sensitivity and range performance of laser radar systems utilizing ordinary Raman scattering, several pulsed laser systems operating in the visible and near ultraviolet regions are compared, and typical characteristics are listed in Table 5.12 [5.20]. The characteristics are rather conservative, so they should be achievable by present state-of-the-art laser engineering.

In Fig. 5.22, evaluation of the minimum detectable concentration of SO_2 as a typical gaseous air pollutant is summarized as a function of range [5.20]. The solid lines represent nighttime operation, while the dotted lines are for daytime operation. To achieve better sensitivity, the

Table 5.12. Typical characteristics of pulsed laser transmitters and relevant physical parameters involved in evaluating the detection capability for SO_2 (as a typical gaseous pollutant) by the laser-Raman radar depicted in Fig. 5.22

	Ruby	Nd-YAG SH	Ruby SH	N_2 gas
Wavelength [nm]	694.3	532.0	347.2	337.1
Output energy [mJ/pulse]	300	30	50	0.8
Repetition rate [pps]	1	100	1	100
Raman wavelength [nm]	754.6	566.7	361.7	350.8
Raman backscattering cross section for SO_2 [cm^2/sr]	8.1×10^{-31}	2.5×10^{-30}	1.5×10^{-29}	1.8×10^{-29}
Quantum efficiency of detector	7×10^{-3}	7×10^{-2}	1.5×10^{-1}	1.5×10^{-1}
Clear sky radiance [$W/nm\text{-}cm^2\text{-}sr$]	5×10^{-7}	2×10^{-6}	1×10^{-6}	9×10^{-7}

photon counting technique is assumed, together with an observation time $\tau_d = 100$ s, a gated range depth $\Delta R = 10$ m, and a visibility $V_M = 10$ km. Other system parameters employed in this analysis are also shown in the insert of Fig. 5.22.

For night observation, sensitivity is limited by shot noise; and the second harmonic of a Nd-YAG laser at 532 nm is found to yield the best sensitivity. On the other hand, performance in daytime is dependent inherently on the background noise, and the second-harmonic beam from a ruby laser at 347.2 nm appears most suitable. This results from a background noise reduction due to the smaller number of laser shots required for the higher-power ruby laser. Moreover, the estimated value of minimum detectable concentration may be decreased by about one order of magnitude by increasing parameters such as the receiver aperture, background noise rejection, or alternatively, laser output power (unless the safety constraint is strict or unavoidable). Since the range resolution of the laser-Raman radar scheme depends on laser pulse duration and gate time, it can be improved by employing shorter-pulse transmitters and a high-speed detection system, but at the cost of sensitivity.

Figure 5.22 shows that a well-designed laser radar system associated with the ordinary Raman scattering technique will have a sensitivity capable of detecting atmospheric pollutants of medium level concentration, such as a few ppm, at a range of a few hundred meters. Although this figure is not sufficiently sensitive to allow the measurement of pollutants normally dispersed in the ambient air, it can possess the practical capability to monitor and analyze a variety of chemicals from stationary emission sources such as smoke stacks and combustion engines. As a matter of fact, this state-of-the-art for sensitivity and range involved in the Raman scattering scheme has been verified, for the most

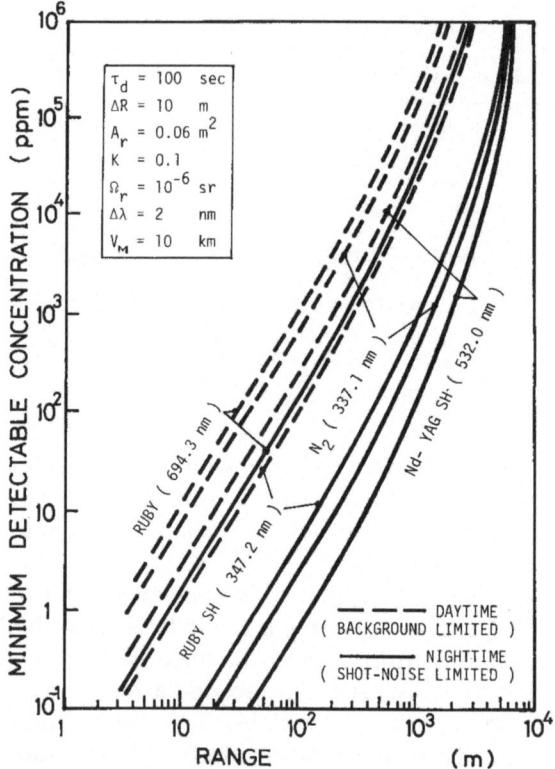

Fig. 5.22. Analytical comparison of minimum detectable concentrations of SO_2 molecules, as a typical example of gaseous air pollutants, as a function of range for the laser radar system utilizing Raman scattering scheme with presently available laser transmitters listed in Table 5.12. (From INABA and KOBAYASI [5.20])

part, by the field measurements, as already described in Subsection 5.4.1.

In order to achieve higher detection sensitivity by detecting the *rotational* Raman spectra of molecular air pollutants (where the cross section is usually about two orders of magnitude larger than that for the vibrational-rotational one), an interferometric filtering method utilizing a very high resolution Fabry-Perot interferometer was proposed by SMITH [5.141]. However, since the rotational lines of most molecules lie within 100 cm^{-1} of the exciting frequency, realization of the expected improvement in sensitivity has been practically difficult due to severe overlaps.

With regard to major molecules present in the ordinary atmosphere, much higher detectivity or longer-range performance can be expected.

Figure 5.23 shows the nighttime detection limit vs. height for several molecular species in the atmosphere, together with their concentration profiles. Included in this figure are not only the major components, such as N_2, O_2, H_2O, CO_2, but also N_2O, for comparison. The second harmonic of a pulsed ruby laser having 1 J output per pulse at 347.2 nm is assumed, in conjunction with a receiving telescope of 0.5 m^2 area (~ 80 cm diameter). Two detection schemes are considered: One is an analog A-scope displaying the return from a single laser shot; the other employs the digital photon-counting technique which accumulates 100 laser shots. These schemes are assumed to have range resolutions of 5 m and 1 km, respectively. The minimum detectable concentration (or the maximum detectable altitude) by these two schemes under night-time operation is obtained at the intersection of the concentration profile for each molecule with the straight line representing the detection limit. We note here that a technique for extending the capability of the laser-Raman approach to measure such profiles of atmospheric constituents, employing a differencing method based on two different portions of the Raman backscatter (monitored simultaneously to subtract the accompanying daytime sky background) was recently proposed by COONEY [5.142].

From Fig. 5.23 it is recognized that atmospheric N_2 and H_2O molecules are detectable up to about 50 km and 9 km, respectively, by the photon counting scheme, which are fairly well in the range confirmed experimentally by recent observations of both species by some laser-Raman radar systems (cf. Subsect. 5.4.1). It will also be of great interest to monitor other molecules, such as O_2 and CO_2, by this remote Raman-scattering scheme. Such data could be obtained up to heights of about 40 km and 15 km for O_2 and CO_2, respectively, by the photon counting operation, and roughly 10 km and 1 km by the analog operation, as predicted by this analysis. This kind of remote measurement should offer novel and valuable information, in real time, of such molecular mixing ratios as oxygen-nitrogen and oxygen-carbon dioxide in the normal atmosphere as a function of altitude, and also appears worthwhile for in situ experimental studies of atmospheric physics, such as water vapor transport and cloud formation, and of atmospheric hygrometry. In addition, since analog detection of Raman signals from atmospheric N_2 molecules appears possible throughout the troposphere, the potential of utilizing this spectral information for remote temperature sensing in three dimensions will also be established in the future. One should stress here that the most important advantage of these laserradar measurements, in addition to their being remote and single-ended, lies in the fact that they do not disturb the region under study, and they can be performed in real time.

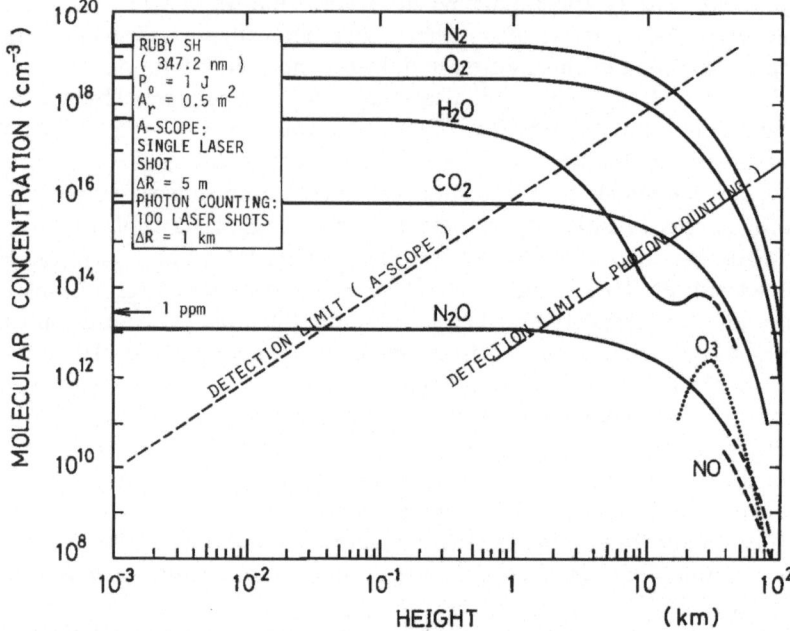

Fig. 5.23. Analytical result of detection capability for several major molecules in the atmosphere by the Raman scattering scheme employing the second-harmonic beam of a pulsed ruby laser for nighttime operation, and their concentration profiles in the standard atmosphere

5.5.3 Detection Capability of Near-Resonance Raman Scattering and Resonance Scattering Schemes in the Lower Atmosphere

For improvement of the scattering efficiency and detection capability of the laser-Raman radar technique, use of the resonance Raman effect appears promising because of resonance enhancement of the scattering cross section, as explained in Subsection 5.2.3. However, this scheme limits the number of constituent species that can be detected with improved sensitivity. In the following analytical discussion of detection sensitivity and range performance, we consider only near-resonance Raman scattering in order to avoid the severe reduction of laser beam power by resonance absorption and to avoid superimposed fluorescence when the laser is tuned to exact resonance. It is also known that almost all molecules possess absorption bands in the ultraviolet region of the spectrum, typically between 200 and 300 nm, and the shortest useful ultraviolet wavelength is limited by the decrease in atmospheric transmittance caused by elastic scattering loss and absorption due to oxygen (below 250 nm). Accordingly, in order to understand the general trends

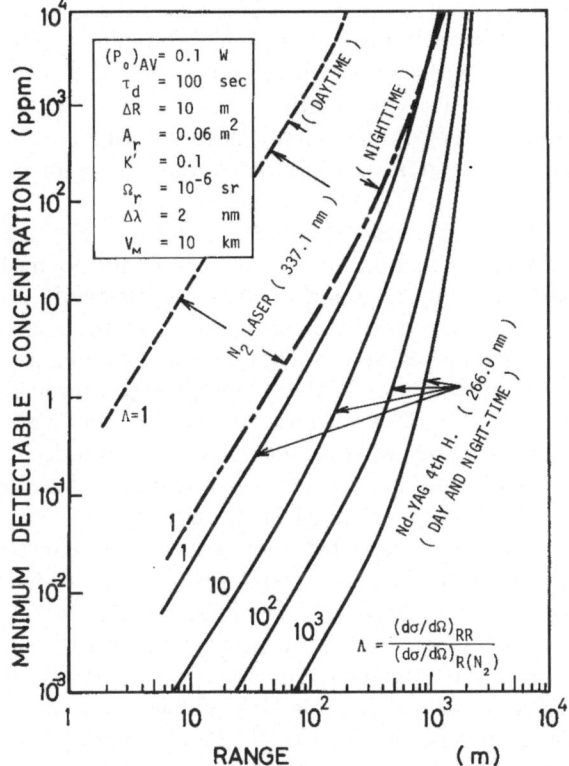

Fig. 5.24. Minimum detectable concentration as a function of range for the laser radar system based on the near-resonance Raman scattering scheme utilizing the resonance enhancement of the scattering cross section. The resonance parameter Λ gives the ratio of differential backscattering cross section of the specific molecule to non-resonant cross section for the N_2 molecule. (From INABA and KOBAYASI [5.20])

in near-resonance Raman scattering, we evaluate and compare system sensitivity and range capability by considering tentatively the N_2 gas laser at 337.1 nm and the 4th harmonic of a Nd-YAG laser at 266.0 nm as possible high-power ultraviolet lasers, although this scheme requires (practically at least) a discretely-tunable laser to match the electronic transition frequency for the individual species to be detected.

In Fig. 5.24 the minimum detectable concentration as a function of range is shown for several different values of the resonance parameter Λ, an enhancement factor of the Raman scattering cross section of the specific molecule of interest (relative to the non-resonant cross section for the N_2 molecule) [5.20]. The average power of these two kinds of pulsed lasers is assumed to be 0.1 W, and the other parameters used for the analysis are the same as those in Fig. 5.22. It is seen that the sensitivity expected by use of the N_2 laser is limited by sky background for daytime operation, while this is not the case for the 4th harmonic of the Nd-YAG laser. This is primarily due to its wavelength being located in the solar blind region, which provides remarkable characteristics during practical operation of the laser radar system irrespective of day- or night-time operation.

It is seen from this result that resonance enhancement of the order of 10^2 to 10^3 in scattering cross section, relative to that for the N_2 molecule, is enough to realize remote detection of minor atmospheric constituents. Trace molecular pollutants at concentrations in the vicinity of 0.01 to 0.1 ppm can be monitored at a maximum range of about a few hundred meters, with a range resolution of 10 m. This inference, applicable to both the day- and night-time observations, may not be too much to expect in view of the fact increases of several orders of magnitude have been observed in cross sections for *resonance* Raman scattering for a few gaseous molecules, as previously mentioned in Subsection 5.2.3.

Recently, ROSEN et al. [5.60] reported their analysis of detection sensitivity and range performance of a laser radar system utilizing resonance Raman scattering for the detection of SO_2, NO_2, and NO molecules in the atmosphere. Their results are based on the listed values of largely enhanced Raman scattering cross sections—experimental ones for SO_2 and NO_2, and a calculated one for NO. The use of a tunable laser matched to the electronic resonance of these molecules with 0.05 mJ output power at 100 pps repetition was assumed to prevent potential eye safety hazards. With the detection parameters of $A_r = 1$ m^2, $\eta = 0.1$, $\Omega_r = 10^{-5}$ sr, and $\Delta\lambda = 0.1$ nm, together with $\tau_d = 100$ s and $\Delta R = 10$ m, and under the condition of S/N ~ 10, they predicted that this scheme will be able to detect pollutant concentrations of SO_2 and NO greater than 10 ppm in smoke plumes at a range of a few hundred meters, even in daytime. This sensitivity appears higher by a factor of about 10 than that achieved recently by HIRSCHFELD et al. [5.114] with ordinary Raman scattering; but the resonance Raman scattering scheme accomplishes this with very modest laser output energy, in compliance with laser safety standards.

With regard to the major molecular constituents in the atmosphere indicated in Fig. 5.23, their remote detection by this resonantly-enhanced scheme through the lower atmosphere is not inherently feasible because their exciting wavelengths are close to or on proper resonance lines, which are mostly in the wavelength range of highly absorptive or significantly low atmospheric transmittance in the ultraviolet. However, the estimated value of resonance Raman scattering cross section for the O_3 molecule excited by its very intense Hartley dissociation band between 250 and 300 nm, could be sufficient large (approximately 10^{-24} cm^2/sr, as reported by PENNEY [5.48]) to allow range-resolved detection of ambient levels of the order of 10 ppb in our environmental air.

Spatially-resolved probing of the stratospheric O_3 layer between about 10 and 35 km altitude, depicted schematically in Fig. 5.23, is likely well within the range of experimental feasibility by means of this resonance Raman scattering scheme using a ground-based laser radar system. For practical application, use of a slightly longer wavelength, near 300 nm, is preferable. This radiation suffers less attenuation and can penetrate to ~ 25 km with a two-way transmittance of $T^2 \sim 0.1$–0.2. As an example, assuming the system parameters of $A_r = 10$ m^2, $K' = 0.1$ and $\eta = 0.2$, the number of detectable photons per Joule of transmitted laser energy backscattered from this altitude was estimated by PENNEY [5.48] to be approximately 1 photon/J. Higher detection sensitivity than this should be expected from elevated ground stations, such as a high-altitude observatory with large receiving optics. Since O_3 plays an essential role in indicating chemical chains and physical processes in the stratosphere, and also for shielding solar radiation below the middle ultraviolet region from the ground, continual observation of this O_3 layer is quite important and valuable.

As discussed in Subsection 5.2.4, the resonance scattering process, regarded as resonantly-enhanced Rayleigh scattering near or inside an isolated absorption line, is expected to offer a very sensitive scheme for remote detection of atomic metal vapors in the lower atmosphere [5.12, 13]. Figure 5.25 illustrates the minimum detectable concentration versus range for some atomic species in the atmosphere under shot-noise-limited conditions (e.g., nighttime observation), evaluated on the basis of scattering cross sections summarized in Table 5.8 and detection parameters indicated in the insert of the figure. As the detectivity may likely be excellent, these measurements could be performed with a single A-scope observation of a single laser shot of 0.1 mJ output power. Also shown in the figure are the analytical curves of Na and Hg atoms for the resonance fluorescence scheme at the same frequencies, for comparison. The minimum detectable concentration for both schemes is usually limited by the presence of intense Mie scattering at the same back-scattered frequency, and hence depends largely upon the visibility.

Improvement of the detection sensitivity for atomic metal vapors by these two schemes should be achievable if high-repetition pulses from tunable dye lasers and their second harmonics are practically available. For instance, with an average output power of 1 mJ and 100 Hz repetition, but with other parameters the same as for Fig. 5.25, and the condition of $S/N \sim 100$, the minimum detectable concentration for resonance scattering and fluorescence is evaluated to be lowered to about 10^{-6} ppb

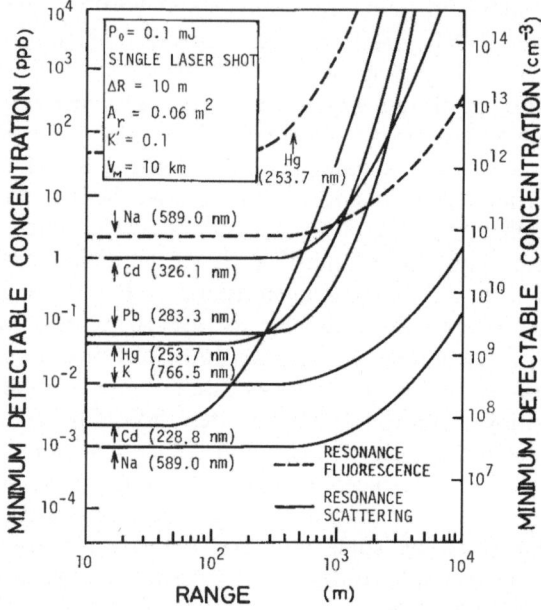

Fig. 5.25. Analytical comparison of minimum detectable concentrations of some atomic metal vapors as a function of range in the lower atmosphere for the laser radar system based on the resonance scattering and resonance fluorescence schemes under the shot noise limited condition

and 10^{-3} ppb, respectively, for Na and K atoms, and 10^{-4} ppb and 10^{-2} ppb for Hg atoms, within the range $R = 1$ km, with an observation time $\tau_d = 1$ s. It should be mentioned that mode-locked dye laser pulses of subnano- or pico-second duration with an oscillation frequency carefully adjusted to the far wing portion of the resonance absorption line at atmospheric pressure (to discriminate from the resonance fluorescence) would permit application of the resonance scattering scheme in the lower atmosphere, but with a trade-off in detection sensitivity [5.12, 13]. Nevertheless, this scheme may still permit range-resolved, remote detection of atomic metal pollutants dispersed in the lower atmosphere within the range of 1 km.

In some situations of range-resolved measurements of air pollutants where high pollutant concentrations are localized, or otherwise medium concentrations are distributed over long distances, the problems of saturation of the pollutant transition and depletion of the transmitted laser beam become important. This is true not only for resonance Raman scattering and resonance scattering, but also for resonance fluorescence, which will be analyzed in the next subsection. A general analytical technique for calculating the actual concentrations detectable by laser radar schemes has been derived by INABA et al. [5.2], and also by BARRETT and BEN-DOV [5.143] by taking into account these effects. KILDAL and BYER [5.29] performed a detailed analysis of these problems in conjunction with the resonance fluorescence and absorption schemes. In order to avoid the complexity associated with these problems, and also to achieve optimum detectivity, the use of transmitting laser wavelengths appropriately off the center of a relevant transition is advantageous. In such cases, and also when pollutants are not so highly concentrated within a thin layer, the system analysis presented in this chapter should be adequate.

5.5.4 Detection Capability of the Resonance Fluorescence Scheme for Ground-Based Monitoring of the Lower and Upper Atmosphere

There should be various situations in which resonance fluorescence can be utilized effectively for spatially-resolved, quantitative probing of atomic and molecular constituents in the atmosphere, within the conditions of applicability pointed out in Subsection 5.2.5. One successful application which has been made was the remote observation of Na and K atoms in the upper atmosphere, as described in Subsection 5.4.2.

In the lower atmosphere, the most favorable situations likely to fit the conditions for remote sensing by resonance fluorescence can be found for some minor molecules such as SO_2, NO_2, and I_2 in the near ultraviolet and visible regions. In Figs. 5.26a–c, the predicted values of minimum detectable concentration as a function of range for this technique are shown for these three molecules [5.78]. We assumed laser characteristics such as $P_0 = 100$ kW, pulse duration $\tau_p = 10$ ns, 100 integrated laser shots, and detection parameters indicated in the insert of each figure. The excitation wavelengths are assumed to be at absorption maxima of 300.1 nm for SO_2, 435.8 nm for NO_2 and 589.6 nm for I_2, for which some spectroscopic parameters related to the differential cross section are listed in Table 5.10. The solid line indicates the case for nighttime operation (limited by shot noise), whereas the dashed line represents daytime operation (characterized by background-limited detection).

As shown in Figs. 5.26a and b, the detection sensitivity for SO_2 and NO_2 molecules was calculated using the optical bandwidth $\Delta\lambda$ of the receiving system as a variable parameter; while that for I_2 in Fig. 5.26c is illustrated only for a fixed $\Delta\lambda = 1$ nm, which provides sufficient detectivity. It is generally found that $\Delta\lambda$ plays an important role in improving sensitivity of the resonance fluorescence technique for remote detection of molecules such as SO_2 and NO_2 with broad fluorescence bands. A wider bandwidth increases sensitivity during both night- and

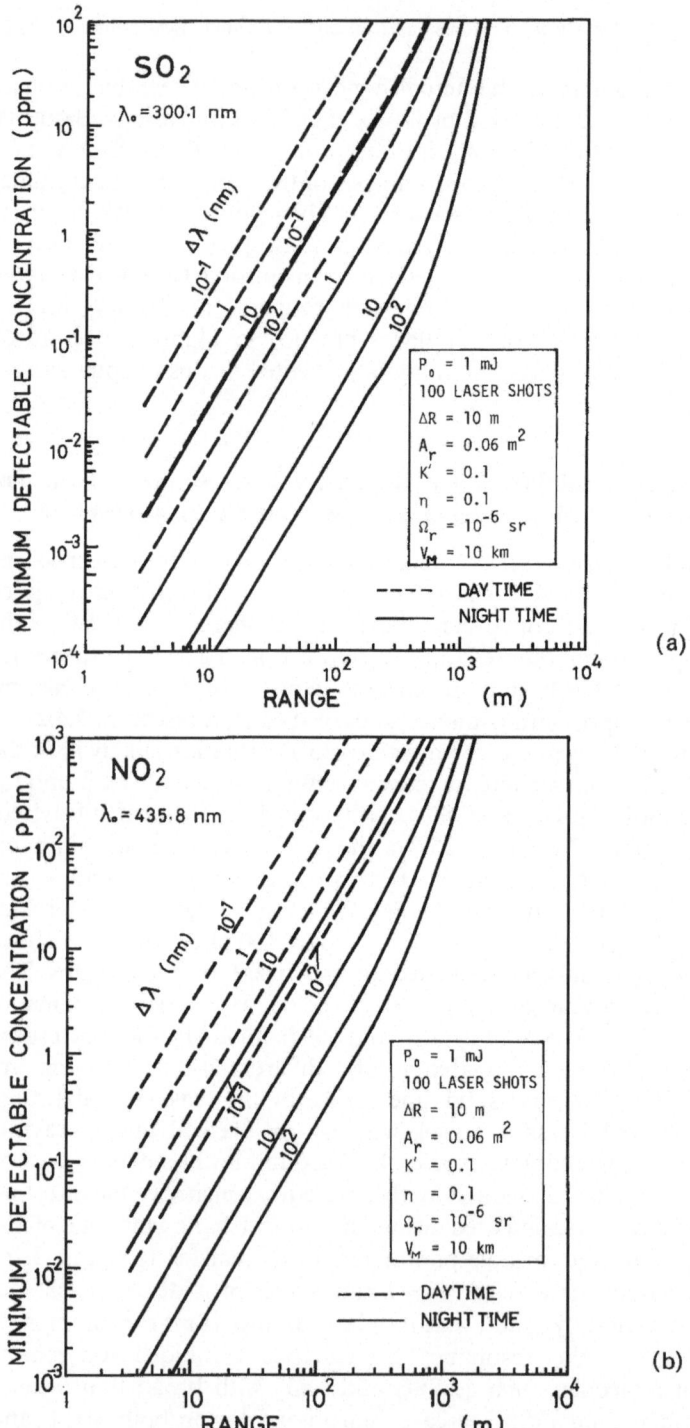

Fig. 5.26a–c. Comparison of calculated results of minimum detectable concentrations as a function of range for the laser radar system operated by the resonance fluorescence scheme employing tunable laser transmitters for (a) SO_2, (b) NO_2, and (c) I_2 molecules

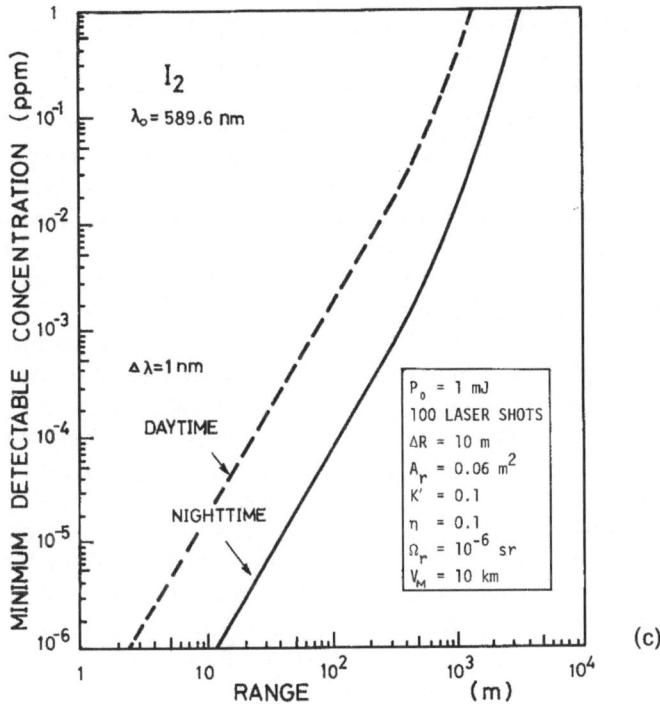

Fig. 5.26c

day-time operations since the signal-to-noise ratio associated with broad fluorescence is proportional to $\Delta\lambda$ and $(\Delta\lambda)^{1/2}$ for the shot noise and background limited detections, respectively, when the detection efficiency specified by $\Delta\lambda/\Delta\lambda_{BF}$ is taken into account (where $\Delta\lambda_{BF}$ is the effective halfwidth of such a fluorescence band). Nevertheless, a compromise should be made between this bandwidth and the background noise reduction to realize the optimum sensitivity.

It is seen from Figs. 5.26a, b, and c that, under nighttime conditions, a minimum concentration of about 1 ppm and 20 ppm should be detectable at a range of 500 m for SO_2 and NO_2, respectively, employing a filter of $\Delta\lambda = 10$ nm; while the sensitivity is a few ppb at the same range for I_2 with $\Delta\lambda = 1$ nm. However, daytime monitoring at a 200 m range may allow us to achieve about 20 ppm for SO_2, 100 ppm for NO_2, and around 10 ppb for I_2. Furthermore, comparing the maximum sensitivity expected from these figures with that for ordinary Raman scattering, resonance fluorescence appears to offer higher detectivity by a few orders of magnitude for molecules relevant to air pollution, and readily allows their detection in the effluents of stationary sources from distances to 1 km. This capability will be comparable to (or slightly better than, in certain practical situations) the near-resonance Raman scattering scheme. If one could be sure that there will be no eye exposure to the beam from the laser transmitter, or no severe restrictions on the system size, then one might be able to use a higher-energy tunable laser or a larger receiver aperture to gain the necessary sensitivity for spatially-resolved measurements of dispersed pollutants. In any event, the highly sensitive, remote monitoring of atmospheric I_2 will be an important application of laser radar technology, as its presence above the sea is known to be related to certain forms of marine biological activity [5.63].

The detectivity and range limitations of the resonance fluorescence scheme have previously been analyzed in detail by KILDAL and BYER [5.29] for CO molecules, using infrared vibrational transitions, and Na and Hg atoms, and SO_2, NO_2 and C_6H_6 molecules through electronic transitions in the visible or ultraviolet. A similar study was performed by MEASURES and PILON [5.14] with regard to remote mapping of SO_2, NO_2, and I_2 molecules. They came to the same basic conclusion: that this scheme for spectroscopic laser radar probing is well suited for the detection of localized or concentrated pollutant sources in the 100 to 1000 m range, but it is marginal for mapping dispersed pollutants. KILDAL and BYER [5.29] also estimated that the fluorescence scheme based on *infrared* molecular transitions is less sensitive by a few orders of magnitude than that accomplished with *electronic* transitions of molecules and atoms.

As already pointed out in Subsection 5.4.2, resonance fluorescence offers detection advantages for upper atmospheric monitoring of atomic and molecular constituents using tunable lasers from the ultraviolet to the near infrared. The earliest proposals on the application of this technique (conventionally called the resonance scattering method) to atmospheric sounding originated in 1964 with three independent groups. The group at Tohoku University [5.144] discussed the possibility of detecting remotely the metastable N_2 molecules, utilizing the near coincidence between its first positive band transition and Nd-glass laser emission; while HIRONO [5.145] suggested use of the Meinel band of the N_2^+ molecule and the resonance transition of the K atom, in combination with a Nd-glass laser and a Raman laser, respectively. Similarly, YOUNG [5.146] remarked that pulsed gas lasers emitting radiation in coincidence with the N_2^+ first negative bands and the N_2 second positive bands as well as the Na resonance doublets may provide useful means for laser radar probing of these chemical species in the upper atmosphere. Later, the measurement of metastable NO molecules employing a suitable frequency-coincident Raman laser through the Ogawa band transition in the near infrared was proposed by NUGENT [5.147]. Although no experimental approach utilizing either a Nd-glass laser, a Raman laser, or a gas laser has yet been reported, the advent of tunable dye lasers led to successful remote detection of Na and K atoms in the upper atmosphere by means of the resonance fluorescence technique, as described in Subsection 5.4.2.

Other upper atmospheric constituents which may possibly be measured with the same scheme by a ground-based laser radar system are listed in Table 5.13, together with some species already mentioned. Table 5.13a summarizes the relevant transitions and resonance wavelengths of some molecular constituents and their expected concentrations and heights. In addition to these species, if any information concerning the location and concentration of molecular constituents such as OH, H_2O, HO_2, H_2O_2, CO, CO_2, NO_2 could be obtained by this remote sounding regime, it would definitely be of importance in aeronomy and photochemistry of the earth's mesosphere or ionosphere. Listed in Table 5.13b are several atomic metal vapors and naturally-existing atoms, such as He and O, excited at metastable states, along with their resonance wavelengths and estimated concentrations. Although the presence of some atomic species (e.g. Na, Li, K, Ca, and Ca^+) has been confirmed by extensive observations of resonance fluorescence of sunlight at twilight [5.148], measurements of the spatial and temporal distributions of these atomic metal vapors by the laser radar technique should be worthwhile to inquire as to their origin in conjunction with meteoric activity and some cosmic materials precipitating onto the upper atmosphere. Furthermore, such a routine observation would provide valuable information on the general atmospheric circulation and wind systems.

As is seen from Table 5.13b, the detection of K atoms is likely to be difficult owing to their small concentration (relative to Na atoms) in the upper atmosphere. It requires the use of a large collecting area, e.g., of 16 m^2 available only at Kingston, Jamaica [5.72]. The estimated values of minimum detectable concentration for a laser radar system with parameters $P_0 = 1$ mJ, $T^2 = 0.5$, $K' = 0.2$, $A_r = 0.2\ m^2$, $\eta = 0.05$, and $\Delta R = 10$ km, with 100 integrated laser shots, are $\sim 20\ cm^{-3}$ at 100 km altitude for Na atoms, $\sim 2 \times 10^3\ cm^{-3}$ at 50 km and $\sim 10^4\ cm^{-3}$ at 100 km for N_2^*, $\sim 3 \times 10^2\ cm^{-3}$ at 100 km for N_2^+, and $\sim 4 \times 10^3\ cm^{-3}$ at 50 km and $\sim 2 \times 10^4\ cm^{-3}$ at 100 km for NO*, respectively [5.125]. Precise assessment of the detectivity and range limitation for most molecular and atomic species shown in Table 5.13 requires knowledge of the individual cross sections and transmitter/receiver characteristics. Further improvement in performance of the ground-based, spectroscopic laser radar system should be pursued compared with that used for the height distribution and its seasonal variation of Na atoms in the upper atmosphere.

Table 5.13. Lists of molecular and atomic constituents in the upper atmosphere which may be detected with the resonance fluorescence scheme by a ground-based laser radar system using a frequency-tunable laser source. (a) and (b) summarize molecular, and atomic and ionic species together with their resonance wavelengths, estimated concentrations and heights, respectively

Table 5.13a

Molecular constituent	Transition		Band system	Resonance wavelength [nm]	Estimated concentration [cm^{-3}]	Expected height [km]
	$B^3\Pi_g - A^3\Sigma_u^+$	(0,0)		$1044 \sim 1054$		
		(1,0)	First	$884 \sim 891$		
N_2^*		(2,1)	positive	$866 \sim 872$	$N_2^* (A^3\Sigma_u^+)$	
		(3,1)		$757 \sim 763$	$\sim 10^8$ (at 100 km)	
	$C^3\Pi_u - B^3\Pi_g$	(0,0)	Second	337		
		(0,1)	positive	358		
N_2^+	$B^2\Sigma_u^+ - X^2\Sigma_g^+$	(0,0)	First	~ 391	10^3	$90 \sim 140$
		(1,0)	negative	~ 358	10^4	200
	$A^2\Pi_u - X^2\Sigma_g^+$		Meinel	$915 \sim 940$		
NO*	$b^4\Sigma^- - a^4\Pi_i$	(1,0)	Ogawa II	$864 \sim 874$	Total NO	$50 \sim 150$
		(2,0)		$786 \sim 794$	$\sim 10^7$	
O_3	$\tilde{D} - \tilde{X}$		Hartley	~ 255	$> 10^{10} (\sim 10^{10})$	60
					$(\sim 10^{12})$	30
$O_2^+ *$	$b^4\Sigma_g - a^4\Pi_u$	(1,0)	First negative	~ 563		

* Excited metastable state.

Table 5.13b

Atomic and ionic constituent	Resonance wavelength [nm]	Estimated concentration [cm^{-3}] or column number density[†] [cm^{-2}]	Estimated concentration relative to Na atom	Expected height [km]
Na	589.0	$10^2 \sim 10^4$ (cf. Fig. 5.6)	1	$80 \sim 110$
	589.6			
K	769.9	$9 \times 10^{7[†]} (\sim 10)$	$< 0.01 \sim 0.1$	$70 \sim 100$
	766.5	(overlapped by O_2 absorption band)		
Ca	422.7		~ 1	
Ca^+	393.4	Highly variable		$100 \sim 200$
	396.8	associated with meteors		
Fe	386.0	$< 10^4$	17	< 700
Cr	425.4		0.2	
Mn	403.1		0.1	
Al*	396.2		1.3	
Li*	670.8	$\sim 6 \times 10^{6[†]} (< 10)$		80
He*	1083.0	$5 \times 10^{7[†]}$		
O*	777.2	$10^2 \sim 10^3$	~ 1	> 100

* Excited metastable state.

5.5.5 Detection Capability of Airborne and Balloon-Borne Probing and Satellite Measurements

Since compact and versatile laser radar systems are available with present state-of-the-art technology, it is only natural to consider the possibility of spectroscopic detection of atomic and molecular constituents in the atmosphere by carrying the instrumentation not only on a high-flying airplane or balloon, but also on an orbiting artificial satellite. In these situations, limitations of laser output power and receiving area could be compensated for (at least partially) by proximity. Also, the laser safety constraint could be avoided unless the beam is directed downward to populated areas.

The minimum concentrations of various and atomic species detectable by airborne or balloon-borne laser radars incorporating the resonance fluorescence scheme was evaluated by PENNEY et al. [5.48]. With a pulsed laser output of 1 J in the visible and ultraviolet region, and system parameters of $K'=0.1, A_r=1$ m^2, T$^2=0.5, \eta=0.2$ for observing a 100 m column at a range of 300 m, for example, the estimated values are 16 ppb for SO$_2$, 16 ppm for NO$_2$, 0.11 ppb for NO, and 0.16 ppb for OH molecules at 20 km altitude, respectively. Spatially-resolved probing of the stratospheric O$_3$ layer can also be made by the same airborne laser radar system, using either the near-resonance Raman scattering or the resonance fluorescence scheme, with a detection sensitivity of about 30 ppb at 20 km. In addition to these molecules, there could be many other species accessible by the fluorescence scheme, such as aromatic hydrocarbons, aldehydes, and various radicals like CN and C$_2$. Numerous atomic species in addition to those listed in Table 5.8 and 5.13b, such as many transition metals and rare-earths (for which large fluorescence cross sections, by reduced quenching, for excitation on resonance lines between the near ultraviolet and the near infrared can be expected) may also be monitored with the excellent detectivity of a few atoms per cm^3 at proximity range (~ 300 m) by an airborne or balloon-borne laser radar system.

Similar performance will be achievable by a rocket-borne laser radar system for real-time measurements of atomic and molecular concentrations in the upper atmosphere. The system, carrying a laser transmitter of 0.1 mJ output and detection equipment with $K'=0.2$, $A_r=0.1$ m^2, T$^2=1, \eta=0.05$ and $\Delta R=1$ m, is estimated to yield a minimum detectable concentration of $\sim 2 \times 10^{-2}$ cm^{-3} for Na, $\sim 10^{-1}$ cm^{-3} for K, ~ 30 cm^{-3} for N$_2^*$, ~ 2 cm^{-3} for N$_2^+$, and 10^2 cm^{-3} for NO* by the resonance fluorescence scheme [5.125].

With respect to satellite measurements of upper atmospheric constituents by the laser radar method, there are severe restrictions of transmitter power and collecting optics, and difficulty of large working distances; whereas advantages are high transmittance and much larger cross section of effectively unquenched fluorescence, compared to the case of the lower atmosphere. If we consider a molecular species for which the fluorescence or resonance Raman scattering cross section has a significantly large value of 10^{-20} cm^2/sr, for example, the estimated minimum detectable concentration in the visible or ultraviolet region becomes about 10^9 cm^{-3} for a laser radar system with $P_0 = 1$ W, $K' = 1$, $A_r = 0.1$ m^2, T = 1, even at the nearest range of $R = 10$ m. This result may indicate that satellite laser radar probing of existing levels of molecular and radical constituents such as OH and NO in the upper atmosphere may be marginal. However, if one could consider the combination of a ground-based laser radar and retroreflectors on an orbiting satellite, the resonance fluorescence and Raman scattering schemes incorporating both backward and forward re-emission along the two-way optical path might offer an alternative method of detection.

In contrast to this, the situation for atomic vapors is much more favorable, and there is a good possibility of monitoring them from a satellite laser radar probe. For many atoms for which the cross section of unquenched fluorescence is of the order of 10^{-12} to 10^{-14} cm^2/sr, we found that the minimum detectable concentration lies in the range roughly 10–10^5 cm^{-3}, depending upon background noise, with a laser radar system of $P_0 = 1$ mW, but with other parameters the same as those used above. PENNEY et al. [5.48] also estimated that a satellite installed with a system of $P_0 = 1$ W, $A_r = 1$ m^2, for observing a 1 km column from an altitude of 250 km, will have the capability of detecting a number of atomic species with minimum detectable concentrations of around 10^3–10^6 cm^{-3} at 200 km, and 10^4–10^8 cm^{-3} at 50 km, respectively. Moreover, several minor ions, such as Mg$^+$, Fe$^+$, and Si$^+$, in addition to Ca$^+$, may also be detected by utilizing their resonance transitions with laser wavelengths less than 280 nm, for which the transmissions are not prevented by absorption due to O_2 and O_3 molecules, as in the case of the lower atmosphere.

References

5.1   H. INABA, T. KOBAYASI, T. ICHIMURA, M. MORIHISA: IEEE J. Quant. Electron. QE-2, 40 (1966)
5.2   H. INABA, T. KOBAYASI, T. ICHIMURA, M. MORIHISA, H. ITO: Electronics and Comm. in Japan 51-B, 36 and 45 (1968)
5.3   R. T. H. COLLIS: Appl. Opt. 9, 1782 (1970)
5.4   R. T. H. COLLIS, E. E. UTHE: Opto-Electron. 4, 87 (1972)
5.5   R. G. STRAUCH, A. COHEN: In Remote Sensing of the Troposphere, ed by V. E. DERR (U.S. Department of Commerce, NOAA, Colorado 1972) Ch. 23
5.6   F. F. HALL, Jr.: In Laser Applications, Vol. 2, ed. by M. ROSS (Academic Press, New York and London 1974) p. 161
5.7   J. M. TAYLOR, H. W. YATES: J. Opt. Soc. Am. 47, 223 (1957)
5.8   G. S. NEWCOMB, M. M. MILAN: IEEE Trans. GE-8, 149 (1970)

5.9   P. L. HANST, J. A. MORREAL: J. Air Pollut. Contr. Assoc. **18**, 754 (1968); also,
      P. L. HANST: Appl. Spectrosc. **24**, 161 (1970)
5.10 P. L. HANST: In *Advances in Environmental Science and Technology*, Vol. 2, ed. by
      J. N. PITTS, Jr. and R. L. METCALF (Wiley-Interscience, New York 1971) p. 91
5.11 R. M. SCHOTLAND: Proc. 3rd Int. Symp. Remote Sensing Environ., Univ. of Michigan,
      Ann Arbor, Michigan, U.S.A. (1964) p. 215
5.12 T. KOBAYASI, H. INABA: Record 11th Symp. Electron, Ion, and Laser Beam Tech-
      nology, ed. by R. F. M. THORNLEY (San Francisco Press, San Francisco, 1971) p. 385
5.13 H. INABA, T. KOBAYASI: Proc. 3rd Biennial Cornell Elect. Eng. Conf., College of
      Eng., Cornell Univ., Ithaca, New York, U.S.A. (1971) p. 73
5.14 R. M. MEASURES, G. PILON: Opto-Electron. **4**, 141 (1972)
5.15 S. A. AHMED: Appl. Opt. **12**, 901 (1973)
5.16 R. L. BYER, M. GARBUNY: Appl. Opt. **12**, 1496 (1973)
5.17 K. W. ROTHE, U. BRINKMAN, H. WALTHER: Appl. Phys. **3**, 115 (1974)
5.18 R. L. BYER: Opt. Quant. Electron. **7**, 147 (1975), and references therein
5.19 T. KOBAYASI, H. INABA: Conf. Abstracts of 1974 Int. Laser Radar Conf., Sendai,
      Japan (1974) p. 108; H. INABA, T. KOBAYASI: Opt. Commun. **14**, 119 (1975); also,
      Opt. Quant. Electron. **7**, 319 (1975)
5.20 H. INABA, T. KOBAYASI: Opto-Electron. **4**, 101 (1972)
5.21 See, for example, J. BEHRINGER: In *Raman Spectroscopy*, ed. by H. A. SZYMANSKI
      (Plenum Press, New York 1967) p. 168
5.22 W. HOLZER, W. F. MURPHY, H. J. BERNSTEIN: J. Chem. Phys. **52**, 399 (1970)
5.23 W. HEITLER: *The Quantum Theory of Radiation*, 3rd ed. (Oxford Univ. Press,
      London 1954)
5.24 D. L. HUBER: Phys. Rev. **158**, 843 (1967); **170**, 418 (1968); **178**, 93 (1969) and B1,
      3409 (1970)
5.25 See, for example, G. HERZBERG: *Molecular Spectra and Molecular Structure*, Vols. I,
      II and III (D. van Nostrand Co., Princeton, N.J. 1950, 1945 and 1967)
5.26 See, for example, G. HERZBERG: *"Infrared and Raman Spectra"*, in *Molecular
      Spectra and Molecular Structure*, Vol. II (1945);
      also S. MIZUSHIMA: In *Handbuch der Physik*, Vols. 26–27, ed. by S. FLÜGGE (Springer,
      Berlin 1958)
5.27 H. INABA, T. KOBAYASI: Digest of Technical Papers, No. 12-1, 1970 Int. Quant.
      Electron. Conf., Kyoto, Japan (1970)
5.28 T. KOBAYASI, H. INABA: Appl. Phys. Lett. **17**, 139 (1970)
5.29 H. KILDAL, R. L. BYER: Proc. IEEE **59**, 1644 (1971)
5.30 G. PLACZEK: *Handbuch der Radiologie*, Vol. VI, Part 2, ed. by E. MARX (Akade-
      mischer Verlag, Leipzig 1934)
5.31 W. F. MURPHY, W. HOLZER, H. J. BERNSTEIN: Appl. Spectrosc. **23**, 211 (1969)
5.32 E. J. STANSBURY, M. F. CRAWFORD, H. L. WELSH: Canad. J. Phys. **31**, 954 (1953)
5.33 B. P. STOICHEFF: Canad. J. Phys. **32**, 630 (1954); **36**, 218 (1958)
5.34 A. WEBER, E. A. McGINNIS: J. Molec. Spectrosc. **4**, 195 (1953)
5.35 J. J. BARRETT, N. I. ADAMS, III: J. Opt. Soc. Am. **58**, 311 (1968)
5.36 A. WEBER, S. P. S. PORTO, L. E. CHEESMAN, J. J. BARRETT: J. Opt. Soc. Am. **57**, 19
      (1967)
5.37 C. M. PENNEY: J. Opt. Soc. Am. **59**, 34 (1969)
5.38 C. M. PENNEY, L. M. GOLDMAN, M. LAPP: Nature **235**, 110 (1972)
5.39 D. G. FOUCHE, R. K. CHANG: Appl. Phys. Lett. **18**, 579 (1971) and **20**, 256 (1972)
5.40 W. R. FENNER, H. A. HYATT, J. H. KELLAM, S. P. S. PORTO: J. Opt. Soc. Am. **63**, 73
      (1973)
5.41 H. A. HYATT, J. M. CHERLOW, W. R. FENNER, S. P. S. PORTO: J. Opt. Soc. Am. **63**,
      1604 (1973)
5.42 D. A. LEONARD: J. Appl. Phys. **41**, 4328 (1970)
5.43 C. J. CHEN, F. WU: Appl. Phys. Lett. **19**, 452 (1971)
5.44 R. L. SCHWIESOW: Am. Inst. Aeron. Astron. J. **11**, 87 (1973)

5.45   J. I. Levatter, R. L. Sandstrom, S.-C. Lin: J. Appl. Phys. **44**, 3273 (1973)
5.46   T. Kobayasi, M. Takemura, H. Shimizu, H. Inaba: IEEE J. Quant. Electron. **QE-8**, 579 (1972)
5.47   H. Shimizu, T. Kobayasi, H. Inaba: Oyo Buturi: Publ. Japan Soc. Appl. Phys. **42**, 889 (1973) (in Japanese)
5.48   See, for example, C. M. Penney: In *Laser Raman Gas Diagnostics*, ed. by M. Lapp and C. M. Penney (Plenum Press, New York 1974), p. 191;
also   C. M. Penney, W. W. Morey, R. L. St. Peters, S. D. Silverstein, M. Lapp, D. R. White: Report for NASA CR-132363, prepared by General Electric Corporate Research and Development, Schenectady, New York (1973)
5.49   M. Jacon, M. Berjot, L. Bernard: Compt. Rend. **273**, Ser. B, 956 (1971)
5.50   P. F. Williams, D. L. Rousseau, S. H. Dworetsky: Phys. Rev. Lett. **32**, 196 (1974)
5.51   D. G. Fouche, R. K. Chang: Phys. Rev. Lett. **29**, 536 (1972)
5.52   R. L. St. Peters, S. D. Silverstein, M. Lapp, C. M. Penney: Phys. Rev. Lett. **30**, 191 (1973)
5.53   M. Berjot, M. Jacon, L. Bernard: Canad. J. Spectrosc. **17**, 60 (1972)
5.54   W. Kiefer, H. J. Bernstein: J. Molec. Spectrosc. **43**, 366 (1972) and J. Chem. Phys. **57**, 3017 (1972)
5.55   P. F. Williams, D. L. Rousseau: Phys. Rev. Lett. **30**, 951 (1973)
5.56   J. Behringer: Z. Elektrochem. **62**, 906 (1958)
5.57   D. G. Rea: J. Molec. Spectrosc. **4**, 499 (1960)
5.58   A. C. Albrecht: J. Chem. Phys. **34**, 1476 (1961)
5.59   P. P. Shorygin: Pure Appl. Chem. **4**, 87 (1962)
5.60   H. Rosen, P. Robrish, O. Chamberlain: to be published
5.61   M. Jyumonji, T. Kobayasi, H. Inaba: Transactions of Tech. Group on Opt. and Quant. Electron., Inst. Electronics and Commun. Engineers of Japan, OQE 73-58 (1973) (in Japanese)
5.62   M. Jyumonji, T. Kobayasi, H. Inaba: Kvantovaya Elektronika (Soviet J. Quant. Electron.) **3**, No. 4 (1976)
5.63   A. J. Moffat, A. R. Barringer: Proc. Symp. on Remote Sensing in Marine Biology and Fishery Resources, College Station, Texas, U.S.A. (1971), p. 98
5.64   C. M. Penney: Conf. Abstracts of 1974 Int. Laser Radar Conf., Sendai, Japan (1974), p. 43
5.65   A. Dalgarno: J. Opt. Soc. Am. **53**, 1223 (1963)
5.66   D. W. O. Heddle: J. Opt. Soc. Am. **54**, 264 (1964)
5.67   C. M. Penney: Phys. Rev. Lett. **14**, 423 (1965)
5.68   J. W. Chamberlain: *Physics of Aurora and Airglow* (Academic Press, New York 1961)
5.69   D. R. Hamilton: Astrophys. J. **106**, 457 (1947)
5.70   S. Chandrasekhar: *Radiative Transfer* (Oxford Univ. Press, London 1950)
5.71   M. Jyumonji, T. Kobayasi, H. Inaba, T. Aruga, H. Kamiyama: Conf. Abstracts of 5th Conf. on Laser Radar Studies of the Atmosphere, Williamsburg, Virginia, U.S.A. (1973), p. 39; also Transactions of Tech. Group on Quant. Electron., Inst. Electronics and Comm. Engineers of Japan, QE 72-58 (1972) (in Japanese)
5.72   F. Felix, W. Keenliside, G. Kent, M. C. W. Sandford: Nature **246**, 345 (1973)
5.73   See, for example, P. Pringsheim: *Fluorescence and Phosphorescence* (Interscience Publishers, New York 1949)
5.74   A. C. G. Mitchell, M. W. Zemansky: *Resonance Radiation and Excited Atoms* (Cambridge Univ. Press, London 1934)
5.75   D. G. Fouche, A. Herzenberg, R. K. Chang: J. Appl. Phys. **43**, 3846 (1972)
5.76   O. Stern, M. Volmer: Phys. Z. **20**, 183 (1919)
5.77   J. I. Steinfeld: Accounts Chem. Phys. **3**, 313 (1970)
5.78   T. Kobayasi, H. Inaba: Conf. Abstracts of 4th Conf. on Laser Radar Studies of the Atmosphere, Tucson, Arizona, U.S.A. (1972), p. 67
5.79   G. W. Bethke: J. Chem. Phys. **31**, 662 (1959)
5.80   C. C. Wang, L. I. Davis, Jr.: Phys. Rev. Lett. **32**, 349 (1974)

5.81 J. A. GELBWACHS, M. BIRNBAUM, A. W. TUCKER, C. L. FINCHER: Opto-Electron. **4**, 155 (1972)
5.82 S. NAKAHARA, K. ITO, T. SAEKI: Digest of Technical Papers, No. 12-2, 1970 Int. Quant. Electron. Conf., Kyoto, Japan (1970)
5.83 H. INABA, T. KOBAYASI: Nature **224**, 170 (1969)
5.84 See, for example: *Dye Lasers*, ed. by F. P. SCHÄFER, Topics in Applied Physics, Vol. 1, (Springer, Berlin, Heidelberg, New York 1973)
5.85 E. g., S. E. HARRIS: Proc. IEEE **57**, 2096 (1969)
5.86 F. ZERNIKE, J. E. MIDWINTER: *Applied Nonlinear Optics* (John Wiley, New York 1973);
      also R. L. BYER: Abstracts of 1974 Int. Laser Radar Conf., Sendai, Japan (1974), p. 57
5.87 P. W. KRUSE, L. D. MCGLAUCHLIN, R. B. MCQUISTAN: *Elements of Infrared Technology* (John Wiley, New York 1963)
5.88 M. BERTOLOTTI, L. MUZÜ, D. SETTE: Appl. Opt. **8**, 117 (1969)
5.89 S. TWOMEY, H. B. HOWELL: Appl. Opt. **4**, 501 (1965)
5.90 R. W. FENN: Appl. Opt. **5**, 293 (1966)
5.91 R. L. BYER: Opt. Quant. Electron. **7**, 147 (1975)
5.92 See, for example, *Laser Raman Gas Diagnostics*, Section II, ed. by M. LAPP and C. M. PENNEY (Plenum Press, New York 1974)
5.93 J. A. COONEY: J. Appl. Meteorol. **11**, 108 (1972)
5.94 T. KOBAYASI, H. SHIMIZU, H. INABA: Conf. Abstracts of 1974 Int. Laser Radar Conf., Sendai, Japan (1974), p. 49
5.95 R. G. STRAUCH, V. E. DERR, R. E. CUPP: Appl. Opt. **10**, 2665 (1971)
5.96 D. A. LEONARD: Nature **216**, 142 (1967)
5.97 J. A. COONEY: Appl. Phys. Lett. **12**, 40 (1968)
5.98 T. KOBAYASI, H. INABA: Opto-Electron. **2**, 45 (1970)
5.99 T. KOBAYASI, H. INABA: Proc. IEEE **58**, 1568 (1970)
5.100 G. S. KENT, P. SANDLAND, R. W. H. WRIGHT: J. Appl. Meteorol. **10**, 443 (1971)
5.101 M. J. GARVEY, G. S. KENT: Nature **248**, 124 (1974)
5.102 See, for example, J. A. COONEY, J. ORR, C. TOMASETTI: Nature **224**, 1098 (1969)
5.103 T. KOBAYASI, M. JYUMONJI, H. INABA: Summaries of 1971 Int. Symp. on Antennas and Propagation, Sendai, Japan, No. 3-IV (1971), p. 259
5.104 S. H. MELFI, J. D. LAWRENCE, JR., M. P. M. MCCORMICK: Appl. Phys. Lett. **15**, 295 (1969)
5.105 S. H. MELFI: Appl. Opt. **11**, 1605 (1972)
5.106 J. A. COONEY: J. Appl. Meteorol. **9**, 182 (1970) and **10**, 301 (1971)
5.107 R. G. STRAUCH, V. E. DERR, R. E. CUPP: Remote Sens. of Environ. **2**, 101 (1972)
5.108 H. SHIMIZU, T. KOBAYASI, H. INABA: Conf. Abstracts of 1974 Int. Laser Radar Conf., Sendai, Japan (1974), p. 25
5.109 T. HIRSCHFELD, S. KLAINER: Opt. Spectra **4**, 63 (1970)
5.110 D. A. LEONARD: Research Report of AVCO Everett Res. Lab., No. 362 prepared for NAPCA, Contract No. CPA 22-69-62 (December 1970)
5.111 S. NAKAHARA, K. ITO, S. TAMURA, M. KANEKIYO, H. INABA, T. KOBAYASI: IEEE J. Quant. Electron. QE-7, 325 (1971)
5.112 S. NAKAHARA, K. ITO, S. ITO, A. FUKE, S. KOMATSU, H. INABA, T. KOBAYASI: Opto-Electron. **4**, 169 (1972)
5.113 S. H. MELFI, M. L. BRUMFIELD, R. W. STOREY, JR.: Appl. Phys. Lett. **22**, 402 (1973)
5.114 T. HIRSCHFELD, E. R. SCHILDKRAUT, H. TANNENBAUM, D. TANENBAUM: Appl. Phys. Lett. **22**, 38 (1973)
5.115 D. A. LEONARD: In *Laser Raman Gas Diagnostics*, ed. by M. LAPP and C. M. PENNEY (Plenum Press, New York 1974), p. 45;
      also Opt. Quant. Electron. **7**, 197 (1975)
5.116 J. J. GROSSMAN, M. MURAMOTO, J. KACIN: Conf. Abstracts of 5th Conf. on Laser Radar Studies of the Atmosphere, Williamsburg, Virginia, U.S.A. (1973), p. 2

5.117 G. KUPER, D. EBELING, F. FRÜNGEL: Conf. Abstracts of the 5th Conf. on Laser
       Radar Studies of the Atmosphere, Williamsburg, Virginia, U.S.A. (1973), p. 119;
       also Conf. Abstracts of 1974 Int. Laser Radar Conf., Sendai, Japan (1974), pp. 40
       and 41
5.118 J. A. SALZMAN, T. A. CONEY: Conf. Abstracts of 5th Conf. on Laser Radar Studies
       of the Atmosphere, Williamsburg, Virginia, U.S.A. (1973), p. 49;
       also J. A. SALZMAN: In *Laser Raman Gas Diagnostics*, ed. by M. LAPP and C. M.
       PENNEY (Plenum Press, New York 1974) p. 179
5.119 J. A. SALZMAN, T. A. CONEY: Conf. Abstracts of 1974 Int. Laser Radar Conf.,
       Sendai, Japan (1974), p. 51
5.120 R. S. HICKMAN, L. H. LIANG: Rev. Sci. Instr. **43**, 796 (1972)
5.121 M. R. BOWMAN, A. J. GIBSON, M. C. W. SANDFORD: Nature **221**, 456 (1969)
5.122 A. J. GIBSON: J. Sci. Instr. **2**, 802 (1969)
5.123 M. C. W. SANDFORD, A. J. GIBSON: J. Atmos. Terr. Phys. **32**, 1423 (1970), A. J.
       GIBSON, M. C. W. SANDFORD: J. Atmos. Terr. Phys. **33**, 1675 (1971)
5.124 T. ARUGA, H. KAMIYAMA, M. JYUMONJI, T. KOBAYASI, H. INABA: Rept. on Ionosphere
       and Space Res. Japan **28**, 65 (1974)
5.125 M. JYUMONJI, T. KOBAYASI, H. INABA, T. ARUGA, H. KAMIYAMA: Proc. of 9th Int.
       Symp. on Space Tech. and Sci., Tokyo, Japan (1971)
5.126 C. J. SCHULER, C. T. PIKE, H. A. MIRANDA: Appl. Opt. **10**, 1689 (1971)
5.127 R. D. HAKE, JR., D. E. ARNOLD, D. W. JACKSON, W. E. EVANS, B. P. FICKLIN, R. A.
       LONG: J. Geophys. Res. **77**, 6839 (1972)
5.128 J. E. BLAMONT, M. L. CHANIN, G. MEGIE: Compt. Rend. **274**, Ser. B, 93 (1972);
       also, Ann. Géophys. **28**, 833 (1972)
5.129 V. W. J. H. KIRCHHOFF, B. R. CLEMESHA: J. Atm. Terr. Phys. **35**, 1493 (1973)
5.130 A. J. GIBSON, M. C. W. SANDFORD: Nature **239**, 509 (1972)
5.131 A. J. GIBSON: J. Phys. E. **5**, 971 (1972)
5.132 A. W. TUCKER, A. B. PETERSEN, M. BIRNBAUM: Appl. Opt. **12**, 2036 (1973)
5.133 E. L. BAARDSEN, R. W. TERHUNE: Appl. Phys. Lett. **21**, 209 (1972)
5.134 See, for example, A. SHULTZ, H. W. CRUSE, R. N. ZARE: J. Chem. Phys. **57**, 1354
       (1972);
       W. M. JACKSON: J. Chem. Phys. **59**, 960 (1973);
       H. W. CRUSE, P. J. DAGDIGIAN, R. N. ZARE: Discuss. Faraday Soc. (London) **53**,
       277 (1973);
       R. H. BARNES, C. E. MOELLER, J. F. KIRCHER, C. M. VERBER: Appl. Opt. **12**, 2531
       (1973);
       C. C. WANG, L. I. DAVIS, JR.: Appl. Phys. Lett. **25**, 34 (1974)
5.135 E. H. EBERHARDT: Appl. Opt. **6**, 161 and 251 (1967)
5.136 E. H. EBERHARDT: Appl. Opt. **6**, 359 (1967)
5.137 Y. SHIMIZU, H. INABA, K. KUMAKI, K. MIZUNO, S. HATA, S. TOMIOKA: IEEE Trans.
       IM-**22**, 153 (1973); also H. INABA, Y. SHIMIZU, Y. TSUJI: Japan. J. Appl. Phys. **14**,
       Suppl. 1.14-1, 23 (1975)
5.138 See, for example, W. K. PRATT: *Laser Communication Systems* (John Wiley, New
       York 1969)
5.139 J. GELBWACHS, M. BIRNBAUM: Appl. Opt. **12**, 2442 (1973)
5.140 M. KERKER: Appl. Opt. **12**, 2787 (1973)
5.141 WM. H. SMITH: Opto-Electron. **4**, 161 (1972)
5.142 J. A. COONEY: J. Appl. Meteorol. **12**, 888 (1973)
5.143 E. W. BARRETT, O. BEN-DOV: J. Appl. Meteorol. **6**, 500 (1967)
5.144 Y. KATO, Y. MORI, H. INABA: Rept. on Ionosphere and Space Res. Japan **18**, 103
       (1964)
5.145 M. HIRONO: J. Rad. Res. Lab. Japan **11**, 251 (1964)
5.146 R. A. YOUNG: Discuss. Faraday Soc. (London) **37**, 118 (1964)
5.147 L. J. NUGENT: Nature **211**, 1349 (1966)
5.148 See, for example, D. M. HUNTEN: Space Sci. Rev. **6**, 493 (1967)
5.149 C. M. PENNEY: IEEE J. Quant. Electron. QE-**11**, 36D (1975)

6. Techniques for Detection of Molecular Pollutants by Absorption of Laser Radiation[1]

E. D. HINKLEY, R. T. KU, and P. L. KELLEY

With 17 Figures

This chapter describes techniques for the detection of gases by differential absorption of laser radiation, with the exception of the lidar technique covered in Chapter 4. Applications include long-path monitoring of gases in the troposphere and stratosphere, *in situ* source monitoring (including chemical process control), and highly specific analyses of gases in collected samples. Long-path monitoring is particularly useful for providing values for the *average* pollutant concentration over long distances to aid in the development of mathematical models for future prediction of pollutant levels. Source monitoring, using this laser technique, can be accomplished without interfering with the effluent stream being analyzed. Finally, unsurpassed specificity may be achieved by tunable laser analysis of effluent samples at reduced pressure, on the basis of a detailed absorption "fingerprint" of the gas under study. As we shall see here, and has been noted in earlier chapters, lasers which are wavelength-tunable are of great importance for monitoring applications. Consequently, part of this chapter is devoted to a state-of-the-art discussion of the capabilities and potentials of continuously-tunable lasers.

Resonance absorption occurs when the wavelength of electromagnetic radiation coincides with one or more characteristic spectral lines of a molecular species. Since the cross-sections for such interactions can be many orders of magnitude larger than those for Raman scattering or resonance fluorescence, discussed in Chapter 5, resonance absorption is generally accepted as the most sensitive optical monitoring technique [6.1]. Several non-laser instruments based on the principle of optical absorption are available commercially for a variety of monitoring applications. These are usually constructed for the detection of one—or at most, a few—specific gases. One of these instruments, the Fourier Transform Spectrometer (FTS), was shown by HANST [6.2] in 1971 to be valuable for multiple-pollutant monitoring of point samples. Since spectra of every pollutant gas present in the sample are recorded simultaneously, the FTS instrument is particularly well suited to studies of atmospheric chemical reactions [6.3]. Other non-laser

[1] This work was supported by the National Science Foundation (RANN), the U.S. Environmental Protection Agency, and the Department of the Air Force.

optical techniques are based on dispersive or correlation spectroscopy [6.4, 5] using either the sun or a high-temperature blackbody source as as the radiant emitter.

For several years the use of lasers to detect gases was limited almost entirely to accidental (usually only partial) coincidences between their emission wavelengths and those of the gas spectral absorption lines. Most notable was the early work of Gerritsen [6.6] in 1966 directed toward the detection of methane using 3.39-μm radiation from a He–Ne laser. With the subsequent development of *tunable* lasers in the visible, ultraviolet, and infrared regions of the spectrum, it has become possible to scan through spectral absorption lines of molecular species, and therefore utilize their relatively large absorption cross sections to the fullest extent possible; and, at the same time, select wavelength regions having minimum potential interferences.

We shall discuss the use of molecular absorption signatures from the far infrared to the ultraviolet region of the spectrum, taking into account limitations imposed by atmospheric transparency, discussed in Chapter 3. In addition, since eye safety must be considered before any laser detection scheme is implemented, a discussion of how safety regulations affect any of the monitoring techniques under consideration will be given near the end of this chapter. As was pointed out in Chapter 2, remote sensing complements the more conventional and generally less expensive point-sampling monitoring techniques; but there will be many cases for which remote sensing will be the only possible technique. It is important, therefore, to explore all such applications fully.

6.1 Absorption Spectra of Molecular Pollutants

6.1.1 Relationship Between Type of Molecular Excitation and Spectral Region

Characteristic molecular absorption spectra occur over a broad portion of the electromagnetic spectrum, from the microwave region where the spectral lines are characteristic of rotational transitions, to the vacuum ultraviolet where they result from outer-shell electronic transitions[2]. The wavelength spread is from 20 cm (or longer) to wavelengths as short as 2.5×10^{-5} cm; and the corresponding energy span is from

---

[2] Shorter-wavelength, inner-shell electronic transitions are generally less useful in identifying a molecular species.

6×10^{-6} eV to 5 eV. This wide range is made possible by various kinds of molecular interactions, as are outlined below:

Microwave and millimeter region—absorption due to rotational transitions. Tunable sources of microwave radiation have been available for many years, and microwave spectroscopy has been very useful for studying the structure of molecules. Absorption measurements must be performed at very low pressures ($\ll$ 1 Torr) in order to avoid the effects of pressure broadening; and this requirement greatly hampers its application to pollutant detection. For example, if we consider the 0–600 GHz region, only about 120 lines are potentially resolvable at atmospheric pressure; and interfering gases reduce the number of possibilities even further. Thus, there is little potential for atmospheric-pressure, long-path monitoring of pollutants using microwaves. Moreover, some molecules, such as CO_2, C_2H_4, and C_6H_6, which lack a dipole moment in the ground electronic state, cannot be detected by microwave absorption. One of the most important pollutants, SO_2, does have a strong dipole moment and has been detected by microwave spectroscopy [6.7]. Some vibrational transitions may also be observed in this region; for example, the inversion spectrum in NH_3.

Far-infrared region (25–500 μm)—absorption due primarily to pure rotational transitions and vibration-rotation bands. There are two main reasons why this region is not very useful for *laser* monitoring: 1) lack of convenient, tunable laser sources (even fixed-frequency laser coverage is sparse); 2) intense water vapor rotational absorption blanks out most of this region, as shown earlier in Fig. 3.1. Water vapor absorption is not as serious in the stratosphere, however, and HARRIES [6.8] has used far-infrared emission lines to monitor stratospheric H_2O, O_3, HNO_3, N_2O, and NO_2.

Middle infrared (2.5–25 μm)—absorption due to fundamental (as well as some overtone) and combination vibration-rotational bands. This is the so-called "fingerprint" region of the infrared, used extensively for chemical analysis and spectroscopic detection of gases. This spectral region is of the order of 4×10^4 atmospheric-pressure-broadened line-widths wide, and strong lines of a number of important pollutant gases can be distinguished from lines of other significant atmospheric absorbers—an important factor for high specificity. Most of this chapter will deal with resonance absorption using vibration-rotation bands in the middle-infrared region.

Near infrared (0.7–2.5 μm)—vibrational overtone and combination bands. This region contains additional overtones and combinations of the fundamental vibration-rotation bands; but since the lines are about 100 times weaker than the fundamentals, it is not usually as useful for detecting trace gases. For cases in which the pollutant burden is high over long atmospheric paths, however, such as carbon monoxide in urban areas, these weaker bands may be preferable for monitoring [6.9] since the fundamental lines may be opaque near line center. Also, for very long paths, such as the planetary limbs, the weaker bands may be preferable for the same reason.

Visible (0.4–0.7 μm)—absorption due to electronic transitions, with vibrational-rotational structure. Only a few molecules have absorption lines in the visible region of the spectrum. Most important of these for atmospheric monitoring is NO_2, for which tunable dye lasers can provide the necessary radiation.

Ultraviolet (0.25–0.4 μm)—absorption due to electronic transitions, with vibrational-rotational structure. All molecules have electronic absorption bands, but only diatomic molecules and small polyatomic molecules have characteristic resolvable structure in the ultraviolet. Some important pollutants, namely SO_2 and O_3, have been detected on the basis of their ultraviolet absorption spectra; however, much of this region suffers from interferences from the more abundant molecules in the normal atmosphere, particularly O_2.

Chapter 2 contains a list of some of the molecular air pollutants of current concern. This list is continually changing, and there will undoubtedly be many additions as new manufacturing processes, products, and sources of energy appear. Two recent additions, for example, are vinyl chloride (C_2H_3Cl) because of its potentially carcinogenic effects [6.10], and the chlorofluoromethanes (widely known as freons) which have been postulated [6.11] to be involved in depleting the ozone (O_3) layer in the stratosphere. The absorption bands of several molecules *normally* present in the atmosphere on a planetary scale have been discussed in Chapter 3. In this chapter we concentrate on the absorption signatures of *trace contaminants* in the atmosphere, many of which have already been analyzed by laser spectroscopy [6.12]. We will focus principally on absorption by vibrational-rotational interactions, since the relevant wavelengths suffer least from interferences by normal atmospheric gases, and thus provide the best opportunity for high detection sensitivity and specificity.

6.1.2 Characteristics of Absorption Lines

We will now discuss some of the properties of absorption lines. An important parameter of an isolated line within the vibration-rotation band of a gas is its integrated intensity:

$$S = \int_{-\infty}^{\infty} \sigma(v)dv , \qquad (6.1)$$

where $\sigma(v)$ is the molecular absorption cross section [cm^2] at wavenumber v [cm^{-1}]. Experimental measurements show that the integrated line intensities are typically 5×10^{-21} cm to 2×10^{-19} cm in the infrared for gases at room temperature. Ideally, for pollutant monitoring using optical techniques, the width of the spectral line should be narrow in order to achieve high specificity, and its optical cross-section large for high sensitivity. On the basis of (6.1), the peak absorption cross-section and linewidth are clearly related for a particular transition, and we will now discuss three of the most common broadening regimes:

Radiatively-Broadened Transitions

The radiative or "natural" linewidth of a molecular transition is determined from the Heisenberg uncertainty principle [6.15] on the basis of the lifetimes of levels involved in the transition. The uncertainty principle may be written in the form

$$\Delta\omega\Delta t \geq 1 , \tag{6.2}$$

where $\omega(=2\pi c v)$ is the circular frequency of the transition and Δt an effective state lifetime. Since Δt is of the order of milliseconds for levels associated with many strong infrared transitions, the corresponding "natural" linewidth is around 1 kHz. The radiative lineshape is Lorentzian and, in the absence of further line splittings, its peak absorption cross section, $\sigma_n(v_0)$, is in the range 2–60×10^{-13} cm^2.

The "natural" lineshape yields the theoretical limit on detection sensitivity and specificity. If gas pressure is sufficiently low and the molecules stationary, the "natural" lineshape of a transition can, in principle, be observed using a very stable, tunable laser. Although low pressures can be achieved, Doppler broadening (to be discussed below) caused by random thermal motion of the molecules is always present, cf. (3.22). Since Doppler broadening is inhomogeneous by nature, nonlinear techniques, or "oven" or "nozzle" beams can be used to reveal homogeneous lineshapes and fine structure which are normally masked. In attempts to observe "natural" lineshapes in the infrared [6.16], the spectral widths have always appeared larger than predicted because of intermolecular collisions or short molecular transit time through the laser beam. However, linewidths as narrow as 12 kHz have been measured [6.17].

As a monitoring technique, the use of fundamental, radiatively-broadened spectra has the potential for high sensitivity and specificity. One obvious application is the detection of virtually every hydrocarbon using their C–H stretch bands in the 3–4 μm region; for even under

Doppler-broadened conditions there are strong overlaps between adjacent spectral lines of molecules having three or more carbon atoms [6.18].

Doppler-Broadened Transitions

At sufficiently low gas pressure the shape and width of a spectral transition is determined by the Doppler effect resulting from random thermal motion of the molecules. Theory predicts a Gaussian contour [6.19]

$$\sigma(v) = (S/\gamma_D)(\ln 2/\pi)^{1/2} \exp[-(v-v_0)^2 \ln 2/\gamma_D^2], \qquad (6.3)$$

where the Doppler-broadened halfwidth is given by

$$\gamma_D = (v_0/c)(2kT \ln 2/m)^{1/2}. \qquad (6.4)$$

In the above expression, m is the molecular mass, c is the velocity of light, k is Boltzmann's constant, and T is the gas temperature in K. If v_0 is in units of cm^{-1}, so is γ_D. It is often convenient to use the terminology "full-width at half-maximum absorption cross section" (FWHM) to characterize spectral linewidth. In wavenumber units, the FWHM is given by $2\gamma_D = \Delta f_D/c$, where Δf_D is the FWHM in frequency units, which may be calculated according to the following expression

$$\Delta f_D \, [\text{MHz}] = (215/\lambda)(T/M)^{1/2}, \qquad (6.5)$$

where λ is the wavelength in μm, and M the gram molecular weight. For a CO line at 4.7 μm, $\Delta f_D = 148$ MHz at room temperature, while for a much heavier molecule such as SF_6, Δf_D (at 10.6 μm) = 30 MHz. The Doppler widths are typically four or five orders of magnitude larger than the "natural" linewidths. Moreover, as pointed out in Chapter 3, temperature variations throughout the troposphere and stratosphere do not exceed a factor of 1.5, which means that the Doppler linewidth of a particular gas will not usually vary by more than 20%.

The development of continuously-tunable lasers has revolutionized spectroscopy, and in recent years Doppler-limited spectra of many gases have been measured [6.12]. Figure 6.1 illustrates the usefulness of this technique in identifying certain distinct spectral features: namely, Lambda-type doubling [6.20] and nuclear hyperfine splitting [6.21] of nitric oxide (NO) spectral lines at room temperature and at 77 K. At pressures below approximately 4 Torr, Doppler broadening is dominant, and the observed linewidth (FWHM) of 129 MHz is very close to the theoretical value. Also, as expected, the 77 K value is nearly

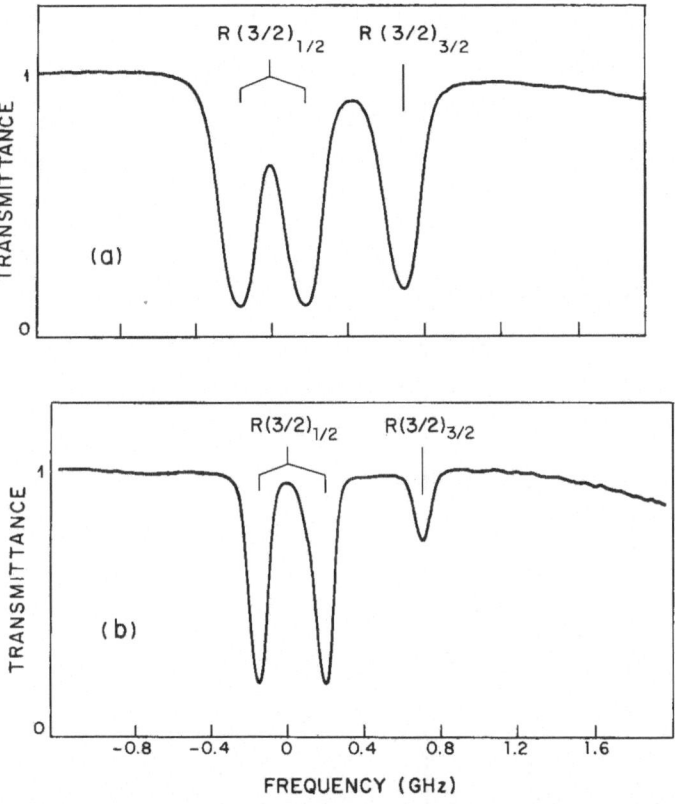

Fig. 6.1a and b. Spectral absorption of $R(3/2)_{3/2}$ and $R(3/2)_{1/2}$ lines of nitric oxide near $1900\,\text{cm}^{-1}$ ($5.3\,\mu\text{m}$), illustrating fine structure due to Λ-type doubling of the $R(3/2)_{1/2}$ line, at (a) 295 K, 4 Torr, and (b) 77 K, 1 Torr. Cell length: 5 cm. (After NILL et al. [6.20])

a factor of two smaller. There has consistently been excellent agreement between the measured and predicted Doppler-broadened linewidths for a variety of gases; and since values for the transition matrix elements are difficult to predict theoretically, we normally rely on data such as shown in Fig. 6.1 for determining absorption cross-sections. It is also worth noting in Fig. 6.1 that there is a strong temperature dependence of the $\Pi_{3/2}$ line intensity relative to that of the $\Pi_{1/2}$ line, arising from the Boltzmann factor described in Chapter 3. This may permit temperature measurements to be made in certain applications. Of course, the Doppler-broadened linewidth can also be used to measure temperature, in accordance with (6.5).

Table 6.1. Peak Doppler-limited absorption cross-sections for several molecules, as measured by laser spectroscopy. In most cases these are typical of strong lines in the bands indicated. The Doppler halfwidths, γ_D, were calculated according to (6.4)

Molecule	Formula	Transition	$\nu_0 \,[\text{cm}^{-1}]$	$\sigma_D \,[10^{-18}\,\text{cm}^2]$	$\gamma_D \,[10^{-4}\,\text{cm}^{-1}]$	Reference
Ammonia	NH_3	$sP(1,0), \nu_2$	948.22	96	14	[6.22]
Carbon monoxide	CO	$P(9), v=1\leftarrow0$	2107.42	63	24	[6.23]
Carbon dioxide	CO_2	$P(35), 01'1\leftarrow01'0$	2305.69	4	21	[6.24]
Ethylene	C_2H_4	In ν_7 band	950.76	81	11	[6.25]
Hydrogen chloride	HCl	$P(4), v=1\leftarrow0$	2798.94	60	29	[6.27]
Methane	CH_4	$R(3), \nu_3$	3057.69	20	47	[6.26]
Nitric oxide	NO	$R(13/2)_{3/2}, v=1\leftarrow0$	1900.52	8.8	21	[6.27]
Nitrous oxide	N_2O	$R(0), \nu_1+\nu_3$	3481.68	0.5	33	[6.18]
Sulfur dioxide	SO_2	$18_{0,18}\leftarrow19_{1,19}, \nu_1$	1139.60	1.8	8.8	[6.29]
Sulfur hexafluoride	SF_6	$P(33), \nu_3$	946.01	50	4.8	[6.30]

Table 6.1 lists a representative set of peak Doppler absorption cross sections in the infrared for a number of important molecules. Values range from 0.5×10^{-18} cm^2 for an N_2O line in the $v_1 + v_3$ combination band, to a high of nearly 1×10^{-16} cm^2 for a strong NH_3 line at 10.5 µm. In some cases tunable laser measurements of the strongest transitions have not yet been made.

For pollutant gas detection, Doppler-limited spectroscopy is most useful in point-sampling applications where the specimen can be reduced in pressure to below 10 Torr, or for *in situ* monitoring in the upper regions of the stratosphere (30 km and above for the infrared). For altitudes between 20 and 30 km (which includes most of the stratosphere), the lines may best be described as convolutions of Gaussian and Lorentzian shapes, known as Voigt lineshapes, to be discussed later. In the next section we will concentrate on the pressure-broadened lineshapes appropriate for pollutant gases at ground level and up to an altitude of approximately 20 km; i.e. pressures from 1 to 0.05 atm.

Pressure-Broadened Transitions

In cases where pressure broadening is dominant, the spectral lineshape is approximately Lorentzian

$$\sigma(v) = (S/\pi)\gamma_L/[(v - v_0)^2 + \gamma_L^2], \tag{6.6}$$

where γ_L is the pressure-broadened half-width. The dependence of γ_L on pressure and temperature may be approximated in some cases by

$$\gamma_L(p, T) = \gamma_L(p_0, T_0)(p/p_0)(T_0/T)^m, \tag{6.7}$$

where $\gamma_L(p_0, T_0)$ is typically the value at 1 atm pressure and 300 K. The value of m according to simple kinetic theory is 0.5 [6.31], but in reality it depends on the particular absorbing and broadening gases. At atmospheric pressure and 300 K, γ_L typically ranges from 0.05 to 0.1 cm^{-1}; however, substantially narrower water vapor linewidths have been observed for high rotational state transitions [6.32].

Figure 6.2 shows the absorption lines of the $R(13/2)_{1/2}$ and $R(13/2)_{3/2}$ transitions of NO at atmospheric pressure (of nitrogen), obtained by HINKLEY and KU [6.33] using tunable diode laser spectroscopy. The peak cross-section for the atmospheric-pressure-broadened $R(13/2)_{1/2}$ line is 6.0×10^{-19} cm^2, which is approximately 15 times smaller than its Doppler-broadened value. The measured pressure-broadened linewidths have not always agreed with theoretical predictions, and attempts are being made to explain the sub-Lorentzian or super-

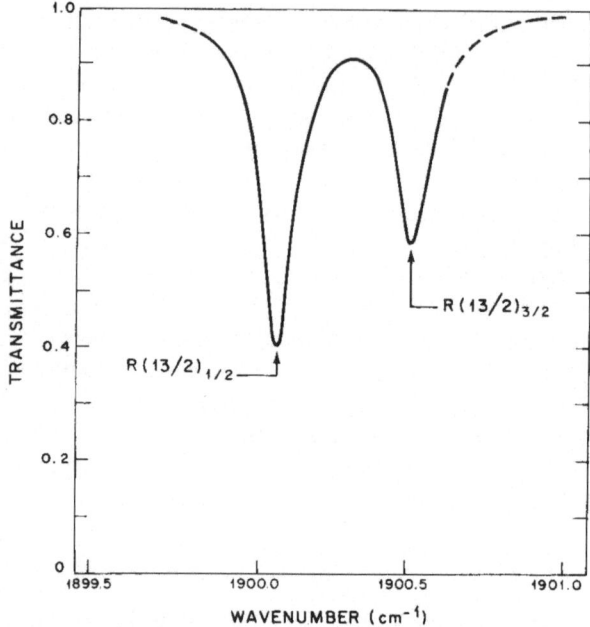

Fig. 6.2. Spectral absorption of $R(13/2)_{1/2}$ and $R(13/2)_{3/2}$ lines of nitric oxide at atmospheric pressure. NO concentration was 2000 ppm in 1 atm N_2; cell length was 30 cm. (After HINKLEY and KU [6.33])

Lorentzian lineshapes observed in the infrared for several gases, since absorption in the wings of spectral lines is very important in determining transmission over long atmospheric paths. Table 6.2 lists absorption parameters for several important gases at atmospheric pressure.

Combined Doppler- and Pressure-Broadened Transitions

For pressures in the intermediate region 0.005–0.05 atm, neither the Gaussian nor the Lorentzian profile accurately describes a spectral line. In this region the lineshape is represented by the Voigt profile, which results from a convolution of Gaussian and Lorentzian lineshapes [6.45], and can be expressed in terms of the real part of the error function of complex argument, $w(z)$ [6.46]

$$\sigma(v) = (S/\Delta v)\, \mathrm{Re}\, \{w(x+iy)\}\,,\tag{6.8}$$

where $\Delta v = (\pi \ln 2)^{1/2} \gamma_D$, $x = (v - v_0)/\Delta v$, $y = \gamma_L/\Delta v$.

Table 6.2. Measured absorption cross-sections for several gases at atmospheric pressure. Where tunable laser scans were made, the pressure-broadened halfwidths are given; but see footnotes for more information on overlapping lines

Molecule	Formula	$\nu\,[\text{cm}^{-1}]$	$\sigma\,[10^{-18}\,\text{cm}^2]$	$\gamma_L\,[\text{cm}^{-1}]$	Reference
Acetylene	C_2H_2	719.9	9.2	—	[6.34]
Ammonia	NH_3	1084.6	3.6	a	[6.35]
Carbon monoxide	CO	2123.7	2.8	0.050	[6.37]
Carbon tetrachloride	CCl_4	793.	4.8	—	[6.34]
Ethylene	C_2H_4	950.	1.7	—	[6.35]
"Freon"-11	CCl_3F	847.	4.4	—	[6.34]
"Freon"-12	CCl_2F_2	920.8	11.	—	[6.35]
Methane	CH_4	3057.7	2.0	b	[6.38]
Nitric oxide	NO	1900.1	0.6	0.060	[6.33]
Nitrogen dioxide	NO_2	22311.	0.2	—	[6.39]
Ozone	O_3	39425.	12.	—	[6.40]
		1051.8	0.9	—	[6.41]
Sulfur dioxide	SO_2	33330.	1.0	—	[6.42]
		2499.1	0.02	—	[6.26]
		1126.	0.2	c	[6.43]
Vinyl chloride	C_2H_3Cl	940.	0.4	—	[6.35]

[a] Total spread (FWHM) of these six NH_3 lines, $sR(5, 0\text{–}5)$, is $0.165\,\text{cm}^{-1}$. For reference, $\gamma_L = 0.079\,\text{cm}^{-1}$ for $aQ(9, 3)$ line at $939.21\,\text{cm}^{-1}$ [6.29]

[b] Total spread (FWHM) of these three CH_4 lines is $0.08\,\text{cm}^{-1}$. For reference, $\gamma_L = 0.0575\,\text{cm}^{-1}$ for the $R(0)$ line at $3028.8\,\text{cm}^{-1}$ [6.38]

[c] Total spread (FWHM) of several overlapping SO_2 lines is $1\,\text{cm}^{-1}$. For reference, $\gamma_L = 0.15\,\text{cm}^{-1}$ for the $(8, 0, 8)\leftarrow(8, 1, 7)$ line at $1148.8\,\text{cm}^{-1}$ [6.44]

The width of a spectral line at intermediate pressure p can be approximated by the following equation

$$\gamma = (\gamma_D^2 + \gamma_L^2)^{1/2} = [\gamma_D^2 + (c_p p)^2]^{1/2}, \qquad (6.9)$$

where c_p is the pressure-broadening coefficient. If we consider the NO molecule, its halfwidth varies from $0.06\,\text{cm}^{-1}$ at sea level to $0.002\,\text{cm}^{-1}$ at 30 km. It is clear that careful observation of lineshape can yield altitude information on the pollutant distribution.

6.2 Lasers for Monitoring Applications

Both fixed-frequency and broadly-tunable lasers are being used and proposed for atmospheric monitoring. The distinction between tunable and fixed-frequency lasers is not as clear as one might first believe:

A reasonable definition of a *broadly* tunable laser is one that is continuously tunable over $1\,\mathrm{cm}^{-1}$ or more and has a total tuning range of hundreds of cm^{-1}; a fixed-frequency laser is one that cannot be tuned over more than a small fraction of a cm^{-1} (e.g., $\sim 0.002\,\mathrm{cm}^{-1}$ for a low-pressure CO_2 laser transition). A fixed-frequency laser may often be step-tuned from one transition to another, but the resulting total *continuous* coverage is typically only $0.1\,\mathrm{cm}^{-1}$ over a $100\,\mathrm{cm}^{-1}$ interval—and for this reason these lasers seldom match perfectly with the absorption lines of molecules to be detected, and sensitivity is therefore reduced. Lasers which fall into the intermediate class between broadly-tunable and fixed-frequency lasers include rare earth and ruby solid state lasers, and the magnetically-tuned He-Ne laser.

Broadly tunable lasers generally have some method of controlling the gross tuning as well as the fine tuning characteristics of the laser either by prechoice of the wavelength range (changing composition in the case of semiconductor lasers, changing dyes in dye lasers) or by changing the operating conditions (tuning to the center of the transition by temperature or magnetic or electric field, or controlling the wavelength dependence of the cavity Q with prisms or gratings). Fine control of the laser wavelength generally involves tuning a single cavity mode with an etalon or by small changes in temperature, magnetic field or electric field. As we discuss individual tunable lasers, the tuning techniques will be described where appropriate.

Lasers can show a considerable variation in linewidth depending on the geometry and number of elements in the cavity as well as on the breadth of the gain peak. Since narrowing the linewidth often involves a considerable complication to the system, it is not advisable to narrow the output beyond some reasonable fraction of the width of the spectral line to be detected (e.g. 1/10 for high specificity and 1/2 for high sensitivity). At sea level spectral lines are typically $0.1\,\mathrm{cm}^{-1}$ wide (FWHM), while in the stratosphere at infrared wavelengths they can be as narrow as $0.002\,\mathrm{cm}^{-1}$. The effective laser linewidths (including frequency jitter) should, therefore, be less than 1.5 GHz and 60 MHz, respectively.

Power levels are also important (both peak and average) as well as pulse duration. Fixed-frequency lasers are generally better than tunable lasers in these characteristics, although there are exceptions.

In volume 2 of this series [6.13] a comprehensive discussion of tunable infrared lasers was given, emphasizing properties related to their application to high-resolution molecular spectroscopy. In the remainder of this section we will highlight the operating characteristics of tunable lasers now available and under development, and concentrate on their applications to atmospheric monitoring.

6.2.1 Dye Lasers

Dye Lasers [6.47] operate by optical pumping from the ground state to an excited singlet electronic state of high rotational and vibrational energy. Subsequently, the vibrational-rotational energy is thermalized by very rapid non-radiative relaxation, so that a population inversion exists, and stimulated emission occurs between low vibrational-rotational energy levels of the first excited singlet electronic state and higher vibrational-rotational levels of the ground electronic state (according to the Franck-Condon principle). Because of the high density of levels of the large dye molecules and the large widths of these levels in solution, very broad, continuous bands of radiation are emitted, typically ~ 50 nm wide. In order to narrow the laser linewidth, yet permit tunability within the broad emission band, highly frequency-selective optical resonators are used. Lasers have been made from dyes of oxazole, xanthene, anthracene, coumarin, acridine, azine, pthalocyanine and polymethine families. Both flash lamps and lasers have served as pumps, and conversion efficiencies of pump to dye of up to 50% have been measured.

cw dye lasers have shown the following characteristics:
1) Power to 10 W, but typically 10–100 mW.
2) Using five or six dyes, operation over the 460–700 nm spectral range has been achieved.
3) Free-running short-term stability of 50 kHz, and long-term stability of hundreds of Hz [6.48].

Pulsed dye lasers have exhibited the following properties:
1) Peak power of 10^8 W.
2) Average power of 60 W.
3) Output wavelength, using a number of dyes, from 0.34 to 1.2 µm.
4) Linewidth as narrow as 35 MHz [6.49].

cw and pulsed dye lasers can be used in the visible and ultraviolet for atmospheric monitoring [6.50–52] of a limited number of pollutants (e.g. SO_2, NO_2, O_3) by differential absorption or fluorescence techniques discussed in Chapters 4 and 5, and have been used to detect Na and Ca above the stratosphere by resonance fluorescence [6.53, 54]. Detection of SO_2 and O_3 requires doubling the dye laser frequency. Doubling and sum- and difference-frequency generation using dye lasers will be discussed in the next section.

6.2.2 Optical Mixers and Parametric Oscillators

Optical mixers and optical parametric oscillators rely on the properties of a nonlinear crystal. In this non-centrosymmetric crystal a significantly large dielectric polarization, quadratic in the strength of an applied laser field, can be induced. The nonlinear polarization can radiate, permitting harmonic generation or sum- and difference-frequency generation when fields at two frequencies are applied. For a substantial buildup of radiation at these new frequencies one must satisfy the "phase matching" condition: The wave vectors of the applied fields and the generated field must have the same relationship as their frequencies, so that, for sum-frequency generation where $\omega_1 + \omega_2 = \omega_3$, then $k_1 + k_2 = k_3$. The usual way to satisfy this condition is to select the propagation direction and crystal temperature so that crystal birefringence offsets the effects of dispersion.

If we have a tunable laser in one region of the spectrum, we can get tunable outputs at shorter or longer wavelengths with sum-frequency and difference-frequency generation. Dye lasers are often the original tunable source. Pulsed difference-frequency generation from the output of a ruby laser plus a dye laser mixed in $LiNbO_3$ has achieved 6 kW of infrared power tunable between 3.1 and 4.5 μm, with a bandwidth of about $10 \, cm^{-1}$ [6.55]. The lower limit can easily be extended to 1.7 μm and the bandwidth can be reduced to less than $1 \, cm^{-1}$. For the most recently developed dye lasers, peak infrared powers of several hundred watts could be generated at repetition rates of up to 30 pulses per second, with long-term stability better than $0.05 \, cm^{-1}$. Mixing in $LiNbO_3$ the output of a stable dye laser with that of a stable argon laser has yielded cw output powers of about 1 μW in the 2–4 μm region, with a linewidth of 15 kHz and continuous tuning of $\sim 1 \, cm^{-1}$. This system has been useful for laser spectroscopic studies of pollutants—particularly in the C-H stretch region of hydrocarbons [6.56].

Longer wavelengths can be generated with other crystals: proustite $(AgAs_2S_3)$ and HgS are usable to about 12 μm. Currently under development [6.57–62] as nonlinear materials in the infrared are I–III–VI$_2$ and II–IV–V$_2$ compounds with the chalcopyrite structure (e.g. $AgGaS_2$, $AgGaSe_2$, $CdGeAs_2$, and $ZnGeP_2$). These phase-matchable materials can be used for sum- and difference-frequency generation as well as in parametric oscillators. Further developments in materials technology could yield materials with improved optical quality that can be widely employed in nonlinear infrared devices.

Step-tuning with nearly continuous coverage of many regions of the middle infrared can be obtained by sum- and difference-frequency generation using the outputs of CO_2, N_2O, CO, and HF lasers. The number of new frequencies is proportional to the square of the number of transitions, and is substantially increased by including all the relatively abundant isotopic species. As an example [6.63], using known laser transitions in the four most abundant isotopes of CO_2 and CO, outputs with an average spacing of the order of 300 MHz are possible near 16 μm; and by increasing the gas pressure to broaden the gain bandwidths, continuous coverage should be possible. KILDAL and MIKKELSEN [6.62] have mixed 97 mW of CO radiation with 1.25 W of CO_2 radiation in a crystal of $CdGeAs_2$, generating 5 μW of power at 13 μm. In addition, KILDAL has irradiated this material with 150-ns CO_2 laser pulses, and obtained 200 mJ of doubled radiation with an optical conversion efficiency of 10%. Both $CdGeAs_2$ and $AgGaSe_2$ should be capable of providing several watts of average power and hundreds of millijoules of pulsed power under similar applications.

Sum and difference frequency generation can also be obtained over a narrow frequency range by modulating gas laser radiation in a nonlinear crystal with a microwave source [6.64, 65]. A tuning range of 200 MHz has been demonstrated, and the technique was used to measure pressure-broadened gain bandwidths for the CO_2 laser. Sum-frequency generation to the ultraviolet has also been accomplished by mixing ruby-laser and dye-laser radiation [6.66]. Between 100 and 200 kW were generated in 7-ns pulses with a 15-cm^{-1} linewidth. This technique, with ammonium dihydrogen phosphate (ADP) as the nonlinear material, can provide an output tuned down to about 235 nm. Second-harmonic generation ($\omega_1 = \omega_2 = \omega_3/2$) of a dye laser in ADP has yielded megawatt pulses with approximately 20 mJ of energy tunable from 280 to 290 nm with a 30 cm^{-1} linewidth [6.67]. With the most advanced dye lasers, this technique could yield kilowatts of peak power at a reasonable repetition rate (> 30 pulses per sec) and long-term linewidths of less than 0.15 cm^{-1}. The use of cesium dihydrogen arsenate (CDA) and its deuterated analog CD*A should increase the average power capabilities of doubling and sum-frequency generation. Four-wave mixing in a gas with two-photon resonance has also been used to generate tunable ultraviolet radiation [6.68]. Four-wave parametric mixing in alkali metal vapors has yielded tunable radiation in the 2–25 μm region [6.69] with 0.2 cm^{-1} linewidth. In this case the average power was very low: 5×10^{-8} W at 2 μm and 5×10^{-11} W at 25 μm.

The optical parametric oscillator (OPO) closely resembles microwave parametric oscillators and amplifiers, and its operation is similar to difference-frequency generation [6.70]. The basic OPO consists of a nonlinear crystal between two wavelength-selective mirrors to form an optical cavity. A laser field at a frequency ω_p (the pump frequency) is applied to the crystal, usually through one of the end mirrors. Initially, the pump radiation mixes with photon noise in the crystal, leading to a buildup of radiation at two frequencies: the signal frequency ω_s and the idler frequency $\omega_i(=\omega_p - \omega_s)$, which are mutually phase-matched for difference-frequency generation with the pump. If losses in the cavity are less than the gain of the buildup process, oscillation occurs. In order to change the phase-matched wavelengths (thereby tuning the oscillator), the indices of refraction of the crystal are varied with temperature, crystal rotation, or electric field.

Optical parametric oscillators operate cw as well as pulsed. Although cw thresholds can be as low as 2.8 mW [6.70], the cw OPO is generally difficult to operate, tends to be unstable, and requires careful geometric design. A singly-resonant LiNbO$_3$ oscillator, pumped with a ruby laser, has yielded peak power up to 340 kW and has had up to 45% conversion efficiency. The average power was as high as 350 mW near 2.1 μm, with a conversion efficiency of 70% when an internal oscillator with a LiNbO$_3$ crystal in a repetitively Q-switched Nd:YAG laser cavity was used [6.71]. In the 9–12 μm region Herbst and Byer [6.72] achieved parametric oscillation by using the 1.8-μm line of a Q-switched Nd:YAG laser as pump and CdSe as the nonlinear material. No single OPO has been tuned over a very wide region because of difficulties with mirror coatings and materials. A commercial device is available, however, that, with changes of mirrors on both the oscillator and pump laser, operates anywhere between 0.55 and 3.5 μm, with peak powers of 80 W to several hundred kilowatts and average powers between 1 and 40 mW (depending on wavelength). Long-term linewidths of less than 0.05 cm^{-1} have been achieved. Although interference techniques have allowed single-mode operation in several systems, long-term stability has not been carefully measured. One recently-proposed and demonstrated scheme for long-term stability is to lock the OPO wavelength to a gas absorption line [6.70]. Linewidths of less than 0.001 cm^{-1} have been demonstrated for an OPO operating at 2.5 μm [6.73, 74].

Remote detection of CO using differential absorption and topographical backscattering targets has been demonstrated [6.75] using an OPO operating at 2.3 μm. This technique will be discussed in detail in a later section.

6.2.3 Semiconductor Diode Lasers

Recombination-radiation semiconductor lasers [6.76–79] operate by stimulating emission across the gap between conduction and valence bands. Population inversion is achieved by electron injection across the band gap either with an electrical current (diode) or by optical pumping or electron-beam excitation. Infrared semiconductor laser materials in the 1–30 μm region include binary compounds such as InAs, InSb, GaSb, PbSe, PbS, PbTe, and pseudo-binary alloys such as $Pb_{1-x}Sn_xTe$, $PbS_{1-x}Se_x$, $Hg_{1-x}Cd_xTe$, $In_xGa_{1-x}As$, and $GaAs_xSb_{1-x}$. Materials such as GaAs, InP, $Ga_xAl_{1-x}As$, $GaAs_xP_{1-x}$, and $GaAs_xSb_{1-x}$ can be used as tunable lasers in the visible and near infrared. We shall discuss here only the lead-salt diode lasers because these devices are most useful in pollution monitoring.

The infrared frequency of lead-salt diode lasers is coarse-tuned by adjusting the chemical composition (i.e. the factor x in $Pb_{1-x}Sn_xTe$ or $PbS_{1-x}Se_x$ crystals), which determines the energy gap of the semiconductor and thus the wavelength of the spontaneous emission. Coarse tuning of the finished lasers can be accomplished by changing the temperature, applied pressure, or magnetic field. Fine tuning can also be obtained by changing these parameters, which affects the index of refraction and thus the cavity-mode frequency.

Most diode lasers emit radiation in several modes whose infrared frequency increases with increased drive current [6.80]. Continuous tuning of a single mode can be several cm^{-1} before a mode jump occurs, but is usually $0.5-1\ cm^{-1}$. The spontaneous emission from these devices tunes of the order of $300\ cm^{-1}$ between 4 K and 77 K [6.81], and the cavity tuning rates for typical lead-salt lasers are about one-third the spontaneous rate. The continuous tuning range of a single mode is quite adequate for scanning a pressure-broadened gas absorption line in the infrared, which is typically about $0.1\ cm^{-1}$ wide.

The magnetic-field dependence of wavelength in diode lasers has been extensively investigated [6.79, 82]. Because of the large mass anisotropy in the lead salts, the magnetic energy levels depend on the orientation of the applied magnetic field with respect to the crystallographic axes. Both the conduction and valence bands have their energies quantized into Landau levels, each of which is split into two spin energy states. Usually, electrons injected into the conduction band thermalize to the lowest available energy state, while holes injected into the valence band thermalize to the highest available energy state before recombining. Fine tuning of the axial modes with magnetic field occurs in a similar fashion to that of temperature tuning. The tuning rate within a mode (and for different modes) can vary with magnetic field; and for a $PbS_{0.82}Se_{0.18}$ diode laser has been measured [6.83] and found to vary from 0.4 to 2.0 MHz/G.

Application of hydrostatic pressure to a semiconductor laser can provide a very broad tuning range for a single device [6.83]. A PbSe diode laser at 77 K has been tuned from 7.5 to 22 µm using hydrostatic pressure up to 14 kB. More recently, the stability of the pressure has been controlled sufficiently to perform Doppler-limited spectroscopy [6.84] using GaAs diodes. Uniaxial pressure tuning of diode lasers can avoid the limitations of low temperature operation; however, in practice it is difficult to keep from crushing the devices. The recent development [6.85] of diode lasers which operate continuously at 77 K or higher will greatly facilitate the usefulness of hydrostatic pressure tuning; and eventually, using this technique, only one or two alloy semi-conductors may be needed to cover the entire 2–35 µm wavelength range.

Double-heterostructure, stripe-geometry lasers have been fabricated by means of liquid phase epitaxy [6.85]. A cw power of 10 mW in four modes was measured at 12 K, and single-mode operation was observed at 77 K with 1.2 mW power. The laser tuned from 10.5 µm at 12 K to 8.2 µm at 80 K—a range of 280 cm^{-1}. In a direct measurement of diode laser linewidth, HINKLEY and FREED [6.86] observed a 54 kHz $(1.8 \times 10^{-6}$ cm$^{-1})$ width for a 0.25-mW device operating at 10.6 µm. It would appear that widely-tunable infrared lasers having narrow linewidths and moderate powers will be available soon.

Diode lasers have already been extensively used in air pollution monitoring applications. Elsewhere in this chapter we shall describe point monitoring, *in situ* source monitoring, and long-path ambient air monitoring using these lasers.

6.2.4 Spin-Flip Raman Lasers

The spin-flip Raman (SFR) laser [6.87–90, 160] is a device that uses a fixed-frequency laser (CO_2, CO, or HF gas laser) to pump a semi-conductor crystal in a magnetic field. The pump-laser photons lose energy when they collide with an electron in the crystal and flip its spin. The down-shifted Raman photon is separated in energy from the pump photon by the change in electron spin energy $g\beta H$, where g is the gyromagnetic ratio of the conduction electron, β is the Bohr magneton, and H is the magnetic field strength. Consequently, the Raman photon frequency depends on magnetic field. At sufficiently high pump power, stimulated emission of Raman photons can exceed losses, and exponential gain and oscillation occurs.

In the most widely studied SFR laser, n-type InSb is the semi-conductor crystal. The laser has been operated in the pulsed mode [6.87] using CO_2 or CO laser pumps, and in the cw mode [6.88] with a CO laser pump. Under cw conditions, short-term stability relative to the CO pump laser of better than 1 kHz has been achieved [6.91]. A hybrid frequency and phase-lock closed-loop system was constructed [6.92] to maintain the output frequency of the SFR laser at a variable frequency offset from that of the CO local oscillator laser. The error signal derived by heterodyning the SFR laser output and stable CO laser output together with the output from a stable microwave oscillator, in a HgCdTe photodetector, has been used to drive small modulation coils around the InSb sample.

The resonance that exists when the pump laser is at a frequency near the bandgap of the semiconductor allows cw operation with low pump threshold power [6.87]. Less than 50 mW of TEM_{00} pump-laser power in an estimated 50–100 μm focal diameter in the crystal has been found sufficient to achieve threshold for the case in which incident pump and Raman beams were colinear; and a maximum output power in excess of 1 W cw was obtained. In this case, over 75% of the pump energy was converted to the first down-shifted Raman mode. At high conversion efficiencies, the beam divergence of the SFR laser is less than 40 mrad, consistent with a TEM_{00} mode divergence from a 250-μm diameter laser cavity. Operation in single axial and single transverse modes has been observed, as well as mode hopping during magnetic-field tuning. About 40% of the overall tuning range is covered, with continuous tuning in each mode of approximately $0.3\ cm^{-1}$. Mode pulling caused by changing the wavelength of the emission peak results in fine tuning, in contrast with the diode laser where a change in refractive index directly tunes the laser wavelength.

Stimulated SFR scattering in InAs has been obtained [6.89] using an HF-TEA laser [$2\rightarrow1$ band, $P(9)$ line at $3385.34\ cm^{-1}$] to pump near the bandgap resonance. Tunable emission was observed from 3347 to $3332\ cm^{-1}$. The output power in the first Stokes component increased monotonically with magnetic field, with conversion efficiencies of more than 10% measured. Stimulated SFR scattering near 10 μm has also been reported [6.90] using a CO_2-TEA laser to resonantly pump an n-type crystal of $Hg_xCd_{1-x}Te$. High quality crystals must be developed in order to make this a useful spectroscopic device for atmospheric monitoring applications.

The InSb SFR laser at 5 μm has been used in conjunction with a spectrophone to observe NO in automobile exhaust and in the atmosphere [6.93]. In addition, stratospheric NO and H_2O were detected with a similar system suspended from a balloon [6.94, 95]. These measurements are described in Section 6.5.

6.2.5 High Pressure Gas Lasers

Molecular gas lasers, such as CO_2, have most frequently been operated at pressures of around 10 Torr. Consequently, the gain bandwidths are essentially Doppler-limited to approximately 50 MHz. It is desirable to increase the range of continuous tuning, and for this higher pressures are needed to broaden the gain bandwidth. However, high pressure discharges in a long laser tube are difficult to maintain, since in addition to the increased bandwidth the deactivation rate is higher, leading to a higher pump threshold. A number of improvements in laser technology have allowed the development of high pressure gas lasers to proceed. These include: 1) transversely-excited atmospheric (TEA) configuration lasers with ultraviolet pre-ionization [6.96]; 2) optically-pumped lasers [6.97, 98]; 3) electron-beam-pumped lasers [6.99, 100]; and 4) capillary lasers [6.101], including pulser sustainer devices [6.102]. The first three of these devices have operated in the 10-atm range where continuous tuning between rotational transitions appears possible. Recently, O'Neill and Whitney [6.103] reported a 15-atm electron-beam-pumped CO_2 laser emitting 0.1 J, 300 ns pulses, in which the 0.2-cm^{-1} wide radiation was tunable over $2.3\,\text{cm}^{-1}$. The capillary system has operated to several atmospheres pressure. While these lasers have great potential for providing tunable infrared radiation, some effort remains in the engineering of them into practical devices. High-pressure gas lasers should have higher peak and average power than any other currently-conceivable tunable infrared source, and they may ultimately be particularly useful in differential absorption systems involving noncooperative target backscatter.

6.2.6 Tunable Laser Summary

In this Subsection we attempted to survey the available tunable coherent sources and indicate future improvements which may be made. By way of summarizing the most promising types and their present and proposed applications for atmospheric monitoring, Table 6.3 is presented. There is considerable overlap in potential applications, of course, but some consideration has been given to the size and portability of the laser type under discussion.

6.3 Signal Detection Techniques

Measurements of trace quantities of pollutant gases by differential absorption require sensitive detection of small amounts (or increments) of transmitted or absorbed laser energy. We shall now consider three techniques available for such measurements: direct photon detection, opto-acoustic detection, and heterodyne detection.

6.3.1 Direct Photon Detection

This technique for detecting infrared radiation (also known as incoherent detection) is the most commonly used of the three to be discussed. In involves direct conversion of incident photons to charge carriers by photoconduction, photodiode operation, or thermal con-

Table 6.3. Properties of tunable lasers and their potential applications to atmospheric monitoring

Tunable source	Wavelength region [μm]	Typical power [W]		Probable monitoring applications
		cw	Pulsed	
Dye laser	0.34–1.2	10^{-1}	10^{8}	Resonance fluorescence (atoms, molecules) Differential absorption/Mie backscattering
OPO	0.55–3.5 ($LiNbO_3$) 1.2–8.5 (Ag_3AsS_3) 8–12 (CdSe)	10^{-2}	10^{5}	Differential absorption/topographic backscatter
Diode laser	1–34	10^{-3}	10	Differential absorption/cooperative reflector Remote heterodyne detection
SFR laser	3 (HF-pumped) 5–6 (CO-pumped) 9–14 (CO_2-pumped)	1	10^{3}	Differential absorption
HPG laser	9–11 (CO_2)	—	10^{5}	Differential absorption/natural reflectors

version. Direct photon detection is used in fundamental laser spectroscopic experiments, in reduced-pressure and *in situ* source monitoring,
and in long-path ambient air monitoring. Moreover, it is used almost
exclusively for all the single-ended monitoring systems described in
Chapters 4 and 5. In order to detect ultraviolet and visible radiation,
nearly ideal (background-limited) photomultipliers and photodiodes
are available. In view of this, and since differential absorption measurements of pollutants are usually made at longer wavelengths, we shall
concentrate on the available types of *infrared* detectors.

For purposes of atmospheric monitoring, infrared detectors may be
described by four basic parameters: spectral coverage, response speed
(f_c), minimum detectable power [often called noise-equivalent-power
(NEP)], and operating temperature. Room-temperature operation is
desirable for field applications, but there are now fast, sensitive detectors
which operate at liquid nitrogen temperature (77 K)—a temperature
which can be used fairly routinely outside the laboratory [6.104].

Room-Temperature Infrared Detectors

Several types of thermal detectors and photoconductors operate at room
temperature, and some of the most common ones are listed in Table 6.4.
We note that although the PbS detector is the most sensitive, its spectral
coverage is limited to below 4 μm. The InSb detector is very fast, but
lacks the sensitivity possessed by some of the others, and also has a
restricted wavelength response. The three thermal detectors have broad
wavelength coverage; and of these the thermister bolometer is least
sensitive and the Golay cell is too slow for many applications. As a
general-purpose, room-temperature detector, the pyroelectric seems to
be the best choice. Although its NEP increases from 4×10^{-10} W/Hz$^{1/2}$
at ~ 10 Hz in direct proportion to the modulation frequency, it still has
sufficient sensitivity at 1 kHz to be useful for some applications involving
pollutant monitoring by differential absorption. For example, KU

Table 6.4. Typical characteristics of several room-temperature infrared detectors. For the
thermal detectors, NEP measured at low frequency (~ 10 Hz)

Detector	Spectral range	Cutoff frequency [Hz]	NEP [W/Hz$^{1/2}$]
PbS	$<4\,\mu m$	1400	0.1×10^{-10} [a]
Golay cell	Complete	40	1.4×10^{-10}
Pyroelectric	Complete	3000	4×10^{-10}
Bolometer	Complete	1000	10×10^{-10}
InSb	$<7.5\,\mu m$	10^6	10×10^{-10} [a]

[a] At peak wavelength

et al. [6.37] have shown that it is possible to detect a 0.3% change in transmission over a 0.6-km atmospheric path. For a pyroelectric detector having an NEP of 3×10^{-7} W at 1 kHz, a laser of power $> 10^{-4}$ W is required to observe this signal change. Clearly, for smaller laser powers or for smaller values of incremental power due to lower molecular pollutant concentration or absorption cross section, the pyroelectric detector will not be sufficiently sensitive. Moreover, from a practical standpoint, a more sensitive detector is usually needed during the initial alignment procedure.

Cooled Infrared Detectors

The most sensitive cooled detectors are extrinsic or intrinsic photo-conductive (PC) or photovoltaic (PV) detectors of various semi-conductor materials [6.105]. Historically, liquid-helium-cooled Ge:Cu, liquid-hydrogen-cooled Ge:Hg, or liquid-nitrogen-cooled Ge:Au have been used for most sensitive detection. More recently, mixed-crystal semiconductors have demonstrated many of the desirable properties of the extrinsic germanium types—and they operate at 77 K. The four most common types of intrinsic semiconductor detectors are listed in Table 6.5. With the exception of the very fast (1 GHz) HgCdTe photo-voltaic detector, generally used for heterodyne detection, these detectors are at least fifty times more sensitive than a pyroelectric, and the de-tection of weak atmospheric absorption lines or low-power laser radia-tion is correspondingly better. For operation in the 3–5 μm region, InSb is the usual choice; whereas in the 8–12 μm region, either PbSnTe or HgCdTe can be used. Finally, it should be mentioned that a *photo-voltaic* detector usually has better sensitivity at low frequencies (less than a few hundred Hz) than a *photoconductive* detector, which has excess noise due to the electrical current.

Table 6.5. Typical characteristics of several liquid-nitrogen-cooled (77 K) infrared detectors. NEP values correspond to optimum wavelengths

Detector	Spectral range	Cutoff frequency, f_c[Hz]	NEP (30° field of view [W/Hz$^{1/2}$]
InSb (PV)	< 5.3 μm	$\sim 10^6$	1×10^{-12}
PbSnTe (PV)	< 14 μm[a]	$\sim 10^7$	7×10^{-12}
HgCdTe (PC)	< 14 μm[a]	$\sim 10^7$	7×10^{-12}
HgCdTe (PV)	< 14 μm[a]	10^9	5×10^{-9}[b]

[a] Determined by crystal composition, typically peaked around 14 μm
[b] As a rule of thumb, $\text{NEP} \gtrsim 10^{-16} \sqrt{A_d/f_c}$ [watts], due to amplifier limitations (A_d is the detector area)

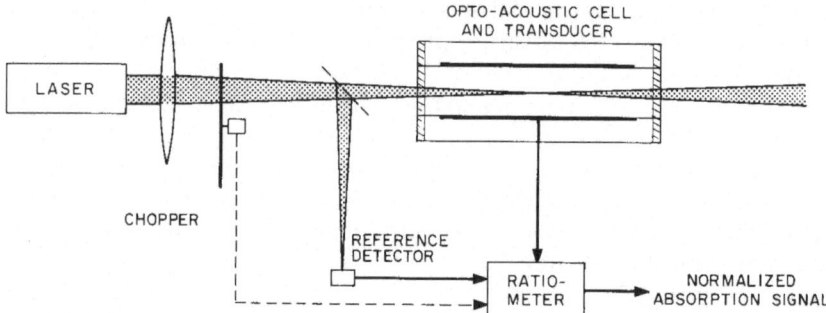

Fig. 6.3. Experimental set-up for air pollution detection by optacoustic absorption measurement. (Adapted from [6.93])

6.3.2 Opto-Acoustic Detection

We turn now from a method based on the measurement of laser *transmission* to one which directly measures *absorption* of laser power—opto-acoustic detection. The opto-acoustic technique uses acoustic (over-pressure) signals to detect the heating of a gaseous sample by absorbed radiation. Although the technique has a long history [6.106], the use of opto-acoustic spectroscopy for air pollutant detection was proposed only recently, by KREUZER [6.107]; and it has been limited primarily to *point sampling* applications using either fixed-frequency or tunable lasers. In addition, because of its excellent sensitivity when used with high-power lasers, opto-acoustic detection is now an important technique for measuring absorption characteristics of trace atmospheric constituents which can affect long-path laser propagation [6.108].

The instrument used for opto-acoustic detection is commonly called a spectrophone [6.109], and consists of a radiation source, modulator, and opto-acoustic detector. Figure 6.3 shows a simplified diagram of a complete spectrophone system using a laser source, reference photon detector, and ratioing electronics. The simplest opto-acoustic detector configuration is a cylindrical tube containing the gas mixture to be analyzed, the ends of which are sealed with windows transparent to the laser radiation. In a typical spectrophone, the laser radiation is mechanically chopped to provide pressure impulses resulting from the absorption of laser energy and its subsequent conversion to translational energy of the gas in the sealed volume. These pressure impulses are detected by a sensitive microphone, manometer, or other pressure transducer located within the cell or connected to it. The electrical signal, synchronously detected by a lock-in amplifier, is proportional to the product of laser power within the cell, gas concentration, molecular absorption cross section, and transducer responsivity. As we shall see, several other factors affect the spectrophone signal and its sensitivity for gaseous pollutant detection.

Each time the chopper blade permits the laser radiation to pass through the opto-acoustic detector, the pressure rises exponentially at a rate determined by the thermal time constant of the cell and thermodynamic properties of the gas—and the same processes govern the decay of pressure when the radiation is blocked [6.110, 111]. We now follow the detailed analysis of ROSENGREN [6.109] to determine the sensitivity of a non-resonant spectrophone for single-pass laser propagation. We assume that the collisional relaxation time of the gas is much shorter than its radiative relaxation time, and that $N\sigma L \ll 1$ (optically thin cell), where N is the density of molecules, σ the cross section per molecule, and L the cell length. The rms value of the first harmonic of the pressure signal is given approximately by

$$p(\omega) = \frac{2^{3/2}\beta P_0 N\sigma L\tau_t}{3\pi V[1+(\omega\tau_t)^2]^{1/2}}, \tag{6.10}$$

where $\beta[=(3/2)(C_p/C_v-1)]$ represents the fraction of laser energy absorbed by the gas and converted into translational energy (C_p and C_v are the specific heats at constant pressure and volume, respectively), P_0 is the laser power within the cell, $\omega/2\pi$ is the chopping frequency, V is the cell volume, and τ_t is the thermal relaxation time.

The thermal relaxation time of the opto-acoustic cell may be given approximately by

$$\tau_t = (V/\pi\xi\delta_0 L)(p/p_0), \tag{6.11}$$

where δ_0 $(=\kappa/\varrho_m C_v)$ is the thermal diffusivity of the gas (in cm^2/s or m^2/s) at a certain pressure p_0, p is the actual cell pressure, and ξ is a parameter related to the laser beam power distribution, and may for a typical situation be given a value of 2.4 [6.109]. The parameters κ and ϱ_m are the thermal conductivity and mass density, respectively.

Ultimately, the minimum detectable pressure signal is limited by noise in the transducer preamplifier, n_e, and noise caused by Brownian motion of the molecules, n_B. Since these noise sources are independent and random, we can use (6.10) to determine N_{min} as a function of noise-equivalent-power of the opto-acoustic cell (NEP)$_{OA}$, electronic bandwidth

B [Hz], signal-to-noise ratio S/N, and other parameters already defined:

$$N_{min} = \frac{(NEP)_{OA}(S/N)\sqrt{B}}{\sigma L P_0}, \tag{6.12}$$

where

$$(NEP)_{OA} = (3\pi/2^{3/2})(V/\beta\tau_t)(n_B^2 + n_e^2)^{1/2}[1 + (\omega\tau_t)^2]^{1/2}. \tag{6.13}$$

Opto-acoustic sensitivity, according to (6.13), appears to be highest at low frequencies; but amplifiers are usually noisiest at low frequencies, so a tradeoff has to be made in order to achieve optimum sensitivity—i.e., the amplifier noise contribution cannot usually be neglected. If the pre-amplifier noise term in (6.13) is rewritten as $n_e = n_0\omega_0/\omega$, where n_0 is the value at a certain angular frequency ω_0, then the optimum chopping frequency can be shown to be given by

$$\omega^* = (n_0\omega_0/\tau_t n_B)^{1/2} \text{ rad/s}. \tag{6.14}$$

Substituting this expression into (6.13) yields the optimum NEP for any pressure

$$(NEP)_{OA}^* = (3\pi/2^{3/2})(V/\beta\tau_t)(n_B + n_0\omega_0\tau_t). \tag{6.15}$$

For a numerical example, consider the detection of a trace gas in air ($\beta = 0.6$, $\delta_0 = 0.26$ cm^2/s) at total pressure p [atm], using a typical opto-acoustic cell [6.109] 20-cm long with a volume of 1.6 cm^3, for which $\tau_t = 0.04p$ seconds. Using a commercial microphone and preamplifier, one can achieve the following parameters [6.109]: $n_B = 0.5 \times 10^{-11}$ atm/Hz$^{1/2}$; $n_0 = 80 \times 10^{-11}$ atm/Hz$^{1/2}$ at $\omega_0 = 2\pi$ rad/s. Evaluating (6.15) for these parameters, we obtain

$$(NEP)_{OA}^* = 4.5 \times 10^{-8}(1 + 0.025/p) \quad [\text{W/Hz}^{1/2}], \tag{6.16a}$$

where the optimum chopping frequency [Hz] is $25/p^{1/2}$, and p is the pressure in atmospheres. At atmospheric pressure the optimum chopping frequency is 25 Hz, and the corresponding $(NEP)_{OA}^*$ is 4.6×10^{-8} W/Hz$^{1/2}$. In the stratosphere at 25 km altitude ($p = 0.012$ atm), the chopping frequency should be increased to around 230 Hz, for which $(NEP)_{OA}^* = 1.4 \times 10^{-7}$ W/Hz$^{1/2}$, or three times higher than the sea-level value.

Using a 1-W laser and 20-cm cell, for a signal-to-noise ratio of 10 at 1-Hz bandwidth, and the optimal NEP, (6.12) becomes

$$N_{min} = (2.2 \times 10^{-8}/\sigma)(1 + 0.025/p) \quad [\text{cm}^{-3}]. \tag{6.16b}$$

Consider the detection sensitivity for NO in the atmosphere at sea level, for which the peak absorption cross section (cf. Table 6.2) is 6×10^{-19} cm^2 for a certain transition. We find that $N_{min} = 3.8 \times 10^{10}$ cm^{-3}, which corresponds to a volume mixing ratio of 1.5×10^{-9}, or 1.5 parts per billion (ppb)[3].

In addition to Brownian and amplifier noise, there are background signals in an opto-acoustic cell due to partially-absorbing windows and absorption in the wings of spectral lines of other species. In addition, there may be noise in the measured data due to laser power fluctuations and variations in the amplifier gain. Attempts to reduce the background signal due to window absorption have included: (a) chopping at a certain rate which results in the establishment of natural acoustic modes within the opto-acoustic cell, with regions of minimum sensitivity (nodes) near the windows [6.113–117]; (b) use of a differential system consisting of two cells in series, with a common window [6.118]; and (c) modulation of either the laser frequency or the spectral frequency of the absorption line to be monitored (rather than chopping the laser radiation) [6.119–121]. A comparison of resonant vs. non-resonant spectrophones has been prepared by DEWEY [6.122], and a detailed analysis of potential interference effects by GELBWACHS [6.123]. Recently, ROSENGREN [6.109] published a very comprehensive treatment of the entire subject of opto-acoustic detector design and evaluation.

One of the most useful features of opto-acoustic detection is wide dynamic operating range. Figure 6.4 shows the dependence between the signal of a resonant spectrophone and C_2H_4 concentration, varying over nearly four orders of magnitude. The data were obtained by ENG [6.115] using the P (14) line of a CO_2 laser at 10.5 μm. The nonlinearity at high concentrations is probably caused by the microphone; for, since $N\sigma L \ll 1$ even at 200 ppm, it could not stem from appreciable laser power attenuation within the cell. The extraneous level noted on Fig. 6.4 corresponds to the total system noise. Extrapolation of the spectrophone signal to the extraneous noise level indicates that 50 ppb of C_2H_4 can be detected (at 336 Torr total pressure) with $S/N = 10$ for a laser power of 1.1 W. Since the absorption cross section of C_2H_4 at the $P(14)$ CO_2 laser transition is 1.34×10^{-18} cm^2 [6.125], the corresponding absorption coefficient is $N_{min}\sigma = 7.4 \times 10^{-7}$ cm^{-1}.

Using a differential spectrophone operating at 1-Hz chopping frequency, with integration times to 2 min, DEATON et al. [6.118] reduced the background signal to a value corresponding to 3.3×10^{-9} cm^{-1} per watt of laser power. This recent development should permit not only the

---

[3] The terms "ppm" and "ppb", often written as "ppmv" and "ppbv", correspond to mixing ratios (by volume) of 10^{-6} and 10^{-9}, respectively.

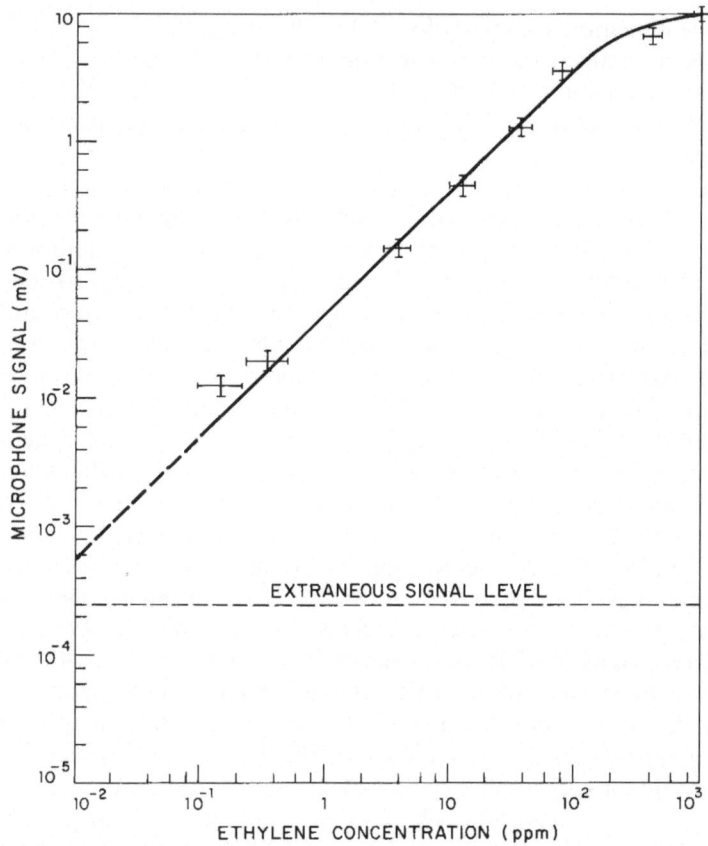

Fig. 6.4. Spectrophone signal vs. C_2H_4 concentration using $P(14)$ line of CO_2 laser (949.48 cm^{-1}) mechanically chopped at 4310 Hz. Laser power was 1.1 W, cell length 15 cm, integration time 10 s, and balance of 336 torr total pressure composed of N_2. (After Eng [6.115])

detection of gases such as C_2H_4 in concentrations approaching 0.02 ppb with fixed-frequency gas lasers, but the use of lower-power (but tunable) semiconductor diode lasers for opto-acoustic detection in the low ppb range.

KREUZER and PATEL [6.93] used the opto-acoustic technique in conjunction with a tunable SFR laser to detect NO in vehicular-exhaust and ambient-air samples to a sensitivity of 10 ppb; and PATEL et al. [6.94] and BURKHARDT et al. [6.95] constructed and launched a balloon-borne SFR laser-spectrophone detection system to monitor NO and H_2O in the stratosphere. These results will be discussed in Sect. 6.5 on applications.

KREUZER et al. [6.124] studied the application of opto-acoustic detection employing CO and CO_2 discretely-tunable lasers to detect a variety of pollutant gases. In an attempt to decouple opto-acoustic cell responsivity from window heating, BONCZYK and ULTEE [6.119] and KALDOR et al. [6.120] used a magnetic field to Zeeman-modulate the frequency of an NO transition, and CHAKERIAN and WEISBACK [6.121] actually placed the spectrophone cell within a CO laser cavity. In another attempt to minimize window noise, DEWEY et al. [6.113] developed an acoustically-resonant spectrophone, and detected n-butane with a 3.39-μm He-Ne laser; and, using a similar setup, SCHNELL and FISCHER [6.125] measured the fundamental absorption parameters of a variety of gases using several different CO_2 laser lines.

On the basis of sensitive measurements already reported, together with ongoing development of low-cost, tunable lasers, the opto-acoustic detection technique will clearly be important in many future pollution-monitoring applications. For localized monitoring it offers certain advantages over the multiple-reflection transmission techniques which often require cooled detectors. In Section 6.4 we will evaluate the sensitivity of the transmission techniques so that a direct comparison can be made with the capabilities of the spectrophone analyzed above.

6.3.3 Heterodyne Detection

In the infrared, heterodyne detection is the most sensitive method for measuring low-power laser radiation. It can also permit passive remote detection of gases by the mixing of radiation from a characteristic emission line with that from a tunable laser, in a wideband infrared detector [6.29, 126–128]. Using the sun as blackbody source, heterodyne detection can be valuable for continuous surveillance of the atmosphere, yielding information as to the pollutant concentration vs. altitude [6.129, 130].

In systems which detect radiation scattered from non-cooperative reflectors, such as atmospheric aerosols or natural terrain at long distances, the return signal is usually very small. Consequently, heterodyne detection has been used [6.131] to detect scattered radiation from CO and CO_2 lasers whose lines were selected to match pollutant gas absorption lines. The use of heterodyne detection in conjunction with low-power semiconductor lasers has been proposed [6.132] for monitoring pollution layers by ground reflection of laser radiation from an airplane.

Because of the rather unique and important applications of heterodyne detection to laser monitoring of the atmosphere, the entire subject is covered in detail in Chapter 7.

6.4 Monitoring by Differential Absorption—Theory

In order to detect the absorbance of a spectral line, hence the concentration of a particular molecular species, differential absorption is used.
The three operating modes most commonly employed are: (i) discrete-
frequency "on-off" absorption; (ii) first-derivative detection; and (iii)
second-derivative detection. A fourth method involves direct line scan
with a tunable laser, and is particularly useful for obtaining spectroscopic
information such as line shapes and relative positions.

   We remind the reader that in the region of the atmosphere between
sea level and the lower stratosphere at around 15 km, infrared spectral
lines of gases are predominantly broadened by collisions with air molecules. Consequently, the lineshapes are Lorentzian with widths (FWHM)
ranging from 0.3 cm^{-1} to slightly below 0.1 cm^{-1}. In the upper portions
of the stratosphere (30 km altitude) and above, where the ambient
pressure is less than 10 torr, the spectral lines are mainly Gaussian in
shape, with widths typically two orders of magnitude smaller than at
sea level. These differences will be considered in the analyses which
follow.

6.4.1 Discrete-Wavelength Monitoring

Consider the transmission of two laser wavelengths λ_1 and λ_2, "on" and
"off" the spectral absorption line to be monitored—i.e. $\sigma(\lambda_1)=\sigma$, and
$\sigma(\lambda_2)=0$, where σ is the peak molecular absorption cross section. If the
laser radiation is transmitted over an atmospheric path of length L to a
remote reflector, the power received back at the transceiver is

$$P_r(\lambda_1)=K'(\lambda_1)P_0(\lambda_1)\exp\left[-2\sigma\int_0^L N(z)dz\right]\exp\left[-2\int_0^L \alpha(\lambda_1, z)dz\right]$$
(6.17a)

$$P_r(\lambda_2)=K'(\lambda_2)P_0(\lambda_2)\exp\left[-2\int_0^L \alpha(\lambda_2, z)dz\right] \tag{6.17b}$$

where $P_0(\lambda)$ is the transmitted power, $K'(\lambda)$ the overall system efficiency,
and $\alpha(\lambda, z)$ the effective extinction due to "wing" absorption from other
constituents and scattering from particles and aerosols. The average
concentration $\bar{N}$ over the path is given by

$$\bar{N}=\frac{1}{L}\int_0^L N(z)dz$$
$$=\frac{1}{2\sigma L}\ln\left[\frac{K'(\lambda_1)P_0(\lambda_1)P_r(\lambda_2)}{K'(\lambda_2)P_0(\lambda_2)P_r(\lambda_1)}\right]$$
$$+\frac{1}{\sigma L}\int_0^L [\alpha(\lambda_2, z)-\alpha(\lambda_1, z)]dz . \tag{6.18}$$

If we assume that the wavelength dependence of the interferences is low enough to make the last term of (6.18) negligible, that the system efficiency is also independent of wavelength over the range of interest, and that $P_0(\lambda_2) = P_0(\lambda_1)$, we obtain

$$\bar{N} = \frac{1}{2\sigma L} \ln \left[P_r(\lambda_2)/P_r(\lambda_1) \right]. \tag{6.19}$$

In the limit where $2\sigma \bar{N} L \ll 1$, we can rewrite (6.19) in the following form

$$\bar{N} \simeq \frac{1}{2\sigma L} \left(\frac{\delta P}{P_r} \right), \tag{6.20}$$

where $\delta P[= P_r(\lambda_2) - P_r(\lambda_1)]$ represents the differential absorbed power collected by the infrared detector system. For background-limited operation, the minimum detectable received power for a given voltage signal-to-noise ratio (S/N) is

$$\delta P_{\min} = \text{NEP} \, (S/N) \sqrt{2B}, \tag{6.21}$$

where B is the system bandwidth. The factor of two under the square root takes into account noise resulting from the two measurements required (one at each wavelength) [6.9]. Combining (6.20) and (6.21), we find

$$\bar{N}_{\min} = \frac{1}{2\sigma L} \left[\frac{\text{NEP} \, (S/N) \sqrt{2B}}{K P_0 T_2} \right], \tag{6.22}$$

where T_2 is the background transmittance, and K takes on the four possible values given in Table 6.6 in terms of the optical geometry. Here A_T, A_{Ret}, and A_R are, respectively, the areas of the transmitter, reflector, and receiver optics. Also, the ξ's are the corresponding optics efficiencies and f is a numerical factor of order unity. Column 1 and row 1 of Table 6.6 refer to the situation when the diffraction-limited laser beam is within the retroreflector and receiver areas, respectively; whereas the converse is true for column 2 and row 2.

As an example of the sensitivity which is possible with background-limited operation in which all turbulence effects, etc. are eliminated in accordance with (6.19), consider the monitoring of atmospheric CO over a 1 km path using a remote retroreflector. From Table 6.6 we note that if the retroreflector and collection optics intercept all of the laser power, $K = \xi$, which is taken to be unity for this illustration. For a detector NEP of 1×10^{-12} W/Hz$^{1/2}$ (cf. Table 6.5), a system bandwidth

Table 6.6. Values for optical geometry parameter K for various conditions of long-path monitoring. A_T, A_{Ret}, and A_R are the areas of the transmitter, retroreflector, and receiver optics, respectively. L is the path length to retroreflector, λ is the laser wavelength, and f is a parameter related to the radial power distribution of the laser beam

	$A_T A_{Ret}/\lambda^2 L^2 > 1$	$A_T A_{Ret}/\lambda^2 L^2 < 1$
$A_{Ret} A_R/\lambda^2 L^2 > 1$	$K = \xi = \xi_T \xi_{Ret} \xi_R$	$K = \xi f A_T A_{Ret}/\lambda^2 L^2$
$A_{Ret} A_R/\lambda^2 L^2 < 1$	$K = \xi f A_{Ret} A_R/\lambda^2 L^2$	$K = \xi f^2 A_T A_{Ret}^2 A_R/\lambda^4 L^4$

of 1 Hz, an absorption cross section of 2.8×10^{-18} cm^2 (cf. Table 6.2), $T_2 = 1$, and $(S/N) = 10$, we obtain, for a 0.1 mW laser, $\bar{N}_{min} = 2.5 \times 10^5$ cm^{-3}, which corresponds to a volume mixing ratio of 1×10^{-5} ppb at sea level. This sensitivity is clearly adequate for CO, whose concentration is generally above 50 ppb, and for most other pollutants as well. It should be stressed, however, that calculations according to (6.22) represent the ultimate in sensitivity only when other factors have been eliminated by suitable techniques. Experimentally, changes in received power equal to 0.003 P_r have been measured over outdoor paths [6.37]. Incorporating this result into (6.20), we obtain $\bar{N}_{min} = 5.4 \times 10^9$ cm^{-3}, or 0.2 ppb. It is hoped that improvements in the signal processing techniques will result in sensitivities closer to the detector limit.

Although long-path monitoring by differential absorption employing a remote reflector has the very important advantage of requiring only low-power laser radiation (less than 0.1 mW, usually), attempts have been made to use topographic targets in order to achieve greater versatility and range resolution by beam steering [6.131, 132]. The sensitivity of long-path monitoring using topographic scattering is the same as before, cf. (6.19), but the power needed is much larger. The transmitter/receiver system efficiency includes the factor $\varrho A/\pi L^2$, which takes into account Lambertian scattering from the target. Here ϱ is the effective target reflectivity and A the receiver area. Fortunately, $\varrho \sim 1$ in the infrared, although it is approximately 0.1 in the ultraviolet and visible regions of the spectrum [6.133]. For this case we have, from (6.22)

$$\bar{N}_{min} = (\pi L/2\sigma\varrho A\xi T_2) \, \text{NEP} \, (S/N)\sqrt{2B}/P_0 \tag{6.23}$$

where $\xi (= \xi_T \xi_R)$ represents the optical system efficiency. As an example, consider the monitoring of atmospheric CO over a 1 km path using a topographic target for which $\varrho = 1$, and a receiver area of 100 cm^2. The other parameters are identical to those used in the above example, except we now assume that the laser power P_0 is 1 W. We find, using (6.23), that $\bar{N}_{min} = 8 \times 10^9$ cm^{-3}, which corresponds to a volume mixing ratio at sea level of 0.3 ppb. Using infrared heterodyne detection, the ultimate sensitivity of this type of monitoring system can be improved since the NEP is typically lower by several orders of magnitude, as shown in Chapter 7.

6.4.2 Derivative Monitoring

As an alternative to measuring gas concentrations by using two discrete wavelengths, modulation of the radiation frequency of tunable lasers can be employed, with the signal being proportional to the *derivative* of the absorption lineshape. We shall confine our attention to the first and second derivatives. As is the case with discrete-wavelength switching, if the modulation rate is rapid compared to atmospheric fluctuations, laser power variations, and changes in amplifier gain, these sources of noise may be reduced or eliminated.

Derivative Monitoring of Doppler-Broadened Lines

For a double-ended system, the transmittance in the vicinity of a Doppler-broadened spectral line is

$$T = \exp(-2\alpha_D L), \tag{6.24}$$

where the absorption coefficient is given by

$$\alpha_D = \bar{N}\sigma \exp[-(\nu-\nu_0)^2(\ln 2)/\gamma_D^2]. \tag{6.25}$$

The ratio of $dT/d\nu (\equiv T')$ to T is

$$T'/T = 4(\ln 2)(\bar{N}\sigma L)(\nu-\nu_0)\gamma_D^{-2} \exp[-(\nu-\nu_0)^2(\ln 2)/\gamma_D^2]. \tag{6.26}$$

Note that since atmospheric turbulence causes fluctuations in T with time, which also occur in the derivatives, it is preferable to normalize the derivative signals with respect to transmission. We anticipate this theoretically, as above. In the limit $\bar{N}\sigma L \ll 1$, $|T'|$ is a maximum for $\nu = \nu_0 \pm \gamma_D/\sqrt{2(\ln 2)}$, in which case, solving (6.26) for $\bar{N}$ yields

$$\bar{N} = \frac{\exp(1/2)}{2\sqrt{2(\ln 2)}\sigma L}\left(\frac{P_r^{(1)}}{P_r^{(0)}}\right)\left(\frac{\gamma_D}{\delta\nu}\right), \tag{6.27}$$

where $P_r^{(0)}(=\bar{T}P_0)$ and $P_r^{(1)}(=T'P_0\delta\nu)$ are the received powers at the zeroth and first harmonic of the modulation frequency of amplitude $\delta\nu$. $\bar{T}$ is the average background transmittance (essentially equivalent to T_2 in the "on-off" analysis above when $\bar{N}\sigma L \ll 1$), and $\delta\nu \ll \gamma_D$ for this analysis to be valid.

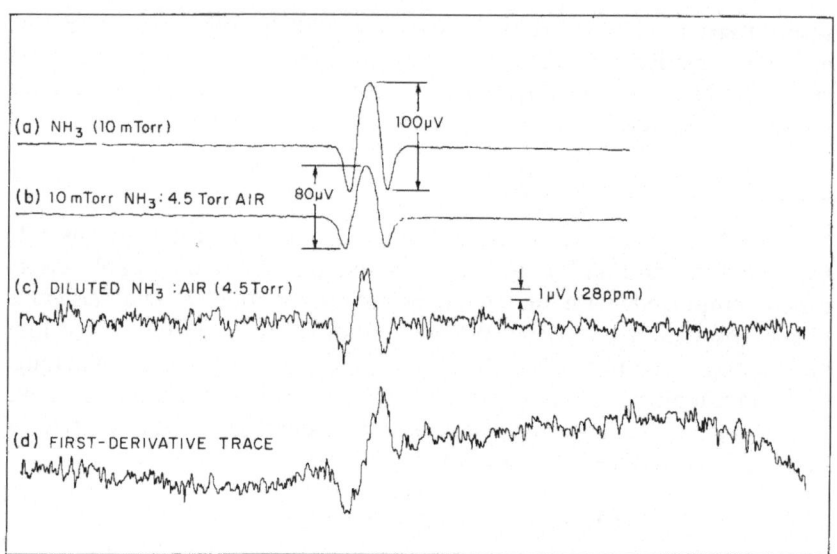

Fig. 6.5a–d. Point-sampling sensitivity measurement taken with a tunable diode laser using a 10-cm-long gas cell and a Ge:Cu infrared detector. (a) Second-derivative trace of NH_3 line at 0.01 Torr pressure; (b) second-derivative trace after 4.5-Torr air was added; (c) second-derivative trace after sequential dilutions with air until the signal at 4.5-Torr total pressure was comparable to noise level; (d) first-derivative trace of (c), illustrating long-term fluctuation effects not present in the second-derivative scans. $\sigma(v_0) = 30 \times 10^{-18}$ cm^2. (After Hinkley [6.29])

The second derivative ratio can be shown to be the following

$$T''/T = 4(\ln 2)(\bar{N}\sigma L)\gamma_D^{-2} \exp\left[-(v - v_0)^2 (\ln 2)/\gamma_D^2\right]$$
$$\cdot \{1 - 2(\ln 2)(v - v_0)^2/\gamma_D^2 + 4(\ln 2)(\bar{N}\sigma L)(v - v_0)\gamma_D^{-2}$$
$$\cdot \exp\left[-(v - v_0)^2(\ln 2)/\gamma_D^2\right]\} . \tag{6.28}$$

When $\bar{N}\sigma L \ll 1$, the principal maximum for the second derivative occurs at line center ($v = v_0$), in which case (6.28) is greatly simplified, and when solved for $\bar{N}$ yields

$$\bar{N} = \frac{1}{4(\ln 2)\sigma L}\left(\frac{P_r^{(2)}}{P_r^{(0)}}\right)\left(\frac{\gamma_D}{\delta v}\right)^2 \tag{6.29}$$

where $P_r^{(2)}[= T'' P_0(\delta v)^2]$ is the received power at the second harmonic of the modulation frequency.

An example of first- and second-derivative detection of an isolated Doppler-broadened spectral line of NH_3 is shown in Fig. 6.5. In trace (a) is the second derivative, T'', taken with a tunable diode laser, of a strong line of pure NH_3 at 10^{-2} torr pressure in a 10 cm cell. Since the absorption strength at line center is $1\,cm^{-1}/Torr$, the transmittance T is 0.90. In trace (b), 4.5 Torr of air was added (equivalent to 2220 ppm NH_3), resulting in slight pressure broadening which reduced the signal by 20%. In order to determine experimentally the minimum detectable concentration, several dilutions of the gas were made until trace (c) was obtained, showing a peak-to-peak noise of approximately 1 µV for a time constant of a few seconds. Since 80 µV represents a partial NH_3 pressure of 10^{-2} Torr, and the signal at line center is directly proportional to gas concentration according to (6.29), the 1-µV value corresponds to 1.25×10^{-4} Torr, or 28 ppm in 4.5 Torr air. Since the observed noise is caused by the infrared detector (Ge:Cu in this case) or amplifier, sensitivity can be improved by using a laser having more than the 10 µW power of that used for this test, although laser-induced photon noise will eventually occur [6.105]. Trace (d) shows the first-derivative signal, exhibiting nearly the same (S/N) as trace (c), but with an added complication of a slow change in signal level over the laser tuning range—caused by a wavelength dependence of laser power.

With arguments similar to those used in arriving at (6.22), we find that the minimum detectable concentration using first-derivative detection is

$$\bar{N}_{min} = \left(\frac{\exp(1/2)}{2\sqrt{2(\ln 2)}\sigma L}\right)\left(\frac{NEP(S/N)\sqrt{B}}{KP_0\bar{T}}\right)\left(\frac{\gamma_D}{\delta v}\right)$$

$$= \left(\frac{0.7}{\sigma L}\right)\left(\frac{NEP(S/N)\sqrt{B}}{KP_0\bar{T}}\right)\left(\frac{\gamma_D}{\delta v}\right) \tag{6.30}$$

when $\bar{N}\sigma L \ll 1$. For $\delta v \approx \gamma_D$ (for which this derivation is only an approximation), the sensitivity is comparable with that for "on-off" monitoring shown in (6.22). A comparison between low-pressure and atmospheric-pressure monitoring sensitivity is made below.

Derivative Monitoring of Pressure-Broadened Lines

As discussed earlier, between sea level and the lower stratosphere, spectral lines are broadened by collisions with air molecules. Consequently, for heights to approximately 15 km, the lineshapes are Lorentzian with halfwidths ranging from $0.15\,cm^{-1}$ to slightly below $0.05\,cm^{-1}$. The transmittance is given by

$$T = \exp(-2\alpha_L L), \tag{6.31}$$

with

$$\alpha_L = \frac{\bar{N}\sigma\gamma_L^2}{(v-v_0)^2 + \gamma_L^2}. \tag{6.32}$$

The first-derivative ratio for Lorentzian lines is

$$T'/T = 4(\bar{N}\sigma L)\gamma_L^2(v-v_0)/[(v-v_0)^2 + \gamma_L^2]^2, \tag{6.33}$$

where σ now refers to the pressure-broadened cross-section (cf. Table 6.2).

When $\bar{N}\sigma L \ll 1$, T' is a maximum for $v = v_0 \pm \gamma_L/\sqrt{3}$, in which case (6.33) may be solved for $\bar{N}$ to yield

$$\bar{N} = \frac{4}{3\sqrt{3}\sigma L}\left(\frac{P_r^{(1)}}{P_r^{(0)}}\right)\left(\frac{\gamma_L}{\delta v}\right), \tag{6.34}$$

where the symbols have their usual meanings.

By a second differentiation of the transmittance, we obtain the following ratio

$$T''/T = \frac{4\bar{N}\sigma L\gamma_L^2}{[(v-v_0)^2 + \gamma_L^2]^2} \cdot \left[1 - \frac{4(v-v_0)^2}{(v-v_0)^2 + \gamma_L^2} + \frac{4\bar{N}\sigma L(v-v_0)^2\gamma_L^2}{[(v-v_0)^2 + \gamma_L^2]^2}\right]. \tag{6.35}$$

As with the Doppler-broadened case, the principal maximum for the second derivative ratio signal is at line center, and the corresponding expression for pollutant concentration is

$$\bar{N} = \frac{1}{4\sigma L}\left(\frac{P_r^{(2)}}{P_r^{(0)}}\right)\left(\frac{\gamma_L}{\delta v}\right)^2. \tag{6.36}$$

The minimum detectable concentration using the first derivative is, for $\bar{N}\sigma L \ll 1$,

$$\bar{N}_{min} = \frac{4}{3\sqrt{3}\sigma L}\left(\frac{\text{NEP}(S/N)\sqrt{B}}{KP_0\bar{T}}\right)\left(\frac{\gamma_L}{\delta v}\right)$$

$$= \frac{0.77}{\sigma L}\left(\frac{\text{NEP}(S/N)\sqrt{B}}{KP_0\bar{T}}\right)\left(\frac{\gamma_L}{\delta v}\right), \tag{6.37}$$

assuming $\delta v \ll \gamma_L$.

In order to appreciate the difference in sensitivity in first-derivative monitoring between pressure-broadened and Doppler-limited lines, we can divide (6.37) by (6.30), realizing that the absorption cross sections are given, respectively, by (6.6) and (6.3)

$$\frac{\bar{N}_{min}(\text{Lorentzian})}{\bar{N}_{min}(\text{Doppler})} = \frac{8(\ln 2)\sqrt{2\pi}}{3\sqrt{3}\exp(1/2)}\left(\frac{\gamma_L}{\gamma_D}\right) = 1.6\left(\frac{\gamma_L}{\gamma_D}\right), \qquad (6.38)$$

where in each case the amplitude of the frequency modulation, δv has been taken to be a fixed fraction, e.g. 10%, of the respective halfwidths γ_L or γ_D.

Consider as an example the difference in sensitivity for detection of CO at atmospheric pressure and at a reduced pressure of 5 Torr (which would minimize interference effects). From Table 6.2, $\gamma_L = 0.05\,\text{cm}^{-1}$, and from Table 6.1, $\gamma_D = 0.0024\,\text{cm}^{-1}$. We assume that the low-pressure lineshape is essentially Doppler-broadened, and the ratio (6.38) becomes 33. However, since only a fraction (5/760) of the molecules remain after evacuation, the effective ratio is 0.22; that is, the molecular detection sensitivity at atmospheric pressure is five times better than it is if the specimen is evacuated to 5 Torr—however, specificity is superior in the latter case by a factor of 21 ($= \gamma_L/\gamma_D$).

Finally, if we compare the sensitivity of "on-off" detection (6.22) with first-derivative detection of a Lorentzian line (6.30), we find the ratio of the corresponding minimum detectable concentrations to be given by $\delta v/\gamma_L$. Although we have stipulated that the frequency modulation index must be small for the mathematical analysis to hold, it is not unreasonable to expect that $\delta v \approx \gamma_L$ gives the maximum sensitivity for derivative monitoring—a sensitivity comparable with the "on-off" technique. Finally, it must be noted that the "on-off" sensitivity was for a 50% duty cycle; and that anything less will have proportionately less sensitivity.

Derivative Monitoring of Lines in the Voigt Regime

In the stratosphere (between ~ 15 and 30 km), the ambient pressure varies from 10 to 90 Torr, such that spectral lines in the infrared are, for the most part, neither fully Lorentzian nor Gaussian in shape, but a convolution of the two—known as the Voigt lineshape. In order to calculate the derivatives of a Voigt lineshape, we must resort to numerical calculations for each particular application. A discussion of the first derivative detection under varying conditions of spectral broadening is given by PENNER et al. [6.140].

6.5 Monitoring by Differential Absorption—Applications

In this section we review the various methods of laser monitoring based on the technique of differential absorption, from an applications standpoint, centering our attention on measurements which have already been performed and from which projections can be made for the future. These applications are designated as follows: point sampling, localized *in situ* monitoring, and long-path monitoring.

Gas lasers tunable to discrete wavelengths in the infrared have been available for several years. Their application to atmospheric monitoring was first proposed by Jacobs and Snowman [6.134] in 1967, and analyzed more fully by Hanst [6.135] in 1970. Subsequently, there have been many investigations of coincidences between normal and isotope gas laser lines and pollutant absorption lines due, in part, to interest in atmospheric monitoring as well as optical communication and high-power laser propagation. Table 6.7 is a summary of some of the best laser lines (in terms of large absorption cross section) for detecting specific gaseous pollutants. It must be understood, however, that for actual monitoring applications in which interfering lines from other gases may also be present, alternative choices may have to be made [6.112].

Table 6.7. Measured absorption cross sections for detection of pollutant gases, at atmospheric pressure, using selected gas laser lines

Molecule	Formula	Laser line	ν [cm^{-1}]	σ [10^{-18} cm^2]	Reference
Ammonia	NH_3	CO_2 R(30)	1084.6	3.6	[6.125], [6.35]
Benzene	C_6H_6	CO_2 P(30)	1037.5	0.09	[6.125]
1,3 Butadiene	C_4H_6	CO P(13)	1609.0	0.27	[6.124]
1-Butene	C_4H_8	CO_2 P(38)	927.0	0.13	[6.124]
Carbon Monoxide	CO	CO_2 P(20) $\times$ 2	2093.8	0.80	[6.136]
Ethylene	C_2H_4	CO_2 P(14)	949.5	1.34	[6.124], [6.125]
Fluorocarbon-11	CCl_3F	CO_2 R(30)	1084.6	1.24	[6.125]
Fluorocarbon-12	CCl_2F_2	CO_2 P(42)	923.0	3.68	[6.125]
Fluorocarbon-113	$C_2Cl_3F_3$	CO_2 P(26)	1041.2	0.77	[6.125]
Methane	CH_4	He–Ne	2948.7	0.6	[6.137]
Methyl Alcohol	CH_3OH	CO_2 P(34)	1033.5	0.89	[6.124]
Nitric Oxide	NO	CO P(11)	1917.5	0.67	[6.124]
Nitrogen Dioxide	NO_2	CO P(14)	1605.4	2.68	[6.124]
Ozone	O_3	CO_2 P(40)	1052.2	0.56	[6.138]
Perchloroethylene	C_2Cl_4	CO_2 P(42)	923.0	1.14	[6.125]
Propane	C_3H_8	He–Ne	2948.7	0.8	[6.137]
Propylene	C_3H_6	CO P(9)	1647.7	0.09	[6.124]
Sulfur Dioxide	SO_2	$C^{12}O_2^{18}$ R(40)	1108.2	0.25	[6.139]
Trichloroethylene	C_2HCl_3	CO_2 P(20)	944.2	0.56	[6.125]

The match between conventional gas laser lines and pollutant absorption lines is seldom ideal; and in some cases it is so poor that the resulting sensitivity is inadequate. With the availability of tunable infrared lasers, it is possible to ideally match pollutant absorption lines and then tune the laser radiation frequency away from line center to provide the "off"-line signal, in accordance with the theory presented in the preceding section. We now consider the applications of tunable and fixed-frequency lasers to pollutant monitoring.

6.5.1 Point Sampling of Molecular Pollutants

Point sampling measurements may be made at ambient pressure, or at reduced pressure for better specificity. A multi-reflection cell can be used to achieve long pathlengths for the laser radiation in confined areas [6.127, 141], to improve sensitivity. Detection of gases in extracted samples employs either a radiation detector to measure laser *transmission*, or a spectrophone for a direct measurement of *absorption*. In addition to these techniques, point sampling also permits the application of an electric or magnetic field to the gas specimen in order to obtain very specific identification of molecular species by modulation of the center frequency of a known spectral line. Several illustrations of point monitoring with lasers will now be given.

Ambient Air Sampling

Both fixed-frequency and tunable lasers have been used to study samples of ambient air. Since trace gases in air are usually present in low concentrations, their detection by the transmission technique may require very long optical paths; and opto-acoustic detection of the absorbed laser radiation appears better suited to this application.

Using a tunable SFR laser from which approximately 15 mW of coherent radiation entered a spectrophone, KREUZER and PATEL [6.93] measured concentrations of NO in air samples taken from various locations. Some of their measurements are shown in Fig. 6.6. Trace (a) is a calibration scan of 20 ppm NO in N_2 at a total pressure of 300 Torr. The lines designated as 1, 5, 6, 8, and 11 are due to NO absorption, while the others are due to water vapor. Trace (b) shows the noise level with the laser radiation blocked—due primarily to Johnson noise in the first amplifier stage. Trace (c) represents a sample of ambient air from their laboratory, indicating approximately 0.1 ppm of NO. The scan for an air sample taken in the vicinity of a busy highway is shown in trace (d), and yields a higher NO concentration of 2 ppm. Trace (e) was obtained for a sample of automobile exhaust, with a NO concentration of over 50 ppm. These results illustrate an important

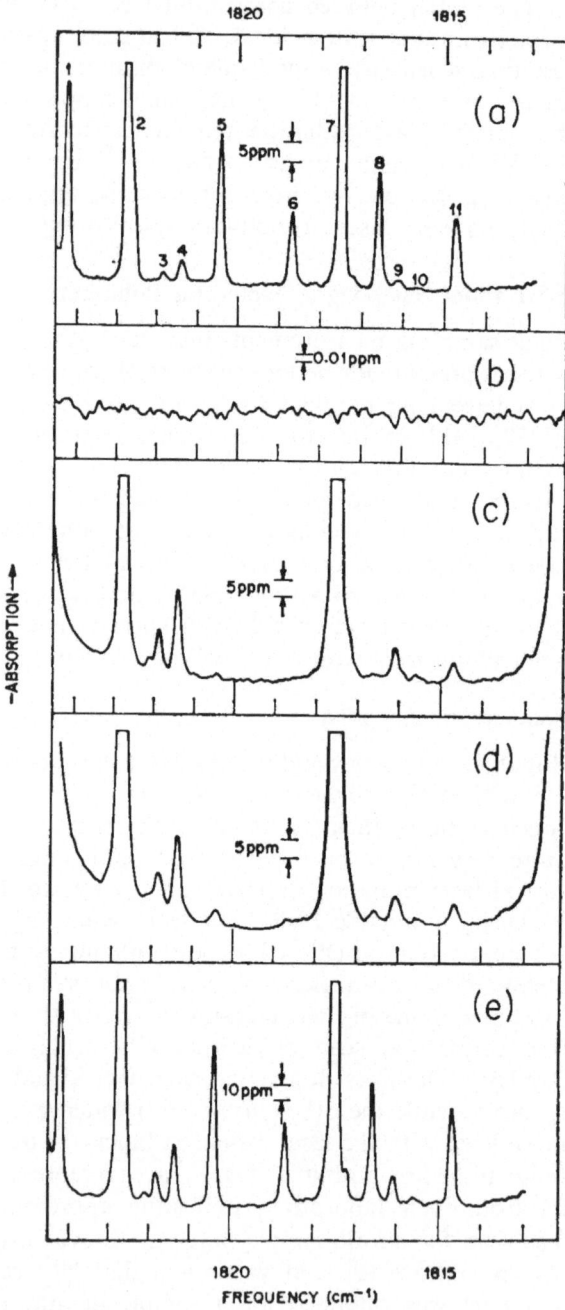

Fig. 6.6a–e

capability of opto-acoustic detection which should, in principle, hold for all other gases as well: that is, this technique can cover a wide dynamic range: 500:1 in this case.

Vehicle Exhaust Sampling

As shown in Fig. 6.6, the opto-acoustic detection technique can be used to measure the concentration of NO, and presumably other gases as well, in samples of automobile exhaust. In another such application, KALDOR et al. [6.120] used a spectrophone in conjunction with a fixed-frequency CO laser to measure the NO content of automobile exhaust samples. In this application a magnetic field was used to Zeeman-shift the spectral line into coincidence with the laser line; resulting in a system with a sensitivity of 3 ppm for NO detection.

Tunable semiconductor diode lasers operating in the infrared have been used to detect CO, NO, C_2H_4, and H_2O in samples of automobile exhaust [6.25, 29, 37, 126, 127, 142] both at ambient and at reduced pressure. In Fig. 6.7 is shown a set of diode laser scans of several C_2H_4 lines in the v_7 band around 10.6 μm [6.126]. The top trace represents 2000 ppm of C_2H_4 in N_2 at a total pressure of 5 Torr, and is used as a quantitative calibration of the signal as well as a qualitative reference signature for the pure gas. The lower three traces represent samples of untreated automobile exhaust from different vehicles. The presence of C_2H_4 is unmistakable because of the identical signatures; and by comparing their amplitudes with that of the calibration trace, a quantitative determination can be made. At this pressure of 5 Torr there is no noticeable interference from any of the other components such as H_2O and other hydrocarbons also present in large quantities. It is clear, however, that for gases such as C_2H_4 which have many closely-spaced spectral lines, it is better to use reduced pressure whenever possible in order to minimize the overlap between adjacent lines and potential interferences.

For diatomic molecules such as CO and NO, also present in large quantity in automobile exhaust, scans of individual lines can be made

---

◄ Fig. 6.6a–e. SFR/spectrophone absorption spectra of various gas samples, in the range 1815–1825 cm^{-1}, at a pressure of 300 Torr. (a) Calibration spectrum of 20 ppm NO in N_2. Lines numbered 1, 5, 6, 8, 11 are due to NO; others due to H_2O. (b) Optoacoustic cell noise when laser was blocked (sensitivity 100 times that of (a), and integration time 4 s). (c) Room air at 21° C, 30% relative humidity (vertical sensitivity same as in (a), integration time 1 s. (d) Air sample from highway in New Jersey. Vertical sensitivity same as (a), integration time 1 s; (e) Automobile exhaust sample. Vertical sensitivity is 0.4 that of (a). Integration time 1 s. Reprinted with permission of the American Association for the Advancement of Science. (After KREUZER and PATEL [6.93])

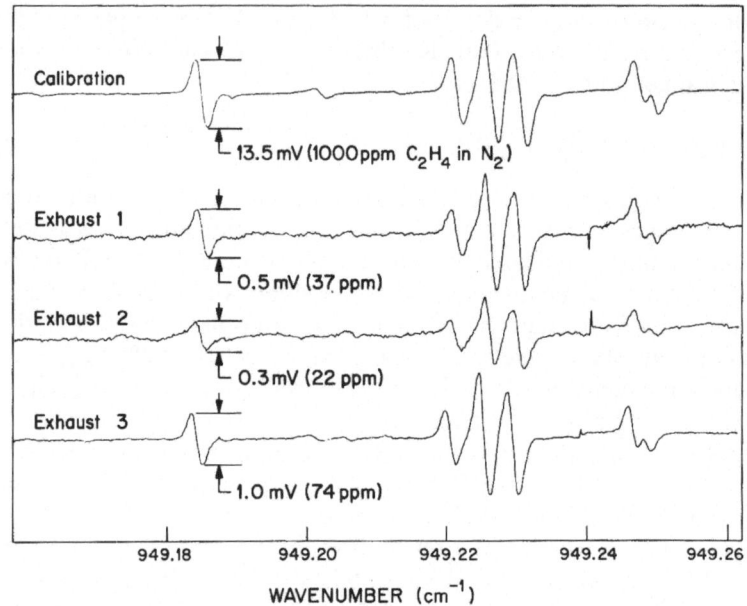

Fig. 6.7. First-derivative spectra of C_2H_4 in the 10.5-μm region obtained using a tunable diode laser. The calibration trace (top) corresponds to 1000 ppm C_2H_4 in N_2 at 5 Torr total pressure. The three exhaust samples from different automobiles were analyzed at different amplifier gain settings, as indicated. Cell length: 30 cm. (After HINKLEY and KELLEY [6.126])

even at atmospheric pressure. KU et al. [6.37] have made laser scans of CO, as shown in Fig. 6.8. Here the $P(7)$ line, centered at 2115.6 cm^{-1}, is scanned by a $PbS_{0.82}Se_{0.18}$ diode laser for exhaust from two vehicles drawn into a 10-cm-long absorption cell. The solid traces represent the actual laser scans of the exhaust, whereas the dotted curves are for calibrated mixtures of pure CO in N_2. Again, because of the lack of significant differences in shape between the solid and dotted curves, we conclude that there is negligible interference from water vapor and the other constituents in this region, at least for the CO concentrations present.

Stack Gas Sampling

In addition to transportation vehicles, another major cause of atmospheric pollution is effluent from power plant and industrial smokestacks. One of the most important pollutants usually present in stack gas is SO_2, which has been measured at reduced pressure by the differential absorption laser technique.

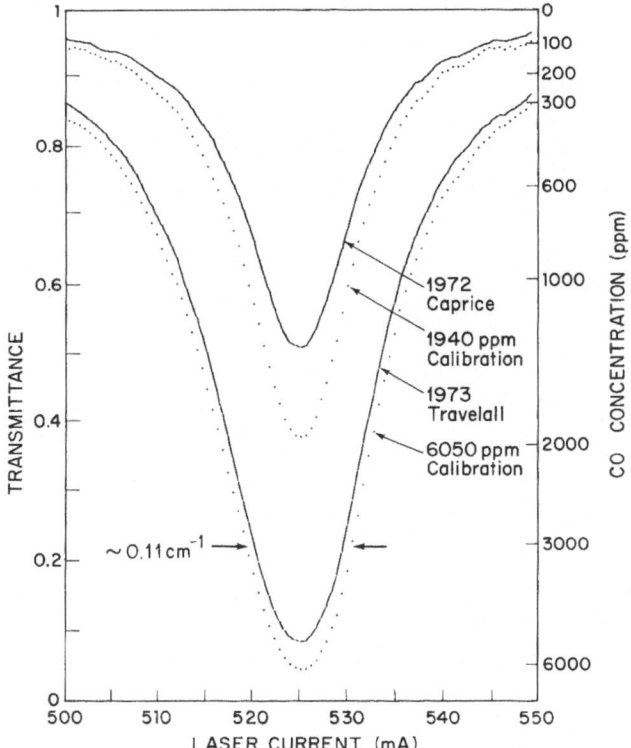

Fig. 6.8. Diode laser scans of the CO $P(7)$ line centered at $2115.6\,\mathrm{cm}^{-1}$ for specimens of untreated automobile exhaust. The solid lines represent data from a 1972 Chevrolet Caprice and a 1973 International Travelall; the dotted curves represent scans of pure CO in N_2 at the indicated mixing ratios. Cell length: 10 cm. (After Ku et al. [6.37])

Using a tunable $Pb_{0.93}Sn_{0.07}Te$ diode laser operating around 8.8 μm, several lines in the v_1 band of SO_2 were scanned by the first-derivative technique [6.127]. In Fig. 6.9 is shown a calibrated measurement of SO_2 in a sample of gas withdrawn from the stack of an oil-fired heating plant. Trace (a) corresponds to the signal from 0.2 Torr pure SO_2, and trace (b) represents the stack sample evacuated to 5 Torr pressure. Comparison between the two traces yields a value for the SO_2 concentration in the sample of 670 ppm, with very high specificity due to the reduced pressure. It is worth noting, moreover, that these are not the strongest SO_2 lines in the v_1 band; and if more appropriate ones were used, the derivative signal could be 20 times larger, resulting in a sensitivitiy of a few ppm for this short path.

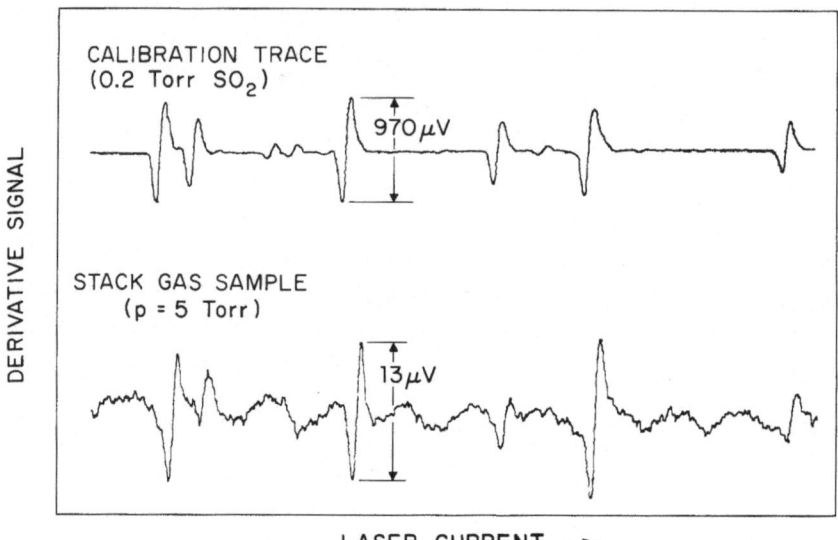

Fig. 6.9. First-derivative detection of SO₂ in a sample of stack gas vs. diode tuning current (laser frequency). Upper trace is a calibration of 0.2 Torr pure SO₂. Lower trace is signal from stack sample evacuated to 5 Torr total pressure. Diode laser wavelength region: 8.8 μm; cell length: 7.3 m. (After Hinkley [6.127])

6.5.2 In situ Localized Monitoring

This type of monitoring does not involve the taking of a sample for "off-line" analysis; and, using the transmittance detection technique, it eliminates entirely the problems associated with monitoring reactive gases such as NO and SO₂ which could be converted during an extraction procedure. In general, however, we consider both the transmittance measurement and opto-acoustic detection to be *in situ* if the ambient pressure is not changed and the measurement is performed in real time.

Vehicle Exhaust Monitoring

Real-time measurements of molecular gases in automobile exhaust effluent have been made using tunable semiconductor diode lasers [6.127]. Figure 6.10 shows results for the detection of C_2H_4 in the exhaust of a 1972 station wagon. The exhaust effluent flowed through a windowless tube 1.15-m long, through which the laser radiation also passed. Monitoring of C_2H_4 was performed for two engine starting sequences: (a) a normal start in which the accelerator was not depressed; and (b) a

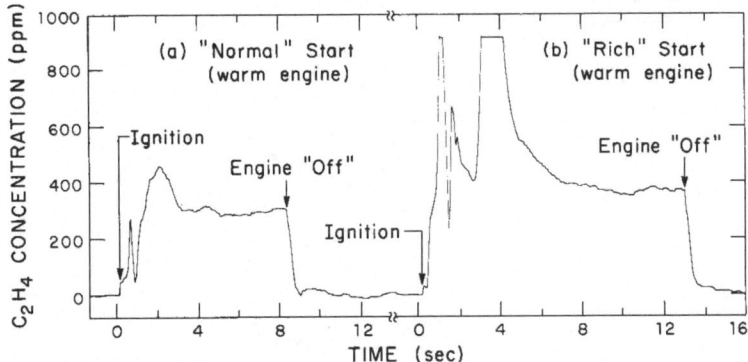

Fig. 6.10a and b. Real-time emissions test for C_2H_4 in exhaust from a 1972 station wagon, using a pulsed diode laser operating at 10.57 μm. (a) represents a normal start of a warm engine, with ignition at time $t = 0$. For (b) the accelerator pedal was depressed fully several times during ignition, to produce a "rich" start, with significantly higher C_2H_4 emission. Cell length: 1 m. (After HINKLEY [6.127])

"rich" start. The measured concentration of approximately 300 ppm during idle was confirmed by checking a sample at reduced pressure by the derivative technique described above. The advantage of this type of measurement is that it can be performed *in situ*, and the response is essentially instantaneous with respect to changes in engine operation.

Stack-Gas Monitoring

The differential absorption technique has been used to monitor the SO_2 content of smokestack effluent with lasers operating in either the infrared or ultraviolet regions of the spectrum. In the infrared, 8.8-μm radiation from a $Pb_{0.92}Sn_{0.08}Te$ diode laser was transmitted across a 5-m wide stack to an infrared detector on the opposite side [6.43]. Although the transmittance was only 10%, limited by particles from coal combustion, the "on-off" wavelength technique described earlier made it possible to monitor the SO_2 concentration in real time with an adequate signal-to-noise ratio. The laser results showed satisfactory agreement with a conventional point monitoring instrument. Using a tunable, frequency-doubled dye laser operating in the ultraviolet from 298 to 309 nm, KUHL and SPITSCHAN [6.143] monitored SO_2 at the exit of a smokestack by directing the beam to a retroreflector located on the opposite side. Their results are illustrated by the scans in Fig. 6.11, where the lower trace in (a) represents absorption by pure SO_2 at 1 Torr pressure in a 15-cm cell. Curve (b) is a similar scan when the laser beam was directed through the stack plume, and (c) shows the back-

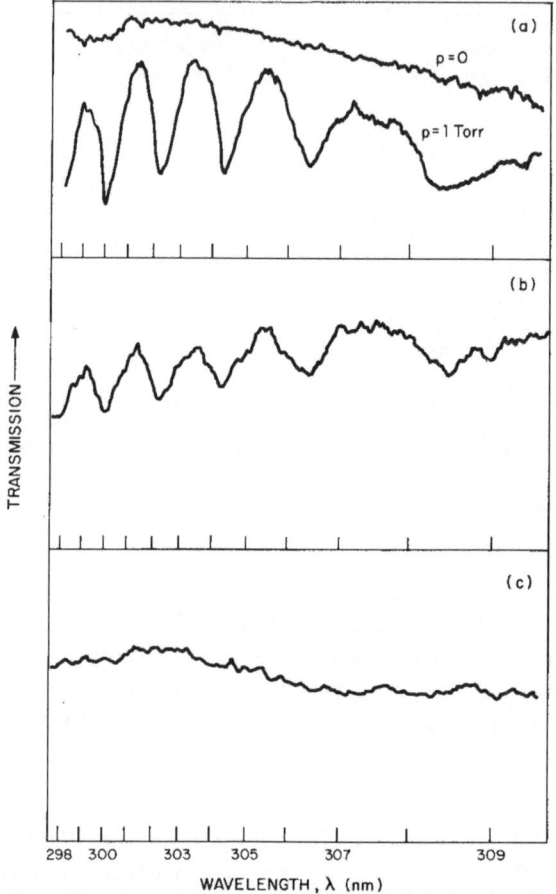

Fig. 6.11a–c. Detection of SO_2 in a smoke stack plume by absorption using a tunable dye laser and remote retroreflector. (a) Laboratory measurements using a 15-cm-long cell filled with pure SO_2 at 1-Torr pressure (many overlapping lines). (b) Measurement in the atmosphere through the stack plume. (c) Measurement in the atmosphere away from the SO_2 emission source. In (b) and (c) the total absorption path was approximately 450 m. (After Kuhl and Spitschan [6.143])

ground tuning signal variation when the laser was directed through the clear atmosphere. Comparison between the stack and calibration scans can be used to determine the SO_2 concentration. Similar studies using a tunable dye laser in the ultraviolet, are being carried out by Hoell and co-workers [6.144].

Tropospheric and Stratospheric in situ Monitoring

The techniques described above can be used for *in situ* monitoring of trace gases in the troposphere and stratosphere using a variety of in-

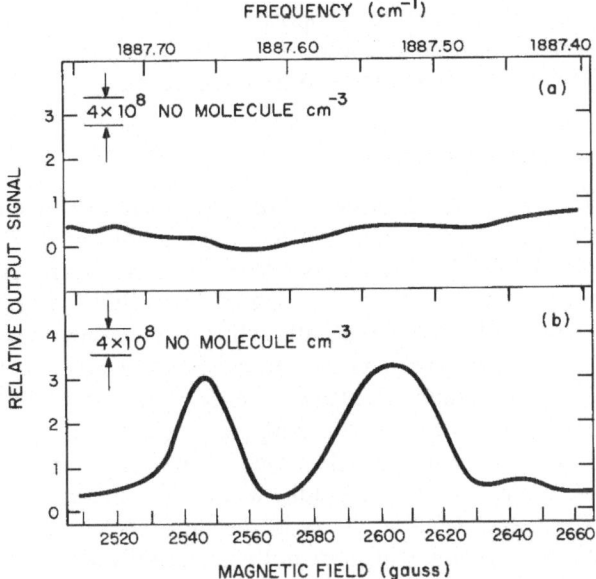

Fig. 6.12a and b. Detection of NO in stratosphere at 28 km altitude, using an SFR laser-spectrophone detector. In (a) is shown the opto-acoustic signal as a function of frequency taken before the ultraviolet sunrise. In (b) a similar scan is made after sunrise. The line at 1887.53 cm^{-1} is wider due to the merging of two Λ-doubled lines (cf. Fig. 6.1). (After PATEL et al. [6.94])

strumentation platforms (e.g. vans, balloons, helicopters, airplanes). Localized measurements *in situ* at sea level have usually been directed toward the determination of absorption of laser radiation through the atmosphere, and have involved both spectrophone and long-path White-cell measurements [6.108]. With regard to higher-altitude measurements, tunable diode lasers have recently been flight-tested on research aircraft [6.145] for eventual application to upper atmospheric detection of gaseous pollutants by employing retroreflectors on the wing tips or tail assembly, or by heterodyne detection of laser radiation reflected from the surface of the earth [6.132].

The first *in situ* measurement of gases in the stratosphere, using laser absorption techniques, was made by PATEL et al. [6.94] in 1974. Using a balloon-borne SFR laser-spectrophone system, concentrations of NO at an altitude of 26–28 km were measured and found to be approximately 1 ppb. The diurnal cycle of NO was also measured, as well as the H_2O concentration at this altitude. Figure 6.12 shows the results of these measurements, during which the SFR laser scanned several NO and H_2O lines. Calibration checks were made automatically during the measurements.

For continuous surveillance of the troposphere and stratosphere, airborne platforms may not be needed if some of the long-path techniques described in Subsection 6.5.3 below prove to be satisfactory.

6.5.3 Long-Path Atmospheric Monitoring

Long-path monitoring of pollutant gases can be performed with fixed-frequency or tunable lasers, employing either a remotely-located detector or a natural or artificial reflector to return the laser beam. As mentioned in Chapter 2, long-path monitoring yields values for the *average* pollutant concentration over the path, which is generally more meaningful for regional models than point samples. Moreover, it can result in considerable saving in manpower and expense needed for certain measurements. Two different schemes for long-path monitoring have emerged during the past few years. One involves the use of discretely-tunable lasers (CO_2, CO, etc.) which are readily available commercially. The other involves the use of tunable lasers in the infrared, visible, or ultraviolet regions of the spectrum, which are not yet as readily available. Because of this difference, we consider fixed-frequency and tunable laser applications to long-path monitoring separately below. Several atmospheric pollutant gases have already been detected by laser long-path techniques, as shown in Table 6.8.

Long-Path Monitoring with Discretely-Tunable Lasers

In the lower atmosphere, where spectral lines are pressure-broadened to 0.1–0.3 cm^{-1}, there is a greater probability of overlap with lines from molecular lasers than there is at higher altitudes. Some of these coincidences are indicated in Table 6.7. Several gases have been detected in the ambient air by the use of a discretely-tunable laser

Table 6.8. List of gaseous pollutant measurements using laser long-path techniques

Pollutant	Laser	λ [μm]	Pathlength [km]	Reference
NO_2	Argon-ion	0.5	3.3	[6.146]
CH_4	He–Ne	3.4	0.3	[6.157]
CO	Diode	4.7	1.0	[6.37]
NO	Diode	5.3	0.22	[6.14, 35]
NO	CO	5.3	0.5	[6.131]
O_3	CO_2	9.5	0.7	[6.149]
C_2H_4	Diode	10.5	0.25	[6.127]
O_3	CO_2	9.5, 10.3	0.08	[6.151]
O_3	CO_2	9.5	0.5	[6.131]

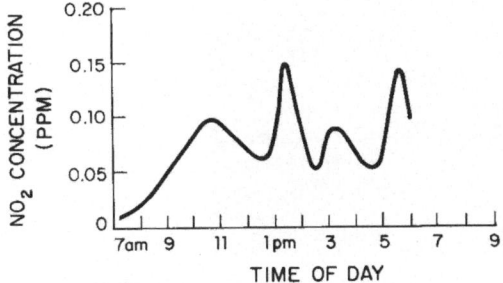

Fig. 6.13. Monitoring of NO_2 concentration in city of Atlanta, using a pulsed argon-ion laser, operating sequentially at 496.5 and 501.7 nm, and a corner-cube retroreflector 3.34 km away. (After O'Shea and Dodge [6.146])

and remote reflector. O'Shea and Dodge [6.146] used two wavelengths (496.5 nm and 501.7 nm) from an argon-ion laser to monitor NO_2 in Atlanta, Georgia, over a 3.34-km path. Their results, shown in Fig. 6.13 for an 11-h period, display the expected diurnal behavior related to commuter traffic, with peaks to 0.15 ppm. Snowman and Gillmeister [6.147] proposed the use of certain discrete lines from a CO_2 laser to monitor ambient concentrations of O_3, NH_3, C_2H_4, and other gases, and Mc Clenny et al. [6.148] confirmed the technique for O_3 with the aid of a conventional point O_3 monitor moved along the path. The use of CO and CO_2 lasers for long-path monitoring was proposed in 1971 by Menzies [6.150]. Long-path measurements of both O_3 (with a CO_2 laser) and NO (with a CO laser) were performed by Menzies and Shumate [6.131]. One unique feature of these measurements was the use of heterodyne techniques to detect diffusely-scattered laser radiation from buildings or natural targets. These measurements are discussed in detail in Subsection 7.3.3. Asai and Igarashi [6.151] detected O_3 over long paths in Tokyo using the $P(14)$ and R(14) lines of a CO_2 laser, at 9.5 and 10.5 µm, respectively, and a HgCdTe detector. They measured O_3 concentrations to 0.2 ppm, with a peak at around 1400 hours.

Long-Path Monitoring with Continuously-Tunable Lasers

Lasers which can be tuned continuously have the advantage of being operated at line center, in the wing of a spectral line, or at another line which may be more suitable for detection of a particular gas. Henningsen et al. [6.75] used an OPO to detect absorption lines in the 2.3-µm overtone band of CO when the laser radiation was scattered by natural foliage from a distance of 107 m. Their result is shown in Fig. 6.14, in which the laser scanned two adjacent rotational lines of CO. Although this measurement does not represent an actual monitoring of

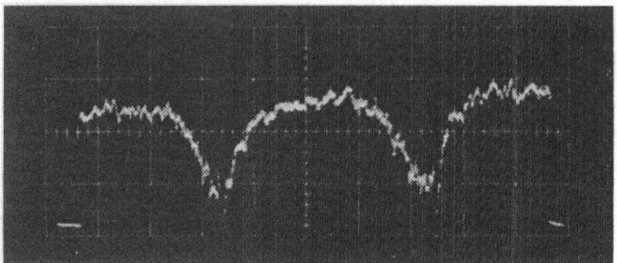

Fig. 6.14. Detection of CO spectral lines in overtone region (2.3 μm) in cell 107-m away using OPO radiation backscattered from a topographic target. Zero transmission is indicated by the small horizontal lines at the beginning and end of the trace. Total CO burden: 18 atm-cm. (After HENNINGSEN et al. [6.75])

atmospheric CO, since the gas was contained in a cell along the path, it does illustrate the feasibility of using such overtone bands when very long paths are needed (as discussed in Subsection 6.1.1.), and of employing natural targets such as foliage and buildings.

The most useful type of tunable laser for long-path monitoring using remote reflectors has been the semiconductor diode laser because it can be made to emit anywhere in the infrared "fingerprint" region [6.152]. Diode lasers have been used to monitor CO, NO, C_2H_4, and H_2O in the atmosphere [6.37, 14, 127, 35]. Although atmospheric turbulence can, for a typical long-path setup, cause the received signal to vary by up to 50% in a few tenths of a second, the use of fast derivative detection (in which the modulation frequency is rapid compared to the time scale of the turbulence) or rapid two-wavelength differential absorption, reduces this effect on the measured pollutant concentration considerably [6.37].

Finally, as part of the U. S. Environmental Protection Agency's Regional Air Pollution Study in St. Louis, Missouri, comparative measurements have been made between long-path monitoring results using tunable diode lasers and data obtained from point-samples taken along the path or by an instrument at either end. One example of this type of comparison is shown in Fig. 6.15, where the solid curve represents the continuous laser readings (taken with a 1-s time constant, but averaged over 10 min for comparison) and the circles represent the localized readings. Even though the path length is relatively long (~ 1 km), fairly good correlation between the two sets of data indicates that the laser calibration procedure was accurate and that the CO concentration was uniform over this rural area during the time the measurements were made. Similar comparative measurements, involving both CO and NO detection over pathlengths to 2 km are currently underway [6.154].

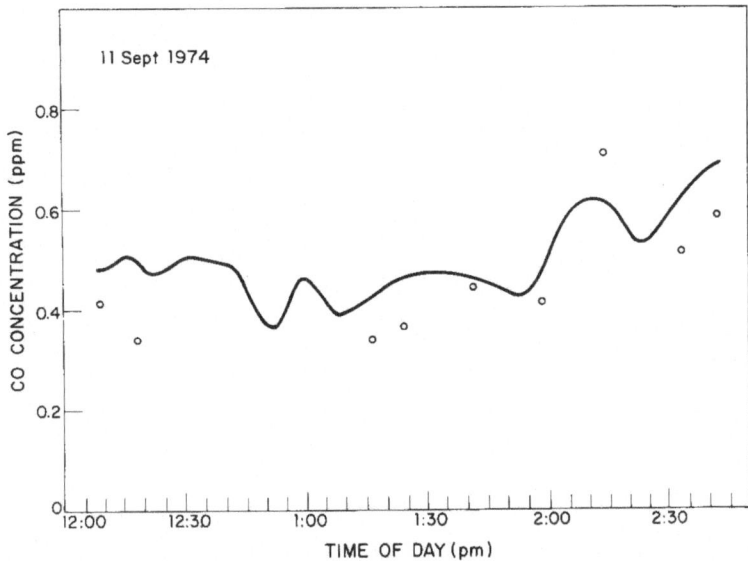

Fig. 6.15. Diode laser monitoring of ambient CO in Granite City, Illinois. Path length to remote retroreflector was 0.75 km. A 1-s integration time was used for original data, but smoothed to a 10-min time for display. Circles represent point-monitor values of CO concentration, using a conventional instrument. (After Ku et al. [6.153])

Vertical Profile Monitoring

MENZIES [6.129] has shown theoretically that it is possible to analyze the shape of a spectral line obtained by vertical transmission of tunable laser radiation through the atmosphere in terms of concentration of the pollutant gas as a function of altitude. Such measurements may be made from either airborne or ground-based platforms. In the former case, NASA (National Aeronautics and Space Administration) is considering the use of airplanes and satellites containing tunable lasers whose radiation is reflected from the earth [6.155], and whose return signal may be detected by heterodyne techniques [6.131].

Two possible configurations for a *ground-based* system for atmospheric monitoring are shown in Fig. 6.16. In (a) the sun serves as an incoherent source of wideband radiation, and the tunable laser is used as a local oscillator in the heterodyne configuration for sensitive, narrow-band detection [6.129, 130, 156]. This technique is effectively single-ended (monostatic) and the optical alignment and laser power requirements are not beyond the present state-of-the-art. In (b) is shown an active (bistatic) system which employs a satellite as a

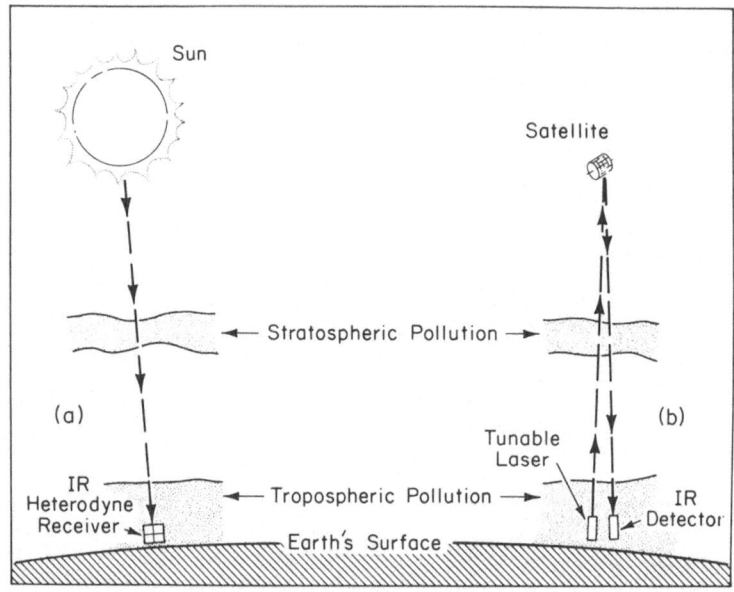

Fig. 6.16a and b. Diagrams for long-path vertical profile monitoring of the atmosphere using (a) passive system with sun as radiation source and tunable-laser heterodyne detection, and (b) an active, ground-based system employing cooperative satellite reflector

cooperative reflector, and an infrared detector which may be used in either the direct or heterodyne mode.

In either of the cases illustrated in Fig. 6.16, the pollutant gas in the stratosphere will have relatively narrow spectral lines; whereas, in the lower troposphere the lines will be considerably wider. Although such a measurement has yet to be made under actual conditions, it has been partially simulated in the laboratory [6.130] using a tunable diode laser and two gas cells, with the results shown in Fig. 6.17. One cell contained a mixture of CO in air at atmospheric pressure to represent 0.12 ppm $(3.1 \times 10^{12} \text{ cm}^{-3})$ over a 1-km path from sea level; the other cell contained CO at low pressure to simulate 0.23 ppm $(2.0 \times 10^{11} \text{ cm}^{-3})$ over a 1-km path in the stratosphere at 25 km altitude[4]. If the pollutants are stratified in similar layers, deconvolution of the spectral lineshape into concentration vs. height data may not be difficult, and the potential economy of this technique could be considerable[5].

---

[4] These concentrations are not unreasonable; see, e.g. [6.37 and 158].

[5] R. T. Menzies very recently measured O_3 by the technique of Fig. 6.16(a).

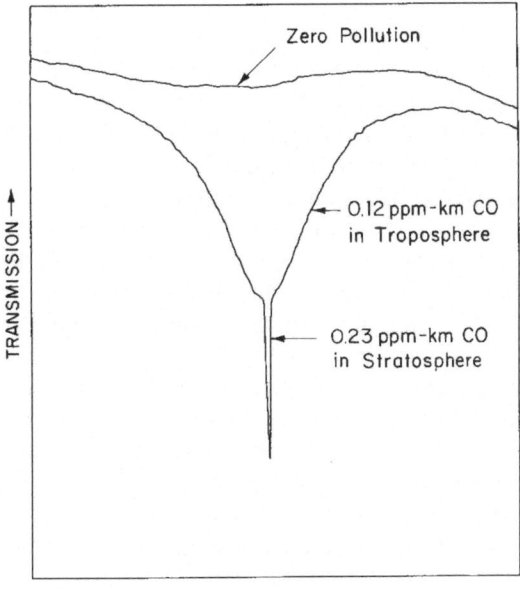

Fig. 6.17. Laboratory simulation of vertical profile monitoring based on technique shown in Fig. 6.16. Diode laser scan of the $P(7)$ CO line of gas contained in two cells—one at low pressure simulating 0.23 ppm CO over a 1-km path at 25 km in stratosphere, using technique (a) of Fig. 6.16; and the second at atmospheric pressure simulating 0.12 ppm between ground level and 1 km altitude. Using the two-way transmission of Fig. 6.16b, the mixing ratios calculated from these scans would be half as great over the same 1-km pathlength increments. (After HINKLEY [6.130])

6.6 Laser Safety

Each of the active laser-monitoring techniques discussed in this volume present potential safety hazards, and may require large-area optics in order to reduce the power and energy densities to safe levels. The danger is greatest when visible laser radiation is used, since it can pass through the eye and injure the retina. For radiation between 0.4 and 1.4 µm, the maximum permissible cw power density is $1\,\mu W/cm^2$ according to recent eye-safety standards [6.159]. In the infrared, beyond 1.4 µm, this limit is increased to $100\,mW/cm^2$, and energy densities of $1\,J/cm^2$. Consequently, in a 20-cm-diameter beam, a cw power of 125 W can be safely transmitted. As we have seen in this chapter, for all applications of long-path sensing using a remote, cooperative retroreflector, the useful power level can be well below this value.

6.7 Conclusion

The absorption of laser radiation can be employed in a variety of ways for pollutant detection and monitoring. Under point-sampling conditions, where sample pressure can be reduced, or during *in situ* measurements at high altitudes, where spectral lines are narrow, very specific and sensitive detection can be accomplished. Short-path *in situ* monitoring of effluents at their sources can be performed; and while not as free from potential interferences at atmospheric pressure, it nevertheless, offers the advantage that the stream under investigation is not disturbed by the measurement. This is particularly important for some of the more reactive species. Finally, long-path monitoring yields information as to the *average* pollutant concentration over a relatively long atmospheric path—an important parameter for the evaluation of various mathematical models being developed for regional pollution prediction. Techniques based on long-path spectral information are being developed to provide continuous surveillance of the troposphere and stratosphere from either airborne or ground-based instrumentation.

Acknowledgements. The authors thank R. J. KEYES for helpful comments on the infrared detector section, and R. T. MENZIES and M. S. SHUMATE for reviewing the opto-acoustic section. This work was supported by the National Science Foundation (RANN), the U.S. Environmental Protection Agency, and the Department of the Air Force.

References

6.1   H. KILDAL, R. L. BYER: Proc. IEEE **59**, 1644 (1971)
6.2   P. L. HANST: In *Advances in Environmental Science and Technology*, Vol. 2, ed. by J. N. PITTS and R. L. METCALF (John Wiley & Sons, New York 1971) Chapter 4. see also [6.28]
6.3   P. L. HANST: Symposium on Long Path Techniques, U.S. Environmental Protection Agency, Research Triangle Park, N. C. (4 February 1975)
6.4   A. R. BARRINGER: J. Opt. Soc. Am. **60**, 729 (1970)
6.5   D. E. BURCH, D. A. GRYVNAK: In *Analytical Methods Applied to Air Pollution Measurements*, ed. by R. K. STEVENS and W. F. HERGET (Ann Arbor Science, Ann Arbor 1974) Chap. 10
6.6   H. J. GERRITSEN: Trans. Am. Inst. Mining Eng. **235**, 428 (1966)
6.7   A. R. BRENNER, S. G. KUKOLICH: Anal. Lett. **6**, 691 (1973)
6.8   J. E. HARRIES: Nature **241**, 515 (1973);
       see also, B. CARLI, D. H. MARTIN, E. PUPLETT, J. E. HARRIES: Nature **257**, 649 (1975)
6.9   R. L. BYER, M. GARBUNY: Appl. Opt. **12**, 1496 (1973)
6.10   See, for example, S. M. FREUND, D. M. SWEGER: Anal. Chem. **47**, 930 (1975) for a list of early references to the vinyl chloride controversy
6.11   See S. C. WOFSY, M. B. MCELROY, N. D. SZE: Science **187**, 535 (1975), and list of references

6.12 For recent surveys of laser spectroscopic measurements, see [6.13] or [6.14]

6.13 H. WALTHER (editor): *Topics in Applied Physics*, Vol. 2: Laser Spectroscopy of Atoms and Molecules. (Springer-Berlin, Heidelberg, New York 1976)

6.14 E. D. HINKLEY: Opt. Quant. Electron. **8**, 155 (1976)

6.15 W. HEISENBERG: Z. Physik **43**, 172 (1927)

6.16 R. G. BREWER: Science **178**, 247 (1972)

6.17 J. L. HALL, C. BORDÉ: Phys. Rev. Lett. **30**, 1101 (1973)

6.18 A. S. PINE: J. Mol. Spectrosc. **54**, 132 (1975); also personal communication

6.19 S. S. PENNER: *Quantitative Molecular Spectroscopy and Gas Emissivities* (Addison-Wesley, Reading, Mass. 1959)

6.20 K. W. NILL, F. A. BLUM, A. R. CALAWA, T. C. HARMAN: Chem. Phys. Lett. **14**, 234 (1972)

6.21 F. A. BLUM, K. W. NILL, A. R. CALAWA, T. C. HARMAN: Chem. Phys. Lett. **15**, 144 (1972)

6.22 E. D. HINKLEY: Phys. Rev. A**3**, 833 (1971)

6.23 K. W. NILL, F. A. BLUM, A. R. CALAWA, T. C. HARMAN: Appl. Phys. Lett. **19**, 79, (1971)

6.24 F. A. BLUM, K. W. NILL: *Laser Spectroscopy*. Proc. Laser Spectroscopy Conf., ed. by R. G. BREWER and A. MOORADIAN (Plenum Press, New York 1974)

6.25 G. P. MONTGOMERY, J. C. HILL: J. Opt. Soc. Am. **65**, 579 (1975); also J. C. HILL: personal communication

6.26 A. S. PINE: personal communication

6.27 K. W. NILL, F. A. BLUM: unpublished

6.28 P. L. HANST, A. S. LEFOHN, B. W. GAY, JR.: Appl. Spectrosc. **27**, 188 (1973)

6.29 E. D. HINKLEY: "*Development and Application of Tunable Diode Lasers to the Detection and Quantitative Evaluation of Pollutant Gases*", Final Technical Report to the U.S. Environmental Protection Agency (September 1971)

6.30 R. S. McDOWELL, H. W. GALBRAITH, B. J. KROHN, C. D. CANTRELL, E. D. HINKLEY: Opt. Commun. **17**, 178 (1976)

6.31 See, for example [6.19]

6.32 F. A. BLUM, K. W. NILL, P. L. KELLEY, A. R. CALAWA, T. C. HARMAN: Science **177**, 694 (1972)

6.33 E. D. HINKLEY, R. T. KU: "*Diode Laser Multi-Pollutant Ambient Air Monitoring*", Annual Report to the National Science Foundation (RANN) (June 1974)

6.34 P. L. HANST, L. L. SPILLER, D. M. WATTS, J. W. SPENCE, M. F. MILLER: J. Air Pollution Control Assoc. **25**, 1220 (1975)

6.35 E. D. HINKLEY, R. T. KU, K. W. NILL, J. F. BUTLER: Appl. Opt. **15**, 1653 (1976)

6.36 R. L. BYER: "*Infrared Differential Absorption for Atmospheric Pollutant Detection*", Report N74-33949, National Aeronautics & Space Admin. (October 1974)

6.37 R. T. KU, E. D. HINKLEY, J. O. SAMPLE: Appl. Opt. **14**, 854 (1975)

6.38 A. S. PINE: J. Opt. Soc. Am. **66**, 97 (1976)

6.39 H. INOMATA, T. IGARASHI: Japanese J. Appl. Phys. **14**, 1751 (1975)

6.40 M. GRIGGS: J. Chem. Phys. **49**, 857 (1968)

6.41 R. T. KU: Unpublished diode laser spectroscopy (1976)

6.42 R. T. THOMPSON, JR., J. M. HOELL, JR., W. R. WADE: J. Appl. Phys. **46**, 3040 (1975)

6.43 E. D. HINKLEY: "*Development of In Situ Prototype Diode Laser System to Monitor SO_2 Across the Stack*"; Final Report to the U.S. Environmental Protection Agency, No. EPA-R2-73-218 (May 1973)

6.44 E. D. HINKLEY, A. R. CALAWA, P. L. KELLEY, S. A. CLOUGH: J. Appl. Phys. **43**, 3222 (1972)

6.45 See L. TRAFTON: J. Quant. Spectrosc. Radiat. Transf. **13**, 821 (1973), and references contained therein

6.46   See Ref. [6.19], p. 32
6.47   See for example, *Topics in Applied Physics,* Vol. 1: Dye Lasers F. P. Schäfer, ed. Top. Appl. Phys. **1**, (Springer-Berlin, Heidelberg, New York 1973)
6.48   J. B. West, R. L. Barger, T. C. English: IEEE/OSA Conf. on Laser Engineering and Applications, Washington, D. C. (May 1975) Paper 5.3
6.49   T. W. Hänsch, I. S. Shahin, A. L. Schawlow: Phys. Rev. Lett. **27**, 707 (1971)
6.50   K. W. Rothe, U. Brinkmann, H. Walther: Appl. Phys. **3**, 115 (1974)
6.51   W. B. Grant, R. D. Hake, Jr., E. M. Liston, R. C. Robbins, E. K. Procter, Jr.: Appl. Phys. Lett. **24**, 550 (1974)
6.52   W. B. Grant, R. D. Hake, Jr.: J. Appl. Phys. **46**, 3019 (1975)
6.53   M. R. Bowman, A. J. Gibson, M. C. W. Sandford: Nature **221**, 456 (1969); A. J. Gibson, M. C. W. Sandford: J. Atmos. Terrest. Phys. **33**, 1675 (1971)
6.54   F. Felix, W. Keenliside, G. Kent, M. C. W. Sandford: Nature **246**, 345 (1973)
6.55   C. F. Dewey, Jr., L. O. Hocker: Appl. Phys. Lett. **18**, 58 (1971)
6.56   A. S. Pine: J. Opt. Soc. Am. **64**, 1683 (1974)
6.57   D. S. Chemla, P. J. Kupeck, D. S. Robertson, R. C. Smith: Opt. Commun. **3**, 29 (1971)
6.58   R. L. Byer, H. Kildal, R. S. Feigelson: Appl. Phys. Lett. **19**, 237 (1971)
6.59   G. D. Boyd, E. Buehler, F. G. Storz, J. H. Wernick: IEEE J. QE-**8**, 419 (1972)
6.60   G. D. Boyd, H. M. Kasper, J. H. McFee, F. G. Storz: IEEE J. QE-**8**, 900 (1972)
6.61   H. Kildal, J. C. Mikkelsen: Opt. Commun. **9**, 315 (1973)
6.62   H. Kildal, J. C. Mikkelsen: Opt. Commun. **10**, 306 (1974)
6.63   A. H. M. Ross: (unpublished calculation)
6.64   V. J. Corcoran, R. E. Cupp, J. J. Gallagher, W. T. Smith: Appl. Phys. Lett. **16**, 316 (1970)
6.65   V. J. Corcoran, J. M. Martin, W. T. Smith: Appl. Phys. Lett. **22**, 517 (1973)
6.66   E. S. Yeung, C. B. Moore: J. Am. Chem. Soc. **93**, 2059 (1971)
6.67   D. J. Bradley, J. V. Nicholas, J. R. D. Shaw: Appl. Phys. Lett. **19**, 172 (1971)
6.68   S. E. Harris, D. M. Bloom: Appl. Phys. Lett. **24**, 229 (1974)
6.69   J. J. Wynne, P. P. Sorokin, J. R. Lankard: In *Laser Spectroscopy,* ed by. R. G. Brewer and A. Mooradian (Plenum, New York 1974) p. 103
6.70   S. E. Harris: Proc. IEEE **57**, 2096 (1969); R. G. Smith: In *Laser Handbook,* ed. by F. T. Arrechi and E. O. Schultz-DuBois (North-Holland Publ. Co., Amsterdam 1972)
6.71   E. O. Ammann, J. M. Yarborough, M. K. Oshman, P. C. Montgomery: Appl. Phys. Lett. **16**, 309 (1970)
6.72   R. L. Herbst, R. L. Byer: Appl. Phys. Lett. **19**, 527 (1971); R. L. Byer: *Digest of Technical Papers,* 7th Intern. Quantum Electronics Conf. Montreal, (May 1972)
6.73   J. Pinnard, J. F. Young: Opt. Commun. **4**, 425 (1972)
6.74   Chromatix Corporation, Mountain View, California has developed an OPO with a similar linewidth
6.75   T. Hennigsen, M. Garbuny, R. L. Byer: Appl. Phys. Lett. **24**, 242 (1974)
6.76   M. I. Nathan: Proc. IEEE **54**, 1276 (1966)
6.77   H. Kressel: In *Lasers,* Vol. 3, ed. by A. K. Levine and A. J. DeMaria (Marcel Dekker, New York 1971) p. 1
6.78   T. C. Harman: In *The Physics of Semimetals and Narrow-Gap Semiconductors* Pergamon Press, New York 1970) p. 363
6.79   I. Melngailis, A. Mooradian: In *Laser Applications to Optics and Spectroscopy,* ed. by S. F. Jacobs, M. Sargent, J. F. Scott, and M. O. Scully (Addison-Wesley, Reading, Mass. 1975) p. 1
6.80   E. D. Hinkley, T. C. Harman, C. Freed: Appl. Phys. Lett. **13**, 49 (1968)

6.81 T.C.HARMAN, A.R.CALAWA, I.MELNGAILIS, J.O.DIMMOCK: Appl. Phys. Lett. **14**, 333 (1969)
6.82 J.F.BUTLER, A.R.CALAWA: In *Physics of Quantum Electronics*, ed. by P.L.KELLEY, B.LAX, and P.E.TANNENWALD, (McGraw-Hill, New York 1966) p. 458; A.R.CALAWA, J.O.DIMMOCK, T.C.HARMAN, I.MELNGAILIS: Phys. Rev. Lett. **23**, 7 (1969); K.W.NILL, F.A.BLUM, A.R.CALAWA, T.C.HARMAN: J. Nonmetals **1**, 211 (1973)
6.83 J.M.BESSON, J.F.BUTLER, A.R.CALAWA, W.PAUL, R.H.REDIKER: Appl. Phys. Lett. **7**, 206 (1965); J.M.BESSON, W.PAUL, A.R.CALAWA: Phys. Rev. **173**, 699 (1968)
6.84 A.S.PINE, C.J.GLASSBRENNER, J.A.KAFALAS: IEEE J. QE-9, 800 (1973)
6.85 S.H.GROVES, K.W.NILL, A.J.STRAUSS: Appl. Phys. Lett. **25**, 331 (1974)
6.86 E.D.HINKLEY, C.FREED: Phys. Rev. Lett. **23**, 277 (1969)
6.87 C.K.N.PATEL, E.D.SHAW: Phys. Rev. B3, 1279 (1971)
6.88 A.MOORADIAN, S.R.J.BRUECK, F.A.BLUM: Appl. Phys. Lett. **17**, 481 (1971)
6.89 R.S.ENG, A.MOORADIAN, H.FETTERMAN: Appl. Phys. Lett. **25**, 453 (1974)
6.90 J.P.SATTLER, B.A.WEBER, J.R.NEMARICH: Appl. Phys. Lett. **25**, 491 (1974) P.W.KRUSE: Appl. Phys. Lett. **28**, 90 (1976)
6.91 C.K.N.PATEL: Phys. Rev. Lett. **28**, 649 (1972)
6.92 S.R.J.BRUECK, A.MOORADIAN: IEEE J. QE-10, 634 (1974)
6.93 L.B.KREUZER, C.K.N.PATEL: Science **173**, 45 (1971)
6.94 C.K.N.PATEL, E.G.BURKHARDT, C.A.LAMBERT: Science **184**, 1173 (1974)
6.95 E.G.BURKHARDT, C.A.LAMBERT, C.K.N.PATEL: Science **188**, 1111 (1975)
6.96 A.J.ALCOCK, K.LEOPOLD, M.C.RICHARDSON: Appl. Phys. Lett. **23**, 562 (1973); J.S.LEVINE, A.JAVAN: Appl. Phys. Lett. **22**, 55 (1973)
6.97 T.Y.CHANG, O.R.WOOD: Appl. Phys. Lett. **23**, 524 (1973); Appl. Phys. Lett. **23**, 182 (1974)
6.98 H.KILDAL, T.F.DEUTSCH: Appl. Phys. Lett. **27**, 500 (1975)
6.99 V.N.BAGRATASHUILI, I.N.KNYAZEV, YU.A.KUDRYATSEV, V.S.LETOKHOV: JETP Lett. **18**, 62 (1973)
6.100 N.W.HARRIS, F.O'NEILL, W.T.WHITNEY: Appl. Phys. Lett. **25**, 148 (1974)
6.101 R.L.ABRAMS: Appl. Phys. Lett. **24**, 304 (1974)
6.102 P.W.SMITH: In *Laser Spectroscopy*, ed. R.G.BREWER and A.MOORADIAN (Plenum, New York 1974) p. 247
6.103 F.O'NEILL, W.T.WHITNEY: IEEE/OSA Conf. on Laser Engineering and Applications, Washington, D.C. (May 1975) Paper 18.9
6.104 For a recent review of infrared detectors for remote sensing, see H.LEVINSTEIN, J.MUDAR: Proc. IEEE **63**, 6 (1975)
6.105 See, for example, R.J.KEYES, T.M.QUIST: In *Semiconductors and Semimetals*, ed. by R.K.WILLARDSON and A.C.BEER (Academic Press, New York 1970) Chap. 8
6.106 For a review of early spectrophone applications, see M.E.DELANY: Sci. Progr. **47**, 459 (1959)
6.107 L.B.KREUZER: J. Appl. Phys. **42**, 2934 (1971)
6.108 P.L.KELLEY, R.A.MCCLATCHEY, R.K.LONG, A.SNELSON: Opt. Quant. Electron. **8**, 117 (1976)
6.109 L.-G.ROSENGREN: Appl. Opt. **14**, 1960 (1975)
6.110 L.-G.ROSENGREN: Infrared Phys. **13**, 109 (1973)
6.111 L.-G.ROSENGREN, E.MAX, S.T.ENG: J. Phys. E (Sci. Inst.) **7**, 125 (1974); also L.-G.ROSENGREN: J. Phys. E: Sci. Inst. **8**, 242 (1975)
6.112 For a recent discussion of interference considerations, see J.SHEWCHUN, B.K. GARSIDE, E.A.BALLIK, C.C.Y.KWAN, M.M.ELSHERBINY, G.HOGENKAMP, and A.KAZANDJIAN: Appl. Opt. **15**, 340 (1976)

294   E. D. Hinkley et al.

6.113 C. F. Dewey, Jr., R. D. Kamm, C. E. Hackett: Appl. Phys. Lett. 23, 633 (1973)
6.114 E. Max, L.-G. Rosengren: Opt. Commun. 11, 422 (1974)
6.115 R. S. Eng: Solid State Research (M.I.T. Lincoln Laboratory) 4, 36 (1974)
6.116 W. Schnell, G. Fischer: Rapport de la Société Suisse de Physique 26, 133 (1975)
6.117 P. C. Claspy, Y.-H. Pao, S. Kwong, E. Nodov: IEEE/OSA Conf. Laser Engineering
       and Applications, Washington, D. C. (1975) Paper 9.10
6.118 T. F. Deaton, D. A. Depatie, T. W. Walker: Appl. Phys. Lett. 26, 300 (1975)
6.119 P. A. Bonczyk, C. J. Ultee: Opt. Commun. 6, 196 (1972)
6.120 A. Kaldor, W. B. Olson, A. G. Maki: Science 176, 508 (1972)
6.121 C. Chakerian, Jr., M. F. Weisbach: J. Opt. Soc. Am. 63, 342 (1973)
6.122 C. F. Dewey, Jr.: Optical Eng. 13, 483 (1974)
6.123 J. Gelbwachs: Appl. Opt. 13, 1005 (1974)
6.124 L. B. Kreuzer, N. D. Kenyon, C. K. N. Patel: Science 177, 347 (1972)
6.125 W. Schnell, G. Fischer: Appl. Opt. 14, 2058 (1975); also personal communication
6.126 E. D. Hinkley, P. L. Kelley: Science 171, 635 (1971)
6.127 E. D. Hinkley: Opto-Electron. 4, 69 (1972)
6.128 M. Mumma, T. Kostiuk, S. Cohen, D. Buhl, P. C. von Thuna: Nature 253, 514
       (1975)
6.129 R. T. Menzies: U. S. Patent 3,761,715, California Institute of Technology (1971)
6.130 E. D. Hinkley: Symp. on Remote Sensing of Environmental Air Pollutants, 1974
       Pittsburgh Conf. on Analytical Chemistry and Applied Spectroscopy, Cleveland,
       Ohio (March 1974)
6.131 R. T. Menzies, M. S. Shumate: IEEE/OSA Conf. on Laser Engineering and
       Applications, Washington, D. C. (1975) Paper 9.2
6.132 R. K. Seals, Jr., C. H. Bair: 2nd Joint Conf. on the Sensing of Environmental
       Pollutants, Washington, D. C. (1973)
6.133 Handbook of Military Infrared Technology, ed. by W. L. Wolfe (U. S. Government
       Printing Office, Cat. 65-62266, 1975)
6.134 G. B. Jacobs, L. R. Snowman: IEEE J. QE-3, 603 (1967)
6.135 P. L. Hanst: Appl. Spectrosc. 24, 161 (1970)
6.136 T. K. McCubbin, Jr.: Air Force Cambridge Research Laboratory Report No
       AFCRL 67-0437 (1967)
6.137 D. N. Jaynes, B. H. Beam: Appl. Opt. 8, 1741 (1969)
6.138 C. Young, R. H. L. Bunner: Appl. Opt. 13, 1438 (1974)
6.139 T. Kobayasi, H. Inaba: Opt. Quant. Electron. 7, 319 (1975)
6.140 S. S. Penner, K. G. P. Sulzmann, H. K. Chen: J. Quant. Spectrosc. Radiat. Transf.
       13, 705 (1973)
6.141 J. J. Ball, R. A. Keller: J. Air Poll. Control Assoc. 25, 631 (1975)
6.142 E. D. Hinkley, H. A. Pike: unpublished
6.143 J. Kuhl, H. Spitschan: Opt. Commun. 13, 6 (1975)
6.144 J. M. Hoell, Jr., W. R. Wade, R. T. Thompson, Jr.: Intern. Conf. on Environmental
       Sensing and Assessment, Las Vegas, Nevada (1975) Paper 10-6
6.145 F. Allario: Personal communication
6.146 D. C. O'Shea, L. G. Dodge: Appl. Opt. 13, 1481 (1974)
6.147 L. R. Snowman, R. J. Gillmeister: Joint Conf. on Sensing of Environmental
       Pollutants, Palo Alto, California (1971) Paper 71–1059
6.148 W. A. McClenny, R. E. Baumgardner, Jr., F. W. Baity, Jr., R. A. Gray: J. Air
       Pollution Control Assoc. 24, 1044 (1974); see also S. E. Craig, D. R. Morgan,
       D. L. Roberts, and L. R. Snowman: "Development of a Gas Laser System to
       Measure Trace Gases by Long Path Absorption Techniques"; Final Report to the
       U. S. Environmental Protection Agency, No. EPA-650/2-74-046a, June 1974
6.149 R. R. Patty, G. M. Russwurm, W. A. McClenny, and D. R. Morgan: Appl. Opt.
       13, 2850 (1974)

6.150 R.T.MENZIES: Appl. Opt. **10**, 1532 (1971)

6.151 K.ASAI, T.IGARASHI: Opt. Quant. Electron. **7**, 211 (1975)

6.152 A.R.CALAWA: J. Lumines. **7**, 477 (1973)

6.153 R.T.KU, E.D.HINKLEY, J.O.SAMPLE, L.W.CHANEY, W.A.McCLENNY: 68th Annual Meeting of the Air Pollution Control Association, Boston, Mass. (1975) Paper 75–56.5

6.154 L.W.CHANEY, W.A.McCLENNY, R.T.KU: 68th Annual Meeting of the Air Pollution Control Association, Boston, Mass (1975) Paper 75–56.6

6.155 F.Allario: EPA Symposium on Long-Path Techniques, Research Triangle Park, North Carolina (February 1975)

6.156 B.CHRISTOPHE, D.CAMUS: IEEE/OSA Conf. Laser Engineering and Applications, Washington, D. C. (May 1975) Paper 9.1

6.157 Z.KUCEROVSKY, E.BRANNEN, K.C.PAULEKAT, D.G.RUMBOLD: J. Appl. Meteorol. **12**, 1387 (1973)

6.158 D.H.EHHALT, L.E.HEIDT, R.H.LUEB, N.ROPER: *Proc. 3rd Conf. on CIAP,* ed. by A.J.BRODERICK and T.M.HARD (U. S. Department of Transportation, Washington, D. C. 1974) pp. 153–160

6.159 *American National Standard for the Safe Use of Lasers* (American National Standards Institute, New York 1973), Z 136.1-1973;
see also, "Laser Products," in Federal Register **39**, No. 172, 4 September 1974 (Dept. of Health, Education, and Welfare)

6.160 H.G.HÄFELE: Appl. Phys. **5**, 97 (1974/75)

7. Laser Heterodyne Detection Techniques

R. T. MENZIES

With 9 Figures

Heterodyne radiometers, using lasers as local oscillators, are useful for detecting weak radiation signals which have a narrow spectral width. These signals can be, for example, thermal radiation from gases, laser radiation which has been scattered from gas molecules or aerosol particles, or laser radiation transmitted from a distant source or reflected from a remote surface. Detection of each of these types of signals is useful in probing the atmosphere.

As we have seen in the earlier chapters, most of the atmospheric pollutants interact with radiation at wavelengths in the infrared, where heterodyne detection has several advantages compared with direct photo-detection. The spectral range of remote infrared sensors is restricted to certain "window" regions described in Chapter 3, spectral areas where strong absorption bands of water vapor and carbon dioxide do not limit transmission. Fortunately, almost all of the pollutants absorb somewhere in the window regions, and interferences from water vapor, carbon dioxide, and other pollutants themselves can be reduced, especially if the monitoring instrumentation has high spectral resolution. The heterodyne receiver responds to a very narrow spectral range which is determined by the local oscillator frequency and the bandwidth of the IF electronics, not by a conventional optical filter or dispersive element. Background radiation and intrinsic photodetector current fluctuations do not represent fundamental noise limitations. The narrow field of view of the heterodyne receiver, determined by spatial phase matching requirements, can be a disadvantage when the signal source is extended. This is especially true in the near-infrared and visible regions.

The application of heterodyne detection to remote monitoring of atmospheric constituents was discussed rather recently [7.1–4]. Since that time, several laboratory demonstrations have supported theoretical performance estimates. At the present time, only a few practical atmospheric measurements have been accomplished using heterodyne techniques. However, the rapid evolution of improved components and the ongoing accumulation of experience in the practical matters of atmospheric monitoring point toward many interesting measurements in the near future.

Heterodyne radiometry principles are discussed in Section 7.1 with an historian's sense of order, beginning with the principles as applied to radio receivers. Heterodyne techniques have been used in this spectral region for several decades, and some of the concepts can be translated directly to the optical wavelength region. We then move to a discussion of the concepts which are unique to the optical region. Included in the discussion are the various sources of noise which must be overcome in order to achieve ultimate sensitivity, and the wavelength dependence of the sensitivity to thermal radiation.

Various applications of passive heterodyne radiometry are dealt with in Section 7.2. Included are discussions about component requirements, examples of experimental pollutant gas detection, and descriptions of possible ground-based and airborne monitoring modes. Before employing the heterodyne radiometer for a specific task, one should ask if it can do the job better than other available radiometers. We have made a brief comparison of this sort in Subsection 7.2.4 with another type of radiometer which has already proved its usefulness in high resolution studies: the scanning Michelson interferometer. Although the treatment of most topics in Section 7.2 is quite general, several specific examples, including the experimental examples, deal with radiometers which employ gas laser local oscillators. Gas lasers have proven to be desirable local oscillators in the sense that quantum-limited heterodyne sensitivity can be achieved when they are used. Other types of lasers should become useful local oscillators in the near future; further discussion of this point can be found in Section 7.4.

Section 7.3 deals with the role heterodyne techniques can play in active monitoring systems, such as those based on differential absorption. Most of the discussion deals with bi-static systems which use a non-cooperative, remote scattering surface to provide a return signal. Results of experimental monitoring of ambient ozone and nitric oxide are related. We also describe how an airborne bistatic system can be used to obtain altitude profiles of an atmospheric constituent.

The heterodyne receiver is restricted to spectral regions in which suitable laser local oscillators exist. We shall discuss in Section 7.4 the present spectral limitations and the future improvements concomitant with the use of tunable laser sources which are currently under laboratory investigation.

7.1 Principles of Heterodyne Radiometry

7.1.1 Thermal Radiation

A radiometer is often used to detect thermal radiation from a source at temperature T. Even if the radiation from the source is generated by non-thermal mechanisms, it can often be approximated as thermal radiation at an equivalent temperature over a limited frequency interval. Thus it is appropriate to present a few basic properties involving thermal radiation. The thermal radiation energy density inside a black-body enclosure at temperature T, according to the Planck formula [7.5], is

$$\varrho(f)df = (8\pi f^2/c^3)\frac{hf}{\exp(hf/kT)-1}df, \tag{7.1}$$

where $\varrho(f)$ is measured in units of $\text{J m}^{-3}\,\text{Hz}^{-1}$, h is Planck's constant $(6.63 \times 10^{-34}\,\text{J s})$, f is the frequency of the radiation in Hz, c is the velocity of light $(3 \times 10^8\,\text{m s}^{-1})$, and k is Boltzmann's constant $(1.38 \times 10^{-23}\,\text{J K}^{-1})$. This expression can be obtained by multiplying the average number of photons in a mode k, with frequency f, $\bar{n}_k = [\exp(hf/kT)-1]^{-1}$, by the photon energy hf, and the number of modes per unit volume in the frequency interval df, which is $(8\pi f^2/c^3)df$. Often a more convenient expression to work with is the intensity or brightness,

$$B(f, T) = (2hf^3/c^2)\,[\exp(hf/kT)-1]^{-1}\,[\text{W cm}^{-2}\,\text{Hz}^{-1}\,\text{sr}^{-1}]. \tag{7.2}$$

The thermal radiation fluctuates inherently, and these fluctuations cause noise in the radiometer. The theory of fluctuations of a Bose gas in thermal equilibrium [7.5, 6] states that the average number of bosons in quantum state k, with energy E_k, is

$$\bar{n}_k = \{\exp[(E_k-\mu)/kT-1\}^{-1}, \tag{7.3}$$

where μ is the chemical potential, and the mean square fluctuation is

$$\overline{(\Delta n_k)^2} = \overline{(n_k-\bar{n}_k)^2} = (kT)\,(\partial\bar{n}_k/\partial\mu) = \bar{n}_k(1+\bar{n}_k). \tag{7.4}$$

A photon is a special kind of boson with $E_k = hf$ and μ equal to zero, thus (7.4) applies to thermal radiation fluctuations. We may sum over a group of G_q adjacent states (in a frequency interval Δf), containing a total of $N = \sum_k n_k$ particles. Because the particles we are discussing are independent and do not interact (in a dynamical sense), the fluctuations of populations in each state k are statistically independent. Thus

$$\overline{(\Delta N)^2} = G_q\bar{n}(1+\bar{n}) = \bar{N}(1+\bar{N}/G), \tag{7.5}$$

i.e., the mean square fluctuation of the sum, N, is equal to the sum of the individual mean square fluctuations. Here $\bar{n}$ is the common value of the adjacent $\bar{n}_k$, and $\bar{N} = \bar{n}G_q$. The quantity G_q for photons is the number of modes per unit volume in the frequency interval Δf, $G_q =$

$8\pi f^2 \Delta f/c^3$. We can thus write (7.5) in more explicit form

$$\overline{(\Delta N)^2} = \bar{N}\{1 + [\exp(hf/kT) - 1]^{-1}\}$$
$$= \bar{N}[1 - \exp(-hf/kT)]^{-1}. \tag{7.6}$$

The radiation transfer equation must be used to determine the meaning of the signal to which a radiometer responds. It can be developed by using Kirchhoff's law in addition to the Planck radiation formula. According to Kirchhoff's law and the principle of detailed balance, a sample of matter absorbs as much as it radiates at each frequency f when it is in thermodynamic equilibrium. If a slab of matter with thickness dz exists between a source of brightness B_s and a radiometer, the change in brightness due to this slab, as seen by the radiometer, is

$$dB_s(f) = B_i(f)\alpha(f)dz - B_s(f)\alpha(f)dz, \tag{7.7}$$

where the measurement is at frequency $f(=cv)$, $\alpha(f)$ is the absorption coefficient per unit thickness, and B_i is the intrinsic brightness of the slab, obtained by the use of (7.2), where T is its kinetic temperature. Equation (7.7), when integrated, with the boundary condition $B(f) = B_s(f)$ when $z = 0$, gives a brightness $B(f)$ at the radiometer [7.7, 8]

$$B(f) = B_i(f)\{1 - \exp[-\alpha(f)L]\} + B_s(f)\exp[-\alpha(f)L], \tag{7.8}$$

where L is the total thickness of the intervening matter. This expression is applicable when the medium is uniform, i.e., the brightness does not vary with z. It is useful as an approximation in many cases.

7.1.2 The Dicke Microwave Radiometer

In the radio and microwave spectral regions, $hf \ll kT$ for atmospheric temperatures, and the thermal radiation formulas simplify to well known forms. The Planck distribution becomes the classical Rayleigh-Jeans expression,

$$\varrho(f)df = (8\pi f^2 kT/c^3)df, \tag{7.9}$$

which can be obtained by giving each mode an average energy equal to kT. A look at (7.3), with $E_k = hf$ and $\mu = 0$, tells us that $\bar{n}_k \gg 1$, and this implies, according to (7.4), that $\overline{(\Delta n_k)^2} \approx \bar{n}_k^2$. The fluctuation energy, or noise energy, per mode is then

$$E_N = hf\left[\overline{(\Delta n_k)^2}\right]^{1/2} = hf\bar{n}_k = kT, \tag{7.10}$$

which is the same as the average energy per mode. As a result, a radio-meter in the radio or microwave region can be characterized by an antenna noise temperature T_A which depends on the background within its field of view. Additional losses in the antenna and transmission line, and other sources of noise in the receiver, will add to this, producing a system noise temperature T_s referred to the antenna terminals [7.8, 9].

The Dicke radiometer configuration [7.10] allows one to measure a temperature change ΔT which is much smaller than T_s, the system noise temperature. In a typical superheterodyne receiver, the incoming radiation within a certain frequency bandwidth is amplified, mixed with a local oscillator to convert it down to a lower frequency region, and further amplified with IF amplifiers. The IF section is the same regardless of the frequency of the signal; the receiver is tuned by tuning the local oscillator frequency. The IF amplifier output power is proportional to $(T_s + \Delta T)B_{IF}$, where B_{IF} is the frequency bandwidth. This IF output power is then detected and filtered to produce a voltage. Since the pre-detection bandwidth is B_{IF} the coherence time of the IF amplifier output waveform is $1/B_{IF}$. This radiation can be pictured as consisting of a sequence of independent noise pulses, arriving at the mean rate of B_{IF} per second. We can apply Poisson statistics to this pulse train and state that the variance in the number counted during an integration time τ is equal to $B_{IF}\tau$. Thus, assuming $\Delta T \ll T_s$, the system noise dc voltage level at the output of the receiver is proportional to $B_{IF}\tau$, the mean count. The root-mean-square (rms) fluctuation in this voltage is $\sqrt{B_{IF}\tau}$. The value of ΔT which equals the rms fluctuation is the minimum detectable signal,

$$(\Delta T)_m = T_s/(B_{IF}\tau)^{1/2} . \tag{7.11}$$

The result of this heuristic argument is the same as that obtained using a more formal approach involving linear transform techniques [7.11]. We have assumed the use of an ideal integrator, whose gain as a function of frequency is

$$G(f) = \sin^2(1/4\pi f \tau)/(1/4\pi f \tau)^2 . \tag{7.12}$$

The equivalent rectangular postdetection bandwidth of this (low-frequency) integrator is

$$B_{LF} = [\int_0^\infty G(f)df]/G(0) = (2\tau)^{-1} . \tag{7.13}$$

The equivalent post-detection bandwidths and integration times for other common low-pass filters are given in [7.8, 7.11].

Gain variations in the predetection part of the heterodyne receiver will degrade the sensitivity, since the dc voltage level due to the system noise cannot be nulled exactly. Output fluctuations due to gain variations are statistically independent from the fluctuations due to the system noise temperature T_s. Thus the actual sensitivity of the radiometer is given by a minimum temperature change $(\Delta T')_m$,

$$(\Delta T')_m = T_s[1/B_{IF}\tau + (\Delta G/G)^2]^{1/2} , \tag{7.14}$$

where $\Delta G/G$ is the fractional gain variation. Often the quantity $1/B_{IF}\tau$ is a very small number, e.g., 10^{-8}. Thus very slight gain variations can drastically reduce the radiometer sensitivity.

The Dicke radiometer minimizes the adverse effects of gain variations by switching the receiver input between the signal source and a reference source which is at a known temperature near the source temperature. The switching frequency is high enough so that the gain variation during one switching period is negligible. In practice, a frequency above 30 Hz is usually adequate. In the Dicke radiometer, the IF signal is amplified and detected, then a narrow-band amplifier which operates at the switching frequency is inserted behind the IF detector. The amplified ac signal is subsequently demodulated in synchronism with the switch and then passed through the integrator.

The sensitivity of the Dicke radiometer is slightly different from (7.11) due to the switching process. In the standard case, the signal power is observed only half the time, which increases $(\Delta T)_m$ by a factor of two. If the post-detection bandpass filter and amplifier respond only to the first harmonic of the square-wave signal, there is a further reduction in efficiency. The fraction of a square wave signal which is contained in its first harmonic is $4/(\pi\sqrt{2})$. Thus the minimum detectable signal becomes

$$(\Delta T)_m = \pi T_s/(2B_{IF}\tau)^{1/2} . \tag{7.15}$$

The degradation due to observing time inefficiency can be reduced by switching the antenna terminals between two Dicke-type radiometers [7.8]. Then the signal source is being observed at any given time by one of the two receivers. Adding the two independent receiver outputs will improve the sensitivity by a factor of $\sqrt{2}$.

7.1.3 Photomixing in the Infrared

There is a physical difference between the standard mixing processes in the radio and optical regions. In the radio region, the nonlinear response of the mixing diodes to the total incident electric field produces the mixing action, but these diodes are not perfect square-law detectors. The response of the charge carriers to the electric field of the radiation contains sizeable terms of other orders. The photo-detectors commonly used in the infrared respond as square-law devices to a much better approximation. The sensitive photodetectors in the infrared respond to radiation by producing charge carriers from bound states. These charge carriers are then free to move in an electric field which exists in the detector material, and which is different from the

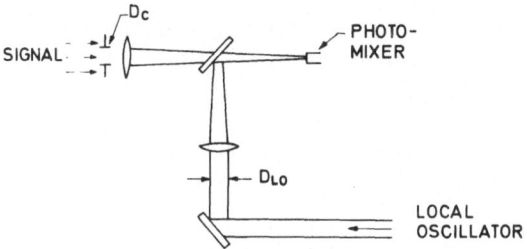

Fig. 7.1. An optical diagram of a heterodyne radiometer. D_c and D_{LO} are diameters of the collecting aperture and LO beam, respectively

electric field of the radiation. When infrared radiation is incident on the photodetector, the probability of producing free charge carriers is pro-portional to the square of its electric field, according to first-order, time-dependent perturbation theory [7.12]. Higher-order (multi-photon) transitions can occur in the photodetector, but at normal local oscillator power levels, these transition probabilities are smaller by several orders of magnitude.

The spatial coherence requirement in photomixing gives the infrared heterodyne radiometer antenna properties, in similarity with radio-wavelength radiometers. The phase fronts of the signal and local oscil-lator must match at the photomixer in order to obtain maximum mixing efficiency. For the heterodyne configuration shown in Fig. 7.1, the focal spot sizes of the signal and local oscillator beams at the photo-mixer should be nearly equal, and they should overlap in order to achieve optimum mixing efficiency. In this case, the receiver field of view will be $\Omega_r \simeq \lambda^2/D_c^2$, where λ is the radiation wavelength to which the radiometer is sensitive. If the local oscillator spot size is increased by reducing D_{LO}, a larger field of view can be obtained while maintaining overlap at the photomixer. However, the mixing efficiency is decreased, for the signal coming from any given direction within the field of view will mix with only a part of the local oscillator, while the entire local oscillator power will induce shot noise in the receiver. (A discussion of noise mechanisms is given in Subsection 7.1.4.) SIEGMAN [7.13] has demonstrated that for any heterodyne receiver configuration, the sensitivity of the receiver can be calculated by assuming it has an integrated effective aperture

$$\iint A_e(\Omega)d\Omega \simeq \lambda^2 , \tag{7.16}$$

where λ is the wavelength of the radiation to which the receiver is sensitive. This relation is equivalent to the well-known antenna theorem at radio frequencies [7.9].

When a heterodyne radiometer is viewing a remote source of thermal radiation and there is no absorption in the intervening path, the received power per unit bandwidth will be

$$dP_r/df = (1/2) \iint B(\theta, \phi) A_e(\theta, \phi)d\Omega , \tag{7.17}$$

where $B(\theta, \phi)$ is the brightness of the source in $[\text{W cm}^{-2}\,\text{Hz}^{-1}\,\text{sr}^{-1}]$ and the direction $\theta, \phi = 0$ corresponds to the axis defined by the local oscillator beam in Fig. 7.1. The factor of 1/2 is due to the fact that the radiometer responds to only one polarization. If the brightness function is angle-independent within the receiver field of view, $B(\theta, \phi)$ is given by (7.2), multiplied by an emissivity factor $\varepsilon(f)$. Then the received power will be

$$P_r = \frac{hf}{\exp(hf/kT) - 1} \int_{f_{LO}-B_{IF}}^{f_{LO}+B_{IF}} \varepsilon(f)df . \tag{7.18}$$

In this expression, T is the source temperature, hf is the quantum energy of the radiation to which the receiver is responding, f_{LO} is the local oscillator frequency, and B_{IF} is the IF bandwidth. The emissivity of the source is equal to its absorptance, the ratio of absorbed radiation intensity to the incident radiation intensity at frequency f. In terms of the extinction coefficient $\alpha(f)$, the emissivity is $\varepsilon_f = 1 - \exp(-\alpha L)$, L being the source thickness. In practice, the frequencies of IF amplification range from a few hertz to several hundred megahertz when the radiometer is viewing thermal radiation from the atmosphere or from hot gases. Then the range of integration is not strictly correct, but it is a good approximation. If a narrow IF bandwidth is used with a non-negligible lower frequency, appropriate modifications to the integral must be made. The emissivity factor is frequency-dependent because often a gas emission line which is being observed might have a line-width comparable to or smaller than B_{IF}. From (7.18) we note that the spatial coherency requirements of the photo-mixing process allow the radiometer to respond to only one mode of the thermal radiation field.

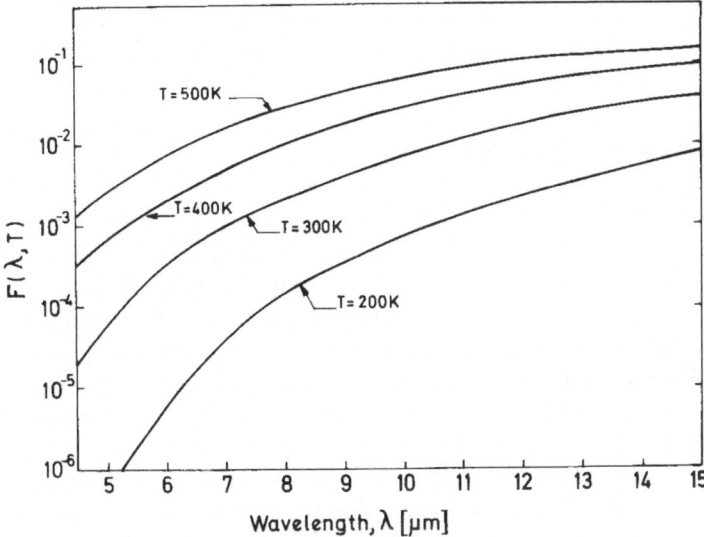

Fig. 7.2. Wavelength dependence of the sensitivity of the heterodyne radiometer to thermal radiation. The function plotted is $F(\lambda, T) = [\exp(hc/\lambda kT) - 1]^{-1}$, where $f = c/\lambda$, and c is the velocity of light, see (7.18, 54) in text. The temperatures indicated are those of the gases within the radiometer's field of view

If the radiometer is viewing a more complicated situation, in which radiation from a remote source is being partially absorbed and re-emitted by a medium with a different temperature, then a radiative transfer equation must be used to determine the received power. In many cases (7.8) can be used. The received power (7.17) will consist of two terms: one due to the remote source, modified by a transmittance factor, and the other due to the intervening medium. In some cases the temperature of the medium might be negligible compared with that of the source, and the re-emitted radiation term can be neglected. This would be the case in viewing solar radiation. On the other hand, if lunar radiation were observed, atmospheric emission would contribute substantially, even if atmospheric absorption amounted to only a few per cent.

Although the frequency dependent expression $[\exp(hf/kT) - 1]^{-1}$ in (7.18) varies slowly enough to be placed outside the integral, it produces a large effect in determining the received signal power for various wavelengths of operation. This factor is plotted in Fig. 7.2 for several temperatures and wavelengths in the infrared. We shall point out later that when the heterodyne radiometer is operating at maximum sensitivity, its noise-equivalent-power (NEP) is proportional to hf. Thus the function plotted in Fig. 7.2 determines the wavelength dependence of the signal-to-noise ratio. It is obvious that the heterodyne radiometer, when viewing thermal radiation, is much more sensitive at the longer wavelengths.

7.1.4 Signal-to-Noise Considerations

The Local Oscillator Fluctuation Limit

In this chapter we are discussing heterodyne detection of radiation which can usually be described as Gaussian in a statistical sense [7.14]. Since neither the signal nor the local oscillator can be accurately described as pure sinusoidal waves, a statistical treatment of heterodyne detection is relevant. We shall use the framework of classical coherence theory to develop the signal and noise expressions in the mixer photocurrent. The development is similar to that of CUMMINS and SWINNEY [7.15].

A short discussion of the fluctuation properties of optical signals is appropriate at this time. When laser radiation is scattered from a diffuse reflector or from aerosols, the scattered radiation intensity at a remote point can be described by the statistical probability distribution of a random Gaussian process [7.14]

$$p(I)dI = (\bar{I})^{-1} \exp(-I/\bar{I})dI ,\qquad (7.19)$$

where the bar denotes an ensemble average or an average over a large number of sampling times t_s, where t_s is shorter than the coherence time of the radiation ($t_s \ll 1/\varDelta f$, the reciprocal bandwidth). When this type of signal is studied by means of a photon-counting experiment, the probability of detecting n photoelectrons of one spatial and polarization mode in time t_s is [7.16, 17]

$$p(n, t_s) = (\bar{n})^n/(1+\bar{n})^{n+1} ,\qquad (7.20)$$

where $\bar{n}$ is the mean count. This is the geometric, or Bose-Einstein distribution. With this type of signal, the variance in the photocount is

$$\overline{(\varDelta n)^2} = \bar{n} + (\bar{n})^2 .\qquad (7.21)$$

This result is equivalent to (7.4), the fluctuation in occupation number of a single mode of thermal radiation. The first term on the right-hand side of (7.21) is called the shot noise term, and the second term is often called the "excess noise" term.

In heterodyne detection analysis, the local oscillator power is usually assumed to be large enough that its fluctuations determine the noise level.

How can its intrinsic fluctuation level be described? Equation (7.21) should not be used to calculate laser fluctuations, since laser radiation is certainly not equivalent to thermal radiation. The properties of laser radiation should ideally be similar to those of a coherent state. When a single mode of radiation is in a coherent state, the probability that it contains n quanta is given by the Poisson distribution [7.18], and the variance in the occupation number is

$$\overline{(\Delta n)^2} = \bar{n}. \tag{7.22}$$

Since photocurrent measurements are phase insensitive, a light source which produces a constant intensity field will also behave statistically according to (7.22). It can be shown that the mean square fluctuation in the photocount number for a stationary field (i.e., its statistical properties are time independent) is given by [7.17]:

$$\overline{(\Delta n)^2} = \bar{n} + \eta^2 \overline{(\Delta U)^2}, \tag{7.23a}$$

$$U(T, t) = \int_t^{t+T} I(t')dt'. \tag{7.23b}$$

where η is proportional to the photodetector quantum efficiency, and U is a measure of the light intensity I integrated over the counting time T. This expression is valid for arbitrary counting times. The constant η relates the (ensemble) average photocount to the (ensemble) average intensity $\bar{n} = \eta \bar{I} T$ when T, the counting time, is much shorter than the coherence time of the light. Equation (7.23a) is a sum of contributions from the fluctuations of classical particles (Poisson behavior) and the fluctuations of classical wave fields. Thus a laser which is amplitude-stabilized will produce shot noise but very little "excess noise". Thermal radiation, according to (7.3) and (7.4), exhibits similar fluctuation properties when $h\nu \gg kT$. On the other hand, when $h\nu \ll kT$, thermal radiation fluctuations can be accurately described in terms of classical fields.

The experiments of FREED and HAUS [7.19, 20] indicate that for a well-designed, single-mode gas laser operating far above threshold, the excess noise due to intensity fluctuations is negligible when compared with the shot noise. A single-mode laser acts as a Van der Pol oscillator, suppressing intensity fluctuations when it is highly saturated. Extraneous noise from, e.g., power supply and plasma instabilities, can cause noticeable fluctuations at certain frequencies. These intensity fluctuations are not fundamental, however, and in many cases these sources of difficulty can be avoided with careful experimental design.

We now present the expressions from classical coherence theory which are useful in determining the power spectrum of the photomixer output current in a heterodyne experiment. If we define a complex analytic function describing the optical field at the photomixer, such that $E^*(t)E(t)$ is the intensity $I(t)$, then according to time-dependent perturbation theory, the rate of photocarrier generation from the photomixer is

$$W^{(1)}(t) = FE^*(t)E(t),$$ (7.24)

where F is a factor containing the quantum efficiency. We assume that spatial coherence exists throughout the active volume of the photomixer, so that the spatial dependence is dropped. The photocurrent at time t is then $i(t) = eW^{(1)}(t)$. The joint probability per unit time that one photocarrier will be generated at time t and another will be generated at time $t + \tau$ is

$$W^{(2)}(t, t + \tau) = F^2 E^*(t)E(t)E^*(t + \tau)E(t + \tau).$$ (7.25)

The averages of these quantities, assuming stationary fields, are

$$\overline{i(t)} = eF\bar{I}$$ (7.26a)

$$\overline{W^{(2)}(t, t + \tau)} = F^2 \bar{I}^2 g^{(2)}(\tau),$$ (7.26b)

where $g^{(2)}(\tau)$ is the normalized second-order correlation function, and $\bar{I}$ is the average intensity of the optical field. (We assume that these random variables are ergodic, so that time averages and ensemble averages are equivalent). The power spectrum of the photocurrent is given by the Wiener-Khintchine theorem

$$P_i(\omega) = (2\pi)^{-1} \int_{-\infty}^{\infty} \exp(i\omega\tau)C_i(\tau)d\tau$$ (7.27)

where

$$C_i(\tau) = \overline{i(t)i(t + \tau)} = e^2 [\overline{W^{(1)}(t)W^{(1)}(t + \tau)}]$$ (7.28)

is the current autocorrelation function, and $C_i(\tau) - C_i^*(-\tau)$. The photocurrent $i(t)$ may be represented by a sequence of photocarrier pulses which are localized at random times t_k, k being a label for the photocarrier:

$$i(t) = e \sum_k \delta(t - t_k).$$ (7.29)

The autocorrelation function $C_i(\tau) = e^2 [\overline{\sum_k \sum_j \delta(t - t_k)\delta(t + \tau - t_j)}]$ can be split into two terms. If $k = j$, we are considering the same photocarrier at t and $t + \tau$; and if $k \neq j$, we are considering distinct photocarriers at times t and $t+\tau$. Then

$$
\begin{aligned}
C_i(\tau) &= e^2\overline{[\sum_k \delta(t-t_k)\delta(t+\tau-t_k)]} + e^2\overline{[\sum_k \sum_j \delta(t-t_k)\delta(t+\tau-t_j)]} \\
&= e^2[\overline{W^{(1)}(t)}]\delta(\tau) + e^2[\overline{W^{(2)}(t,t+\tau)}] \\
&= e^2 F\bar{I}\delta(\tau) + e^2 F^2 \bar{I}^2 g^{(2)}(\tau) .
\end{aligned}
\tag{7.30}
$$

Suppose the optical field incident on the photodetector is a constant-amplitude field with random phase modulation. This is a suitable model for a single-mode gas laser far above threshold [7.16]. Then the complex field function is

$$
E(t) = E_0 \exp\{-i[\omega_0 t + \phi(t)]\} .
\tag{7.31}
$$

This field is characterized by the autocorrelation function

$$
C_E(\tau) = \overline{E^*(t)E(t+\tau)} = \bar{I}g^{(1)}(\tau) ,
\tag{7.32}
$$

where $g^{(1)}(\tau)$ is the normalized first-order correlation function. The optical spectrum of the field is given by the Fourier transform of $\bar{I}g^{(1)}(\tau)$,

$$
\begin{aligned}
I(\omega) &= (\bar{I}/2\pi) \int_{-\infty}^{\infty} \exp[i(\omega-\omega_0)\tau] \\
&\quad \cdot \overline{\{\exp[i\phi(t)]\exp[-i\phi(t+\tau)]\}}dt .
\end{aligned}
\tag{7.33}
$$

This correlation function will decay with decay time τ_c, producing an optical power spectrum with half-width equal to $1/2\pi\tau_c$. The photocurrent spectrum is given by (7.30), where $g^{(2)}(\tau)$, using (7.25) and (7.26b), is found to be equal to unity. Thus the photocurrent spectrum is

$$
P_i(\omega) = e\bar{i}/2\pi + \bar{i}^2\delta(\omega) .
\tag{7.34}
$$

Since $P_i(\omega)$ is defined for both positive and negative values of ω, and it is symmetric about $\omega=0$, we can combine positive and negative frequency parts to obtain the power spectrum for positive frequencies only. Then

$$
P_i(\omega)|_{\omega\geq 0} = e\bar{i}/\pi + \bar{i}^2\delta'(\omega) .
\tag{7.35}
$$

[The integral of $\delta'(\omega)$ over all frequencies $\omega\geq 0$ is equal to unity]. We see in this case that the photocurrent power spectrum contains a shot noise term and a dc term. No excess noise term appears even though the optical power spectrum, given by the Fourier transform of $\bar{I}g^{(1)}(\tau)$, has a finite width due to phase fluctuations. This result can be attributed to the lack of phase sensitivity in the photodetection process, and it is equivalent to the statement made earlier that a constant intensity source produces no excess fluctuation term in a photocount experiment.

If the optical field incident on the photodetector is composed of a random Gaussian signal field and a constant-amplitude local oscillator field, then we are considering a typical heterodyne detection process. The photocurrent autocorrelation function in this case is

$$C_i(\tau) = e^2 F \delta(\tau) \left[\overline{E^*(t)E(t)} \right] + e^2 F^2 \left[\overline{E^*(t)E(t)E^*(t+\tau)E(t+\tau)} \right], \quad (7.36)$$

where $E(t) = E_S(t) + E_{LO}(t)$. Let $E_{LO}(t)$ be described by (7.31). Then when the first term of the right-hand side (rhs) of (7.36) is expanded, the two non-zero terms are equal to

$$e^2 F \delta(\tau) (I_{LO} + \bar{I}_S). \quad (7.37)$$

When the second term on the rhs of (7.36) is expanded, there are only six non-zero terms. Three of these terms are time-independent, and their sum is

$$e^2 F^2 (I_{LO}^2 + 2 I_{LO} \bar{I}_S). \quad (7.38)$$

The three time-dependent terms are

$$e^2 F^2 I_{LO} \exp(i\omega_0\tau) \{ \overline{\exp[-i\phi(t)] \exp[i\phi(t+\tau)]} \} \left[\overline{E_S^*(t)E_S(t+\tau)} \right]$$
$$+ e^2 F^2 I_{LO} \exp(-i\omega_0\tau) \{ \overline{\exp[i\phi(t)] \exp[-i\phi(t+\tau)]} \} \left[\overline{E_S(t)E_S^*(t+\tau)} \right]$$
$$+ e^2 F^2 \left[\overline{E_S^*(t)E_S(t)E_S^*(t+\tau)E_S(t+\tau)} \right]. \quad (7.39)$$

We have assumed that the phase of the local oscillator and the amplitude of the signal field are independent random processes, so that their averages can be taken separately. If $I_{LO} \gg \bar{I}_S$, we can neglect the second terms of (7.37) and (7.38) and the third term of (7.39). Then the current autocorrelation function becomes

$$C_i(\tau) = e i_{LO} \delta(\tau) + i_{LO}^2$$
$$+ i_{LO} \bar{i}_S \left[\exp(i\omega_0\tau) \{ \overline{\exp[-i\phi(t)] \exp[i\phi(t+\tau)]} \} g_S^{(1)}(\tau) \right.$$
$$+ \exp(-i\omega_0\tau) \{ \overline{\exp[i\phi(t)] \exp[-i\phi(t+\tau)]} \} g_S^{(1)}(\tau) \right]. \quad (7.40)$$

The power spectrum of the photomixer current, using (7.27) and the convolution theorem, is given by

$$P_i(\omega)|_{\omega \geq 0} = (e i_{LO}/\pi) + i_{LO}^2 \delta'(\omega) + (i_{LO} \bar{i}_S/\pi) \left[F(\omega) \otimes S(\omega) \right], \quad (7.41)$$

where

$$F(\omega) = \int_{-\infty}^{\infty} \exp[i(\omega_0 - \omega)\tau] \, \overline{\{\exp[-i\phi(t)] \exp[i\phi(t+\tau)]\}} d\tau, \quad (7.42a)$$

$$S(\omega) = \int_{-\infty}^{\infty} \exp[i(\omega_0 - \omega)\tau] g_S^{(1)}(\tau) d\tau, \quad (7.42b)$$

and $\otimes$ indicates the convolution. In certain cases the signal correlation function $g_S^{(1)}(\tau)$ equals $[\exp(-i\omega_S\tau)] [\exp(-\gamma|\tau|)]$. In these cases, $S(\omega)$ is a Lorentzian function centered at (circular) frequency $\omega = |\omega_S - \omega_0|$

$$S(\omega)|_{\omega \geq 0} = (2\gamma)/[(\omega - |\omega_S - \omega_0|)^2 + \gamma^2]. \quad (7.43)$$

This spectrum will be convoluted with the local oscillator field's spectrum, according to (7.41). If the local oscillator is a stable laser, its spectrum will normally have a width γ of several tens of kilohertz [7.21].

Thus the photocurrent power spectrum of the heterodyne receiver contains a shot noise term, equal to $2ei_{LO}B$, where B is the frequency bandwidth, a dc term equal to i_{LO}^2, and a mixing term whose integrated power equals $2i_{LO}\bar{i}_S$. Note that with the assumption of a constant amplitude local oscillator, there is no excess noise term. Using the relation

$$\bar{i}_S = (e\eta/hf)P_S, \quad (7.44)$$

where P_S is the signal power and η is the photomixer quantum efficiency, we find that the value of P_S for which the mixing term power equals the shot noise power is

$$(P_S)_{min} = (hf/\eta)B_{IF}. \quad (7.45)$$

This is the noise-equivalent-power (NEP) for an ideal heterodyne receiver with IF bandwidth B_{IF}. If the receiver is used as a Dicke radiometer, then its NEP is (7.45) multiplied by the radiometer factor $\pi/(2B_{IF}\tau)^{1/2}$, as in (7.15). The Dicke heterodyne radiometer NEP is thus:

$$NEP = (\pi\sqrt{2}) (hf/\eta) (B_{IF}/\tau)^{1/2}. \quad (7.46)$$

The expression (7.46) for the NEP of a heterodyne radiometer must be multiplied by a factor of 2 if the photomixer is a photoconductor. In a photoconductor, the generation and recombination processes are related through the principle of detailed balance. The total fluctuations are the same as if there were independent generation and recombination rates, each having full shot noise [7.22].

Let us return briefly to the effects of non-fundamental local oscillator noise, caused by power supply instabilities, for example. These intensity fluctuations are usually at frequencies below a few kilohertz, and if the IF bandpass is at much higher frequencies, they will not contribute to the IF noise. In the Dicke radiometer mode, the chopping frequency should be chosen so that it does not overlap ripple or oscillation frequencies. According to (7.41), the shot noise is proportional to the local oscillator current. The minimum detectable signal is nearly $(B_{IF}\tau)^{-1/2}$ times this noise level. If phase-sensitive detection is used in the heterodyne radiometer, then heterodyne sensitivity will be degraded by a local oscillator power fluctuation $\Delta P_{LO}/P_{LO} \geq 1/(B_{IF}\tau)^{1/2}$, which is within the frequency interval $\Delta f \simeq 1/\tau$ about the chopping frequency, provided the fluctuation is nearly in phase with the signal. Occasionally this criterion is met in practice, but local oscillator fluctuations within the pass-band of the tuned amplifier preceding the synchronous detection circuit are large enough to cause saturation or overload. It is evident that one must pay careful attention to these extraneous sources of noise in order to eliminate their effects on heterodyne sensitivity.

Additional Noise Sources

The local oscillator power must be large enough so that the shot noise (or generation-recombination noise) which it induces is much larger than other sources of noise in the radiometer. These other sources of noise might be:
1) shot noise induced by the radiation from the signal source within the optical bandwidth,
2) excess noise induced by the same signal source radiation,
3) shot noise induced by background radiation,
4) thermally induced shot noise within the photomixer,
5) Johnson noise in the photomixer, load resistor, and IF amplifier circuit,
6) low frequency (current) noise due to photomixer contact and surface effects.

These noise sources are common in direct photodetection at infrared wavelengths, and detailed descriptions of their origins and characteristics can be found elsewhere [7.22–26].

Normally the total optical power from the signal sources is much lower than the local oscillator power, even if the spectral bandpass of the optics is large and the photomixer responds over a large range of wavelengths. Then noise sources 1) and 2) can be dismissed immediately. In solar heterodyne radiometry this may not be true, especially at wavelengths in the visible or near-infrared regions. In this case, a combination of apertures and optical filters might be necessary to reduce the total solar power incident on the photomixer. If the sun is the signal source, then for infrared wavelengths the excess noise term in (7.21) cannot be neglected when compared with the shot noise term. The solar brightness temperature is around 5000 K in the infrared [7.27]. From (7.3), we find that at 5 μm wavelength, $hf/kT \simeq 0.5$ and $\bar{n}_k^2 \simeq 1.5\,\bar{n}_k$. The average occupation number of each mode is large enough so that classical wave interference effects become important.

The effects of background radiation and thermal shot noise can be minimized by using a small photomixer. If the quoted value of D_λ^* for a particular photodetector is limited by either of these two noise sources, then it can be used to determine the local oscillator power necessary to dominate them. The noise-equivalent-power due to local oscillator fluctuations is $(P_{LO}/hf)^{1/2}(hf)$, where hf is the quantum energy at the local oscillator frequency. The noise-equivalent-power due to background radiation or thermally-generated photocarrier fluctuations is equal to $\sqrt{A}/D_\lambda^*$, where A is the photodetector cross-sectional area [7.22–26]. Thus the necessary local oscillator power is dictated by the following relation

$$P_{LO} > A/[hf(D_\lambda^*)^2]\,, \tag{7.47}$$

where $f = c/\lambda$. Consider a photomixer with a 14-μm long-wavelength cutoff, used in a heterodyne radiometer at 10 μm wavelength. If its value of D_λ^* is limited by radiation fluctuations from a 300 K background in a 2π steradian field-of-view, then $D_\lambda^* = 3 \times 10^{10}$ cm Hz$^{1/2}$/W [7.23–25, 28]. If the photomixer cross-sectional area is 10^{-2} cm^2, then, from (7.47), P_{LO} must be greater than 5×10^{-4} W. Much smaller photomixers can be used at 10 μm wavelength, since focal spot areas with optics of moderate size are less than 10^{-3} cm^2.

When a cooled infrared photomixer is used, the chief source of noise which must be overcome by the local oscillator is often the IF pre-amplifier noise. If the photomixer is copper-doped germanium, for example, the thermally induced generation-recombination noise is negligible at the low temperature of operation [7.25]. The Johnson noise power of the load resistor can be effectively reduced by cooling it along with the photomixer. However, when large IF bandwidths are required, an IF amplifier with 50 ohm input impedance is convenient, if not necessary. The large mismatch in this case between the high impedance extrinsic photoconductor and the amplifier causes inefficient transfer of IF power. Thus the equivalent input noise of the amplifier looks very large, and large local oscillator powers (30–100 mW) are necessary to achieve near quantum limited heterodyne performance. If a smaller IF bandwidth is adequate, the mismatch can be reduced by using a high input impedance preamplifier and a coil which resonates with the parallel capacitance.

For heterodyne applications which require large IF bandwidths, high-speed $p-n$ junction $Cd_xHg_{1-x}Te$ photomixers require the least amount of local oscillator power to overcome other noise sources. Studies in the 10 μm region indicate that nearly quantum-limited performance can be achieved over a 1 GHz bandwidth with a local oscillator power of a few milliwatts [7.29]. The dynamic impedance of the junction can be as low as 100 ohms at a suitable operating point on the voltage-current curve; thus much better impedance matching to wideband IF amplifiers can be achieved. The $p-n$ junction devices are not hindered by slow photocarrier transit times, which is another performance limitation associated with the use of high-speed photoconductors. Recent advances in the fabrication of $Cd_xHg_{1-x}Te$ photodetectors have produced uniform spectral response over large regions from 2 to 13 μm, with low noise operation at temperatures around 100 K [7.30, 31].

We conclude that a few milliwatts of single-mode local oscillator power are sufficient to achieve quantum-limited performance, even in very wide bandwidth heterodyne radiometers. Many infrared lasers can meet this requirement. In practice, the local oscillator stability requirements are often more difficult to meet.

7.2 The Passive Heterodyne Radiometer

7.2.1 Gas Emissivities in the Infrared

Pollutant gases, which we define here as all atmospheric constituents except nitrogen, oxygen and the noble gases, almost always have absorption bands in the infrared if they exist in molecular form. Gases radiate at their resonant absorption frequencies, and since thermal radiation is strongest in the infrared for atmospheric temperatures, this is the ideal spectral region for viewing pollutants via their thermal radiation.

In Subsection 3.1.2 the absorption characteristics of a single spectral line in the vibration-rotation band of a gas were described. The strength of a line is described in terms of its integrated intensity

$$S = \int_{-\infty}^{\infty} \sigma(v)dv, \tag{7.48}$$

where $\sigma(v)$ is the molecular absorption cross-section [cm^2] at wavenumber v. The quantum-mechanical expression for intensity of a single transition from one molecular state to another is given by (3.24) in terms of transition center frequency v_{ij}, dipole matrix element R_{ij}, lower level degeneracy g_j, and lower level fractional number density N_j/N. This latter term may be written as

$$N_j/N = (g_j/Q)\exp(-E_j/kT),\qquad(7.49)$$

where E_j is the lower level energy and Q the total partition function. Substituting (7.49) into (3.24) yields

$$S_{ij} = (8\pi^3 v_{ij}/3hQ)|R_{ij}|^2[1 - \exp(-hcv_{ij}/kT)]\exp(-E_j/kT).\qquad(7.50)$$

One can notice from (7.50) that S is a function of the gas temperature. Often the factor $[1 - \exp(-hcv_{ij}/kT)]$, which takes into account stimulated emission from molecules in the upper level of the transition, can be neglected since $hcv_{ij} \gg kT$ for many transitions.

The shape of an infrared absorption, when collisions are the predominant cause of broadening the resonance (pressures $\gtrsim 0.1$ atm), is described by the Lorentzian lineshape

$$\sigma(v) = \frac{S}{\pi}\frac{\gamma_L}{(v - v_0)^2 + \gamma_L^2},\qquad(7.51)$$

where γ_L is the line halfwidth as defined by (3.18). The dependence of γ_L on pressure and temperature can generally be expressed as

$$\gamma_L = \gamma_L^0(p/p_0)(T_0/T)^m,\qquad(7.52)$$

where p_0 and T_0 are usually 1 atm and 300 K, respectively. The value of m, according to simple kinetic theory, is 0.5 (cf. (3.20) and [7.7]); but, in reality, it depends somewhat on the absorbing gas. At 1 atm ambient pressure and 300 K ambient temperature, γ_L ranges from 0.01 to 0.10 cm^{-1}, depending on the gas absorption line. At lower background pressures, Doppler broadening plays a more important role, and the lineshape is best described by a Voigt profile. The Voigt lineshape is the convolution of a Gaussian (Doppler) lineshape and a Lorentzian lineshape [7.7], which can be expressed in terms of the real part of the error function for complex arguments, $w(z)$ [7.32]

$$\sigma(v) = (S/\gamma')\,\mathrm{re}\{w(x + iy)\},\qquad(7.53)$$

where $\gamma' = \gamma_D(\pi\ln 2)^{1/2}$, $x = (v - v_0)/\gamma'$, $y = \gamma_L/\gamma'$, γ_D being the Doppler halfwidth defined earlier by (3.22).

Line intensities for typical infrared absorption lines which are suitable in remote sensing range from 5×10^{-21} to 2×10^{-19} cm^{-2}/molec-cm^{-3}. At temperatures around 300 K and ambient pressures near one atmosphere, corresponding absorption coefficients range from 1 to 50 atm^{-1} cm^{-1}. At higher temperatures, the gas molecules populate many more energy levels. Thus the absorption spectrum is populated with a greater number of lines, and the intensities of the strong lines are diminished.

The sensitivity of a heterodyne radiometer to a radiating gas within its field of view may be computed by using the expressions for the received power (7.18), and the noise-equivalent-power (7.46). The resulting signal-to-noise ratio (S/N) is

$$S/N = (\sqrt{2}/\pi)\eta(\tau/B_{IF})^{1/2} \left[\exp(hf/kT) - 1\right]^{-1} \int_{f_{LO} - B_{IF}}^{f_{LO} + B_{IF}} \varepsilon(f) df. \quad (7.54)$$

This expression applies to back-biased $p-n$ junction photomixers. The signal-to-noise ratio is a factor of two less when a photoconductor is used. As mentioned earlier, if the radiation originates from a pollutant gas of small optical depth, $\varepsilon(f) \simeq \alpha(f)L$, where $\alpha(f)$ is its extinction coefficient, and L the total pathlength through the gas. Using Fig. 7.2, we find that a heterodyne radiometer with $\eta = 0.5$, $\tau = 10$ s, $B_{IF} = 10^9$ Hz, $\lambda = c/f = 10$ μm will respond with $S/N = 1$ to an amount of radiating gas at 300 K for which $\bar{\varepsilon}(f) \simeq \bar{\alpha}(f)L \simeq 10^{-2}$. The bar denotes an average over the range of integration in (7.54). If the radiometer is responding to a gas emission line for which $\bar{\alpha}(f)L \simeq 20$ atm^{-1} cm^{-1}, then the minimum detectable quantity of gas is 5×10^{-4} atm^{-1} cm^{-1}, or 5 parts per million in a one-meter path. This is typical of the sensitivity to pollutant gases which radiate near 10 μm.

The high resolution capability of the heterodyne radiometer is very important in remote sensing. If the heterodyne IF bandwidth is 1 GHz (0.033 cm^{-1}), the spectral resolution is nearly the same as typical linewidths for gases in the lower atmosphere. For decreasing background pressure in the pressure-broadened (Lorentz) regime, the absorption line narrows; for a fixed concentration of absorber, its absorption coefficient near line center is inversely proportional to background pressure (see Subsection 7.2.5). As the IF bandwidth is decreased to match the decreasing line half-width, the heterodyne radiometer sensitivity to the absorber, for a given mixing ratio times path depth, decreases as $(B_{IF})^{1/2}$. This decrease is quite small when compared with other radiometers, where at best the sensitivity is proportional to the linewidth. High resolution is valuable not only for sensitivity, but also in minimizing interference. There are many spectral regions in the so-called atmospheric windows where several pollutant absorption bands overlap. LUDWIG has illustrated the severity of this problem in a recent feasibility study of remote sensing from airborne platforms [7.33]. His examples, which represent by no means a complete list, point out the need not only for high-resolution sensing instruments, but also for accurate tables of line strengths and frequencies for all atmospheric constituents.

7.2.2 Component Requirements

Local Oscillator

A glance at (7.54) will evoke the obvious primary requirement of the local oscillator: it must have an output frequency which overlaps a suitable emission line of the pollutant of interest. It is important to have lasers available which emit at wavelengths in the atmospheric window regions, and, more specifically, at wavelengths which allow the detection of a single pollutant with a minimum of interference from other pollutants.

The local oscillator must also have the stability necessary for the heterodyne radiometer to achieve its maximum sensitivity. This is an especially important requirement for passive detection of thermal radiation. A substantial increase in noise-equivalent-power above that given in (7.46) will seriously hamper the passive radiometer's effectiveness as an atmospheric sensing device. The local oscillator amplitude-stability requirements were discussed in Subsection 7.1.4. Equations (7.41–43) can also be used to assess the effects of local oscillator phase fluctuations and frequency drift. Phase fluctuations are not a serious problem unless they produce a broadening of the "beating" spectrum which is larger than the IF bandwidth. Frequency drift of the local oscillator should not be significant compared with the half-width of the thermal emission line of the pollutant.

We stated previously that single-mode gas lasers can be constructed to meet the local oscillator stability requirements. One drawback of gas lasers is their limited tunability. On the other hand, lasers which exhibit greater spectral flexibility have not as yet met the necessary stability requirements. Advances in the technology of gas, injection diode, and other solid-state lasers will certainly improve matters. At present, many atmospheric measurements can be conducted with the relatively fixed frequency gas lasers, since numerous spectral coincidences between pollutant absorption lines and gas laser lines have been found [7.2, 34–36].

Figure 7.3 portrays the spectral regions in which CO_2 and CO gas laser emission frequencies exist, along with an overview of several pollutant absorption bands. Suitable spectral overlaps have been found with most of the pollutants shown; one obvious exception is carbon monoxide.

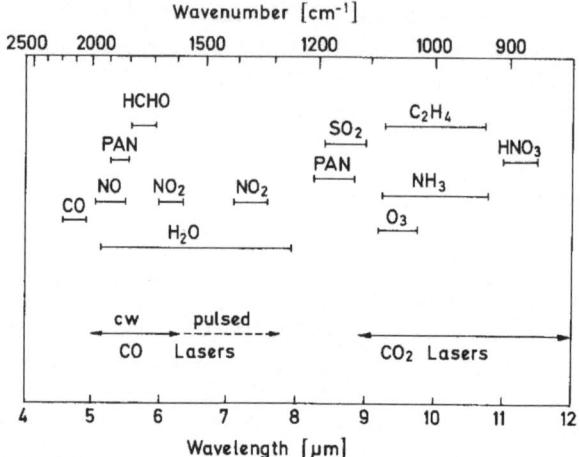

Fig. 7.3. A spectral plot of regions in which various pollutants absorb and CO_2 and CO lasers emit

The oxides of nitrogen react with other combustion products in the formation of photochemical smog [7.37, 38]. On a global scale, they participate in aerosol formation, which is being studied in order to determine its impact on the global radiation budget [7.39]. Oxides of nitrogen in the lower stratosphere can affect the ozone belt [7.40], and measurements of nitric oxide in this region have recently received considerable attention [7.41–43]. Formaldehyde is a product of photochemical smog, and it is also involved in the methane oxidation chain, which is now considered as the primary source of carbon monoxide in the troposphere [7.44–45]. The peroxyacyl nitrates (PANs) are also formed in photochemical smog, and they cause severe plant damage and eye irritation [7.46]. Ammonia and sulfur dioxide react to form a toxic aerosol in urban environments [7.47], and which has also been observed in the stratosphere [7.48]. Nitric acid vapor has been found in the stratosphere [7.49], where it plays an important role in the photochemical reactions which maintain the relative mixing ratios of the stratospheric constituents [7.50]. Ozone is a well-known end product of photochemical smog and an important stratospheric constituent. It is now believed that ozone formation in the troposphere is predominantly due to the methane oxidation chain [7.51].

Several stratospheric trace constituents, which do not appear in Fig. 7.3, have received considerable attention lately, and a number of them have absorption bands in the spectral region where CO_2 lasers emit. There is concern about the propellant solvents for aerosol dispensers (Freons) which, when transported into the stratosphere, are decomposed by solar ultraviolet radiation into odd chlorine, a potent catalyst for the destruction of ozone [7.104]. Major components in the chlorine reaction scheme are HCl and ClO. The spectra of the Freons and ClO are currently under investigation in the 9–12 μm region with laser sources.

Photomixer

The photomixer in a passive heterodyne radiometer must have the necessary spectral response, high quantum efficiency, and high frequency response. Pollutants in the lower troposphere have emission line half-widths around 2 GHz, and to maximize the integral in the S/N expression (7.54), the IF bandwidth should be nearly the same. Copper-doped germanium photoconductors respond with high quantum efficiency over a wide range of wavelengths. When they are compensated with antimony donors, they are capable of flat frequency response in excess of 1 GHz [7.25]. Mercury-cadmium telluride $p-n$ junction photomixers have also been fabricated with flat frequency response to 1 GHz [7.52, 31]. These have been the commonly used photomixers in passive heterodyne radiometry.

The concept of heterodyne conversion gain is useful in determining the factors involved in making a passive heterodyne radiometer achieve quantum-noise-limited sensitivity [7.53, 29]. When shot noise due to background radiation and thermal effects is not important, the noise equivalent power of the heterodyne receiver can be expressed as

$$\text{NEP} = hfB_{\text{IF}}/\eta + k(T_M + T_{\text{IF}})B_{\text{IF}}/G, \tag{7.55}$$

where T_M and T_{IF} are the photomixer temperature and effective input noise temperature of the IF amplifier, and G is the heterodyne conversion gain. The first term in (7.55) is twice as large when the mixer is a photo-conductor. The conversion gain is normally defined as the ratio of the available IF output power at a (circular) frequency ω, to the input signal power, P_S, at a frequency $\omega_S = \omega_{\text{LO}} + \omega$,

$$G = \overline{i^2(\omega)}R_0/4P_S. \tag{7.56}$$

Here R_0 is the mixer IF output resistance, and $\overline{i^2(\omega)}$ is the mean square IF photocurrent at frequency ω

$$\overline{i^2(\omega)} = (\eta e/hf)^2 (P_{\text{LO}}P_S)(\tau/t_r)^2/(1 + \omega^2\tau^2). \tag{7.57}$$

The quantity hf is the local oscillator quantum energy, τ is the (principal) photocarrier lifetime, and t_r is the photocarrier transit time.

If the heterodyne conversion gain is defined as in (7.56), then it is independent of the local oscillator power when a high impedance photoconductor, such as Ge:Cu, is used. The expression for mixer output resistance in this case is

$$R_0 = (hf/\eta e P_{LO})(\ell^2/\mu_h \tau),\qquad(7.58)$$

where ℓ is the interelectrode spacing and μ_h is the hole mobility. The input impedance of the wideband IF amplifier is near 50 ohms. R_0 is of the order of one kilohm for Ge:Cu slabs with $\ell = 0.25\,\text{mm}$, when 50 mW of local oscillator power is incident. This presents a large mismatch, and the actual IF output power is far from that given by (7.56). The analysis in [7.53] accounts for this mismatch by using a value for T_{IF} in (7.55) which is equal to $(R_0/4R_{IF})(T_{IF})_{\text{actual}}$.

A slight modification in the definition of heterodyne conversion gain for the case of the Ge:Cu photoconductor leads to a clearer picture of the dependence of heterodyne radiometer sensitivity on various factors. We shall define G as

$$G = \overline{i^2(\omega)} R_{IF}/P_S,\qquad(7.59)$$

where R_{IF} is the IF amplifier input impedance (nominally 50 ohms). In this case, T_{IF} in (7.55) is the IF amplifier equivalent input noise temperature. Using (7.56, 57), we find that G is proportional to P_{LO} and to the square of the ratio τ/t_r. Since τ is of the order of 10^{-10} s and fixed, it is important that the Ge:Cu photomixer have a very short transit time. Mixer capacitance is another consideration when small mixers are fabricated. The length of the mixer is limited by the infrared absorption coefficient of the material, which is near $6\,\text{cm}^{-1}$ for a typical copper concentration of $7 \times 10^{15}\,\text{cm}^{-3}$ [7.25]. A Cu:Ge mixer of 3 mm length and square cross section has a material capacitance of 0.4 pF.

Limitations on the achievable values for τ/t_r in a high-speed Ge:Cu mixer make the values of G in (7.59) much less than unity. If we consider, for example, a photomixer with $\ell = 0.25\,\text{mm}$, and apply a voltage of 10 volts, then, using a value for $\mu_h \simeq 2 \times 10^4\,\text{cm}^2/\text{volt}\cdot\text{s}$, we obtain a transit time $t_r \simeq 3 \times 10^{-9}$ s. (The applied field cannot be much greater than 400 volt/cm in practice because severe low-frequency noise often appears. There has also been evidence of a drop in hole mobility at higher fields.) Thus $(\tau/t_r)^2 \simeq 10^{-3}$. If 100 mW of local oscillator power at 10 μm wavelength is incident on the mixer, and $\eta = 0.5$, then from (7.59) we obtain a heterodyne conversion gain,

$$G \simeq 0.125 = -9\,\text{dB}.\qquad(7.60)$$

If $T_M \simeq 10$ K, and $T_{IF} \simeq 300$ K ($kT_{IF} \simeq 0.025$ eV), then at 10 μm wavelength ($hf = 2 \times 10^{-20}$ J $= 0.125$ eV) the noise-equivalent power, according to (7.55), is

$$\text{NEP} \simeq 1.25\,(2hf/\eta)B_{IF}\,. \tag{7.61}$$

We use the factor of two in the quantum noise expression because the mixer is a photoconductor. Under these conditions, the theoretical sensitivity limit is nearly achieved.

The high speed mercury-cadmium telluride photodiodes which have recently been used in infrared photomixing experiments [7.29, 31, 52, 54] can be designed such that the equivalent output resistance is around 100 ohms at the operating condition, thus producing a good match to the IF amplifier. The heterodyne conversion gain, using definition (7.56), is [7.29]

$$G = (P_{LO}/2G_D)(\eta e/hf)^2/[1+(\omega/\omega_c)^2]\,, \tag{7.62}$$

where G_D is the incremental shunt conductance of the photodiode, and ω_c is the rolloff frequency,

$$\omega_c = G_D^{1/2}/C_D R_s^{1/2}\,, \tag{7.63}$$

C_D being the junction capacitance and R_s the series resistance. Similarities between (7.62) and the equivalent expression for the Ge:Cu photoconductor are obvious. The only significant difference is the extra factor of $(\tau/t_r)^2$ in the photoconductor gain. This equals 10^{-3} for nearly optimum operating conditions when a Ge:Cu photomixer with $\tau \simeq 10^{-10}$ s. is used. Thus, much reduced local oscillator powers can be sufficient to achieve quantum-limited sensitivity with the photodiode. For one which has a quantum efficiency near 0.5, values of P_{LO} near a few tenths of a milliwatt should be sufficient. Recent experimental studies of photodiode mixer capabilities [7.29] and heterodyne detection of blackbody radiation [7.54] have indicated that $P_{LO} \simeq 1$ mW is sufficient to achieve nearly quantum-limited sensitivity with mixers whose quantum efficiency is near 0.25.

7.2.3 Experimental Heterodyne Detection of Gases

A heterodyne radiometer, with a CO_2 laser local oscillator and a 600 MHz IF bandwidth, has been used to remotely detect several laboratory samples of gaseous pollutants at ambient temperatures [7.55, 56] at the author's Laboratory. These gases were ozone O_3, ammonia NH_3, ethylene C_2H_4, and sulfur dioxide SO_2. Detection of carbon dioxide at a temperature of 450 K was also accomplished with the CO_2 laser local oscillator.

For detection of all the gases except SO_2, a stable laser with Invar cavity spacers and polished granite end plates was used as the local oscillator. The laser was operated in a sealed-off mode. A grating in the Littrow configuration served as one end reflector, and the 10% transmission output mirror was bonded to a cylinder made of piezo-electric ceramic for fine-tuning capability. The amplitude stability of this laser met the requirements discussed in Subsection 7.1.4; commercial dc power supplies with good stability were used. The laser frequency drift was very small; after tuning it to a line center, the laser would remain there for hours, and turning it off and on again would not affect it.

For detection of SO_2, a commercial grating-tuned laser with the rare isotope $C^{12}O_2^{18}$ was used. Laser operation with this isotope can take place at shorter wavelengths, and we used the shortest available output wavelengths, $R(40)$ and $R(42)$ of the 9.2 μm band, to interact with SO_2. These lines are at 1107.948 cm^{-1} and 1108.923 cm^{-1}, respectively [7.57]. The same laser was also used to detect ozone. Substantial frequency drift was common with this laser when it was first turned on; after about one hour, equilibrium was reached, and the output frequency was stable at the center of a particular line after proper adjustment of the piezo-electric transducer (PZT). The beat frequencies observed when the two lasers were mixed had spectral widths of the order of 10^5 Hz when averaged over a several second period.

A CO laser, operating near 5.2 μm, has also been used as a local oscillator for heterodyne detection of a nitric oxide NO laboratory sample. The NO was detected while the gas was at a temperature of 390 K. The CO laser was also sealed off and stabilized with Invar rod spacers. The discharge tube was cooled with a dry-ice/methanol closed-cycle system which provided an outer wall temperature around 200 K. Line tuning was accomplished with a Littrow-mount diffraction grating, and a PZT was used for fine frequency tuning.

A high-speed, copper-doped germanium photoconductor was used as the mixer. We started with antimony donor material. Copper was evaporated onto this material, and, after indiffusion, the Cu concentration was about 7×10^{15} cm^{-3}. The copper and antimony concentrations gave the material an infrared absorption coefficient near 6 cm^{-1}, and a photocarrier lifetime near 10^{-10} s. (See [7.25] for information about Ge:Cu material parameters.) Several mixer elements were fabricated. Pieces were sawed, then etched and polished on their optical faces. They were then each mounted between two platinum strips with indium solder. A fine platinum lead wire was also attached to one of the strips in the same process. The high-voltage breakdown characteristics of each assembly were studied at liquid He temperature. Sizeable variations were observed, probably due to slight variations in the mounting process. We found that a quick dip in either the germanium etch or HF solution, followed by cleaning with high purity methanol, improved the breakdown characteristics of some assemblies. The mixer used for most of the gas detection experiments had dimensions $0.3 \times 0.4 \times 3.5$ [mm].

The local oscillator and thermal radiation signals were combined using a sodium chloride wedge beam splitter, and both beams were then focussed onto the mixer. When a local oscillator power of 40 mW was incident on the mixer, its dc resistance dropped to about 2000 ohms. (The resistance did not appear to be a strong function of applied voltage below the breakdown level.) The beat frequencies between 10 and 600 MHz were amplified by a 50 ohm input amplifier chain. The IF signal from the mixer was passed through a strip-line-mounted dc blocking capacitor, into the first IF amplifier, a low-noise, commercially available unit. The equivalent input noise temperature of the IF amplifier chain was approximately 250 K. The IF gain varied no more than 2 dB over the entire frequency range. When a bias power of 40 mW was applied to the mixer, the quantum noise induced by the local oscillator was enough to raise the total noise level to 3–5 dB above that of the IF amplifier chain over the 10 to 600 MHz region. These were the normal operating conditions of the heterodyne radiometer. At higher bias voltages, $1/f$ noise due to mixer contact effects became evident; occasionally it could be observed on a spectrum analyzer out to 100 MHz before breakdown occurred.

Most of the detected gases were individually placed inside a cell 15 cm long. When NO, SO_2, C_2H_4, and NH_3 were detected, small amounts were leaked into the cell, and then the cell pressure was brought up to atmospheric pressure by adding N_2. The ozone was detected in this cell in an O_2 background at one atmosphere. The CO_2/N_2 mixture was detected in a 75 cm long cell.

A block diagram of the experimental configuration is shown in Fig. 7.4. A reflecting chopper was used to allow the receiver to alternately view the gas cell and a cold, black surface inside a small dewar filled with liquid nitrogen. A mirror placed behind the gas cell directed the receiver field of view to the same cold reference surface. Various integration times were used, ranging up to 30 s. In each case the signal-to-noise ratio was computed using the "rule of thumb" estimate that the rms noise level is one-sixth of the peak-to-peak fluction observed over at least thirty integration times. Occasionally, the output noise levels were measured with the mixer bias off, the local oscillator power off, or the IF amplifiers off to make certain that the radiometer was

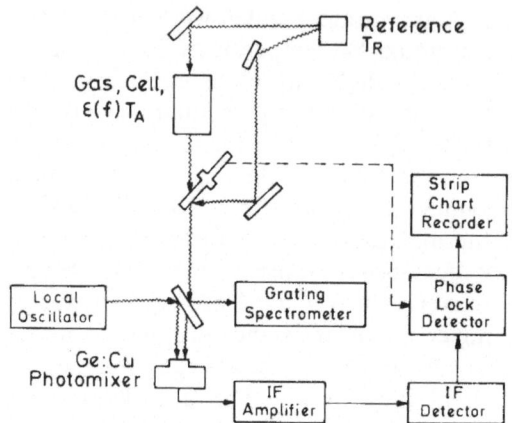

Fig. 7.4. A block diagram of the passive heterodyne gas detection experiment. $\varepsilon(f)$ is gas emissivity at temperature T_A

behaving properly and operating near its quantum-noise-limited sensitivity.

The heterodyne radiometer sensitivities to the various gases are shown in Table 7.1 along with a listing of laser local oscillator lines and the gas absorption coefficients at the indicated temperatures and frequencies. The minimum detectable amounts shown in the third column (in units of concentration times path depth) give a signal-to-noise ratio of unity when a time constant τ of ten seconds is used. As stated previously, the IF bandwidth is 600 MHz. This corresponds to a spectral resolution of $\Delta v = 2B_{IF} = 0.04 \text{ cm}^{-1}$. When the 80% transmission of the optics is taken into account, the experimental sensitivities are very close to the theoretically predicted values, assuming η is near unity.

The sensitivities in Table 7.1 show that detection of ambient concentrations in the parts per billion (ppb) range is possible with passive measurements if path lengths near one kilometer are available. This would be the case for vertical sensing if the pollutants were uniformly mixed under an inversion layer of one kilometer thickness. Then, according to Table 7.1, the minimum detectable concentration of ozone, for example, would be 2 ppb. The sensitivity to SO_2 is not as good, and without further improvements this technique would be inadequate for monitoring ambient concentrations of this gas. The sensitivity to room temperature nitric oxide emission near 5 μm is not nearly as good as the sensitivity to gases which emit at the longer wavelengths, as Fig. 7.2 depicts. However, the sensitivity to NO at elevated temperatures is adequate for monitoring concentrations in certain stationary source emissions. CO_2 monitoring can also be accomplished

Table 7.1. Measured sensitivities of fixed-frequency laser lines to pollutant gases. The gases were at 298 K, except NO, which was at 390 K, and CO_2, which was at 450 K. The background gases were either N_2 or O_2 (see text). The band designations I and II refer to the upper and lower of the mixed CO_2 (10°0, 02°0) states, respectively. (See Subsec. 4.3.2 for alternate definitions of sensitivity)

Gas	Sensitivity [atm-cm]	Laser line	Absorption coefficient [atm-cm]$^{-1}$	Wavelength [μm]
Nitric oxide	10^{-2}	$^{12}C^{16}O$:7 − 6, $P(15)$	3.5	5.19
Sulfur dioxide	10^{-2}	$^{12}C^{18}O_2$:00°1-II, $R(40)$	0.55	9.02
Ozone	2×10^{-4}	$^{12}C^{18}O_2$:00°1-II, $P(40)$	14	9.50
	2×10^{-4}	$^{12}C^{16}O_2$:00°1-II, $P(14)$	13	9.50
Ethylene	5×10^{-5}	$^{12}C^{16}O_2$:00°1-I, $P(14)$	30	10.53
Ammonia	10^{-4}	$^{12}C^{16}O_2$:00°1-I, $P(32)$	17	10.72
Carbon dioxide	2×10^{-1}	$^{12}C^{16}O_2$:00°1-I, $P(20)$	0.015	10.59

at the same time, since normal CO_2 levels in fossil-fuel burning plants are near 10%. The simultaneous monitoring capability would be very helpful if CO_2 concentrations cannot be obtained by other means.

Improved sensitivity to SO_2 and NO is possible using other local oscillator wavelengths which can be obtained with CO_2 and CO lasers. The $C^{12}O_2^{18}$ laser has been operated in a cw, sealed-off mode at R-branch lines as high as $R(58)$ in the 9.2 μm band [7.57]. If this line, at 1116.043·cm^{-1}, were used as the local oscillator frequency, the sensitivity improvement would be at least a factor of two. The 9–8 band, $P(9)$ CO laser line at 1900.050 cm^{-1} [7.58] overlaps the stronger $R(13/2)_{1/2}$ nitric oxide line. The sensitivity improvement at this frequency would be a factor of 4 at 300 K and about 2.5 at 400 K.

7.2.4 A Comparison with the Michelson Interferometer-Spectrometer

A radiometer of the scanning Michelson interferometer type is widely recognized as a valuable instrument for high-resolution spectroscopy [7.59]. It is worthwhile to make a brief comparison of the passive heterodyne radiometer with this type of instrument. Comparisons depend to a large extent on the operational circumstances; in this particular example we assume that both radiometers are sensing small spectral features in 300 K background radiation.

The signal-to-noise ratio of a scanning Michelson interferometer is [7.60]

$$(S/N)_{INT} = g_M(I_v \Delta v)(D_\lambda^*)(A\Omega\Omega_d\tau)^{1/2}, \tag{7.64}$$

where g_M is the overall transmission factor of the instrument, $I_v \Delta v$ is the flux per unit solid angle in the spectral interval Δv centered at wavenumber v, D_λ^* describes the photodetector noise, A and Ω are the collection area and solid angular field of view of the instrument, Ω_d is the photodetector field of view, and τ is the time required to make the interferogram. A maximum value for η near 0.1 can be assumed [7.60]. The angular field of view is dictated by the spectral resolution of the instrument. To achieve an apodized resolution Δv [cm^{-1}], the mirror in the scanning arm of the interferometer must move a distance $x = 1/2\Delta v$. During this motion, the total phase shift deviation between the on-axis ray and the off-axis ray at the maximum allowed half-angle must be less than π radians. This implies an effective angular diameter of $(\lambda/x)^{1/2} = (\Delta v/v)^{1/2}$. In order to obtain high sensitivity, large collecting optics and a small detector with high D_λ^* should be used. The detector size is limited by the relation $A_d \Omega_d \simeq A\Omega$, where A_d is the detector area.

In order to compare the sensitivities of the two instruments, we shall make use of the heterodyne radiometer NEP expression (7.46) and rewrite it in terms of a signal-to-noise ratio similar to (7.64)

$$(S/N)_{HET} = (2/\pi)\eta_b(I_v \Delta v/hv^3)(\tau/c^3 \Delta v)^{1/2} . \qquad (7.65)$$

We have here made use of the property that the integral of the effective area over the field of view is λ^2, see (7.16). The ratio of the two sensitivities is

$$\frac{(S/N)_{HET}}{(S/N)_{INT}} = \frac{(8/\pi)(\eta_b/hv^3)}{(\Omega_d)^{1/2}g_M(Ac^3\Delta v)^{1/2}(\Delta v/v)^{1/2}D_\lambda^*} . \qquad (7.66)$$

Observe that higher resolution (small $\Delta v/v$) and longer wavelengths of operation favor the heterodyne radiometer. Let us choose an operating wavelength region near 10 μm, with $\Delta v = 0.03$ cm^{-1}. Assume the interferometer photodetector D_λ^* equals 10^{10} cm Hz$^{1/2}$ W^{-1}. Then, using the values $g_M = 0.1$, $\Omega_d = 0.5$, $\eta_b = 0.5$, $A^{1/2} = 20$ cm, we obtain a ratio $(S/N)_{HET}/(S/N)_{INT} = 10$. For $\Delta v = 0.01$ cm^{-1}, the ratio becomes 30. It must be pointed out that, during the integration time τ, the interferometer has produced a spectrum of resolution Δv which covers a wide spectral range (equivalent to many resolution elements), while the heterodyne radiometer has responded only to a single resolution element. Thus, in this range of spectral resolution, the interferometer has a definite advantage when a wide spectrum is required, as opposed to a radiance reading at a few selected intervals. Even if a continuously-tunable local oscillator were available, which would allow the heterodyne radiometer to operate analogous to a radio frequency spectrum analyzer, its capabilities as a sensitive radiometer near 10 μm would not necessarily be as good as those of a large, well-designed interferometer spectrometer. If 0.01 cm^{-1} resolution at 10 μm is required, then in this example, the heterodyne radiometer would have the advantage only for spectra containing less than 10^3 resolution elements, since S/N is proportional to $\tau^{1/2}$.

7.2.5 Ground Monitoring Applications

A ground-based heterodyne radiometer, operating as a passive sensor of thermal radiation, can be useful for several applications. It can be used to view the sun, moon, or nearby planets, measuring atmospheric transmittance at specific wavelengths with high spectral resolution [7.61–62, 101–102]. Thermal radiation from ambient pollutants can be measured directly. Emissions from stationary sources can also be monitored from remote locations.

To discuss solar absorption measurements, we shall consider a modified form of (7.8), neglecting the re-radiation term. The incremental absorption of radiation at wavenumber v due to a gas at altitude z is given by $\sigma[v, p(z), T(z)]N(z)\sec\psi\, dz$ when the sun angle with respect to the zenith is ψ, where $p(z)$ and $T(z)$ are pressure and temperature, and $N(z)$ is the number density of absorbing gas molecules. The total optical thickness (defined in Section 3.1) is:

$$\tau(v, \psi) = \sec\psi \int N(z)\sigma(v, z)dz\,, \qquad (7.67)$$

where the integral is over the entire atmospheric thickness in the vertical direction. (We are ignoring the curvature of the earth and its atmosphere; as a result, this formulation does not apply for ψ near 90°.) The intensity measured by a radiometer with spectral resolution Δv_i at v_i will be

$$\begin{aligned} I_i(\psi) &= I_S(v_i) \int_{\Delta v} U(v) \exp[-\tau(v, \psi)]dv \\ &= I_S(v_i) \exp[-\bar\tau(v_i, 0)\sec\psi]\,, \end{aligned} \qquad (7.68)$$

where $I_S(v)$ is the solar intensity at the top of the atmosphere, varying relatively slowly with frequency, and $U(v)$ is the instrumental response function when it is centered at v_i. If one plots $\ln I_i(\psi)$ vs. $\sec\psi$, the slope gives the *zenith* atmospheric absorption over the spectral interval Δv_i.

A high-resolution solar absorption spectrometer with either GHz fine-tuning ability or selected multifrequency operation can be used to obtain altitude profiles of pollutants. From (7.51) and (7.52), we note that at sensing frequencies in the wing of a line, where $v - v_0 \gg \gamma$, the molecular absorption cross-section $\sigma(v)$ is proportional to the background pressure. At the line-center frequency, $\sigma(v_0)$ is inversely proportional to pressure. Thus by taking measurements at various values of $v - v_0$, the relative contribution of the upper-altitude molecules to the total absorption can be varied, and an altitude profile can be computed. The profile should be quite accurate up to the altitude for which Doppler broadening becomes predominant, if an atmospheric temperature profile is known with ± 5 K accuracy at each altitude. A radiometer with resolution near 15 MHz is desirable in order to obtain accurate values for the upper-altitude concentrations. The heterodyne radiometer signal-to-noise ratio in a solar-looking mode can be obtained using (7.54), with $\varepsilon = 1$ and $T = 5000$ K in the 10 μm region. An instrument with a $B_{IF}\tau$ product of 5×10^8 would produce a S/N ratio near 40 dB when looking at the sun through a perfectly transmitting atmosphere [7.61]. The subject of determining altitude profiles via high resolution spectral absorption measurements will be discussed further in Subsection 7.3.4.

The heterodyne radiometer sensitivities to pollutant gases shown in Table 7.1 indicate the capability to monitor stationary source emissions. The heterodyne sensitivity to SO_2 at wavelengths near 9 μm is better at high temperatures than at 300 K. If the SO_2 temperature were 400 K, for example, the sensitivity would be 2.5 times better. With realistic improvements, it should be possible to achieve a limiting sensitivity for 400 K SO_2 of 10 parts per million (ppm) in a one-meter path. Water vapor thermal emission in this spectral region would be equivalent to 20–40 ppm of SO_2, assuming a 12% concentration is present. Water vapor contributions to the emission at each measurement frequency should be determined by measurements at additional wavelengths.

7.2.6 Airborne Monitoring Applications

We shall discuss two modes of operation for the airborne passive heterodyne radiometer: limb observations and downward-looking, or nadir observations. In the limb observation mode, which we shall discuss very briefly, the radiometer points toward the horizon and observes either thermal emission or transmitted solar radiation. In the downward-looking mode, the radiometer observes upwelling radiation from the earth's surface and the intervening atmosphere. We restrict our discussion to the longer infrared wavelength region where upwelling thermal radiation predominates over scattered solar radiation.

Useful information about the atmospheric temperature profile and the concentration profiles of certain pollutants can be obtained with radiometers looking down from satellites, high-altitude aircraft, or balloons. The temperature profile, along with a value for the pressure at some known altitude (usually at the surface), allows the meteorologist to calculate an atmospheric pressure profile, which is of great value in studying weather patterns. The temperature profile is also a vital piece of information when radiometers are used to determine profiles of atmospheric pollutants. Temperature profiles are obtained by observing radiation from molecules which are uniformly mixed with a known mixing ratio. The molecules CO_2 and O_2, for example, are uniformly mixed at least up to 100 km altitude [7.63].

The ability to remotely measure temperature and pollutant profiles with downward-looking radiometers depends on an understanding of the intricacies of the radiative transfer equation. A discussion of this topic can be found in HOUGHTON and TAYLOR's review paper [7.63]. Consider an atmospheric slab at temperature T which contains a thickness dz of absorber in the vertical direction. At a wavenumber v, where the molecular absorption cross-section is $\sigma(v)$, Kirchhoff's law states that under conditions of local thermodynamic equilibrium, the emitted intensity per unit spectral interval from this slice in the upward direction will be $N(z)\sigma(v)B(v, T)dz$, where $B(v, T)$ is the Planck brightness function given by (7.2). The portion of this radiation which reaches the top of the atmosphere is

$$\zeta_z(v) = \exp[- \int N(z)\sigma(v)dz] , \tag{7.69}$$

where the integral is taken between the slab and the top of the atmosphere. Integrating over all such slabs, the total intensity per unit spectral interval $I'(v)$ which reaches the top is

$$I'(v) = [\int B(v, T)N(z)\sigma(v)dz] \exp[- \int N(z)\sigma(v)dz]$$
$$= \int_0^1 B(v, T)d\zeta(v) . \tag{7.70}$$

A convenient independent variable to use is $y = -\ln p$ where p is the pressure [atm]. The variable y would be equivalent to altitude in an isothermal atmosphere. In this case,

$$I'(v) = \int B[v, T(y)] (d\zeta/dy)dy$$
$$= B[v, T(y)]\kappa(y)dy . \tag{7.71}$$

The intensity per unit spectral interval $I'(v)$ is thus pictured as a weighted average of the blackbody intensity, with the weighting function being $\kappa(y) = d\zeta(v)/dy$. Thus far we have not considered the contribution to $I(v)$ due to the upwelling radiation from the earth's surface. The measured radiance at wavenumber v is also affected by the spectral resolution of the radiometer. The total radiance in a resolution element Δv is

$$I(v)|_{\Delta v} = \int_{\Delta v} U(v)dv \int_0^\infty B[v, T(y)]\kappa(y)dy$$
$$+ \int_{\Delta v} U(v)\zeta_0(v)B[v, T(0)]dv, \tag{7.72}$$

where $\zeta_0(v)$ is the transmission between the surface and the top of the atmosphere, and $U(v)$ is the instrument's spectral response profile. The upwelling thermal radiation at v is continuously absorbed and re-emitted, and the more strongly it is absorbed as it travels upward, the higher must be the level from which some of it can reach the instrument. One must measure radiation at frequencies in the wing of an absorption line in order to observe the lower atmosphere, and at line-center frequencies in order to observe the upper atmosphere. If the absorber is uniformly mixed, then the weighting functions $\kappa(y)$ are nearly independent of temperature over the normal range of atmospheric temperature profiles.

The problem to be solved if one desires a temperature profile is: Given certain observations $I(v)$ for various v, what is $T(y)$, assuming the absorber gas is uniformly mixed? Certain sounding frequency regions can be found in the v_2 band of CO_2 near 15 μm which produce weighting functions that peak at various altitudes. Since the weighting functions each have a finite width, only a limited number can be found which do not overlap significantly. Significant overlap implies that the observations at the various sounding frequencies are not independent. The techniques for solving the inverse of (7.72), to obtain a temperature profile, are numerous and complex, even if scattering and cloud effects are neglected. Due to the finite width of the weighting functions and the inevitable existence of noise in the radiance measurements, all of the techniques produce temperature profiles which are limited in altitude resolution to about 4 km in the lower troposphere and about 16 km in the stratosphere [7.63].

In order to sense temperatures within the upper atmosphere, the radiometer spectral resolution must be equal to or better than a Doppler-broadened linewidth, i.e., about 0.001 cm^{-1} (30 MHz). A low-spectral-resolution radiometer can be used to obtain data in the lower altitude region because for any spectral interval containing several absorption lines most of the intensity which reaches the instrument originates from the wings of the lines. Upper-altitude sensing, however, can be achieved only with instruments which combine high sensitivity with spectral resolution high enough so that response only to line center frequency regions is possible.

The problem of obtaining pollutant profiles (assuming the temperature profile is known) is more difficult unless a great deal of *a priori* information is available. If the radiance in (7.72) arises from a particular pollutant molecule whose mixing ratio is to be determined, the brightness functions are known, and the weighting functions $\kappa(y) = d\zeta/dy$ become functionals of $N(y)$. The mixing ratio profile is determined once $N(y)$ is known, assuming the necessary pollutant line parameters are known beforehand. The set of equations of form (7.72) for various radiance observations are manifestly nonlinear in this case, and iteration techniques must be used to obtain a solution [7.64]. The weighting functions $\kappa(y)$, and the altitudes at which they peak, depend on the unknown pollutant profile itself. For this reason, several extra sounding frequencies are needed in order to be prepared for variations in the profile.

In order to assess the capabilities of a heterodyne radiometer in both temperature and pollutant profile determination, it is appropriate to introduce the concept of minimum detectable temperature change. From the signal-to-noise expression (7.54), one can obtain $(\Delta T)_{min}$ by calculating the "normalized" signal change $\Delta S/N$, caused by a temperature change ΔT, and setting the resulting expression equal to unity. We'll assume that $\varepsilon(v)$ is unity, so that $(\Delta T)_{min}$ applies to radiation from an effective blackbody at temperature T. The result is

$$(\Delta T)_{min} = (\pi/2\eta)\,(1/B_{IF}\tau)^{1/2}\,[\exp(hf/kT) - 1]^2$$

$$\cdot\,[(hf/kT^2)\exp(hf/kT)]^{-1}. \tag{7.73}$$

This is a rather complicated function of wavelength $(\lambda = c/f)$, and temperature. The dependence on the other factors is comparatively straightforward. We have plotted $(\Delta T)_{min}$ versus wavelength, for various values of T, in Fig. 7.5. The radiometer properties which are assumed for the left-hand scale are $B_{IF}\tau = 10^{10}$, $\eta = 0.5$. This would typically correspond to a 1 GHz IF bandwidth and a 10 s integration time. These parameters are appropriate for monitoring pollutants in the troposphere from slowly moving platforms. The parameters assumed for the right-hand scale might correspond to $B_{IF} = 50$ MHz, $\tau = 2$ s. These are more appropriate for upper-altitude monitoring. If a heterodyne radiometer such as this were used to measure radiation from the 15 μm v_2 band of CO_2, in order to obtain temperature profiles, the temperature resolution at a typical stratospheric temperature of 250 K would be $\Delta T_{min} = 1$ K.

A downward-looking heterodyne radiometer, using a CO_2 laser, should be capable of obtaining profiles of a few important pollutants. CO_2 lasers operate in the 9–12 μm window region, a most favorable

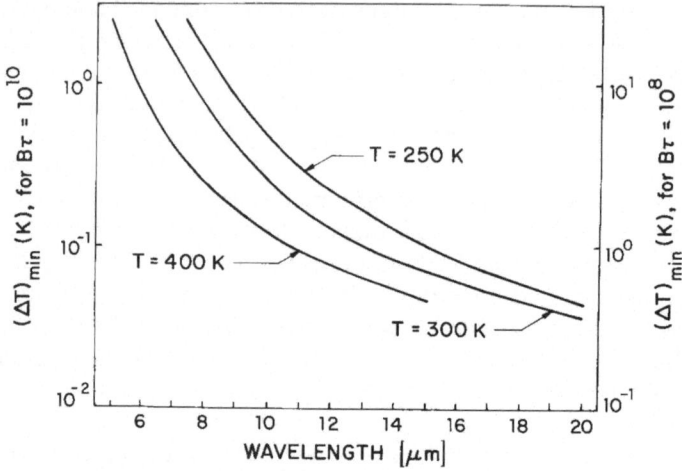

Fig. 7.5. Minimum detectable temperature change $(\Delta T)_{min}$ as a function of wavelength of operation, see (7.73) in text. The photomixer quantum efficiency assumed is $\eta = 0.5$

region for observations of the less abundant pollutants. Ludwig et al., have calculated expected changes in upwelling radiance at various wavelengths due to the presence of certain common air pollutants [7.65]. Two basic meteorological conditions can be distinguished in the troposphere: the existence of a temperature inversion with a height near 1 km, which traps pollutants underneath; and the typical falloff in temperature with increasing altitude at a nearly constant lapse rate, for which pollutant concentrations are lower but extend to higher altitudes. In the first case, a typical signal increase of a few percent at selected wavelengths can be caused by the high-temperature, trapped pollutants at normally-occurring concentrations. Sulfur dioxide, benzene, ethylene, and ammonia are typical examples which affect the 9–12 μm region. In the second case, a signal decrease of a few percent is common. Tropospheric ozone detection is also possible if measurements are made from a platform below the stratospheric band. The plots in Fig. 7.5 are useful in determining heterodyne radiometer sensitivity to signal change. For small ΔT around $T = 300$ K, the signal change in percent is nearly equal to ΔT in K. Thus a radiometer with $B_{IF}\tau = 10^{10}$ should be capable of detecting signal changes of a few tenths of one percent in the 9–12 μm region. The heterodyne radiometer is useful in minimizing interference problems because it is a high resolution device, and because measurements can be made at several selected wavelengths to quantitatively determine the effects of known interfering species. A heterodyne radiometer using a CO_2 laser is currently being developed under a NASA

Advanced Applications Flight Experiment program, and it will most likely be used to monitor one of the above-mentioned pollutants in limb and nadir experiments [7.66].

The downward-looking, passive mode of observation is, in general, insensitive to pollutants below 1 km altitude. The low-altitude pollutants exist at nearly the same temperature as the earth's surface; thus they are "invisible" to the radiometer. This mode of observation is better suited for higher altitude measurements in which atmospheric temperatures differ significantly from surface temperatures.

Limb measurements from high-altitude platforms are a useful means of studying the stratosphere. The long pathlengths through the atmosphere increase the sensitivity to pollutants with small mixing ratios. Limb observations are generally restricted to the stratosphere because of extensive cloud cover in the troposphere. Solar absorption spectra, taken with airborne radiometers, have been useful in detecting trace constituents which exist with mixing ratios near 10^{-9} [7.41, 57]. The usefulness of the sun-looking mode is limited to surveys, due to the space and time restrictions involved. The limb emission mode is more flexible.

Limb emission measurements require sensitive radiometers with high spectral and spatial resolution. Spectral resolution of the order of a typical Doppler-broadened linewidth is optimum. The Doppler half-width cf. (3.22) in wavenumbers, is

$$\gamma_D = (3.58 \times 10^{-7})v_0 (T/M)^{1/2} , \qquad (7.74)$$

where v_0 is the wavenumber of line center, T is the temperature in K, and M is the molecular weight of the molecule. For a typical 10 μm transition ($v_0 \simeq 1000$ cm^{-1}), $\gamma_D \simeq 10^{-3}$ cm^{-1}. A narrow field of view is necessary for good altitude resolution. These requirements mesh well with the properties of heterodyne radiometers.

7.2.7 Turbulence Effects on Passive Heterodyne Radiometry

The effects of atmospheric turbulence on light propagation[1] and coherent receivers is an immense subject. This discussion is confined to a few concepts which are particularly applicable to the spatial coherence requirement of heterodyne radiometry.

The heterodyne receiver responds optimally to a signal which is spatially coherent over the collection aperture. When radiation passes through a turbulent atmosphere, refractive index inhomogeneities distort

---

[1] This subject will be the topic of a forthcoming volume under the editorship of J. W. STROHBEHN.

the phase fronts. These distortions become increasingly important as the collection aperture diameter increases, but the time-averaged signal-to-noise ratio is not significantly affected if the collection diameter is less than a certain value, called the coherence diameter [7.68]. The coherence diameter is defined in terms of the wave structure function, $D_s(d)$, which is a measure of the statistical variations of the propagating wave's phase and log-amplitude at two points separated by a distance d. Based on the Kolmogoroff theory of turbulence, which predicts a spatial correlation of turbulence which decreases in proportion to the 2/3 power of spatial separation, it is possible to show that the wave structure funtion is proportional to $k^2 d^{5/3}$, where $k = |k| = 2\pi/\lambda$ [7.69]. The coherence diameter d_c is defined in terms of the relation [7.68]

$$D_s(d) = 6.88 \, (d/d_c)^{5/3} . \tag{7.75}$$

Thus the coherence diameter increases with increasing wavelength of the radiation. The structure function is also proportional to the integral of the structure constant C_n^2 over the length of the propagation path. For more detailed discussion, see Subsection 3.6.1.

The results of coherence diameter calculations indicate that turbulence is a limitation at infrared wavelengths only for large systems and long pathlengths. Reference [7.68] contains nomographs which can be used to estimate the coherence diameter over a range of wavelengths and propagation pathlengths, for both horizontal and vertical paths. At wavelengths above 5 μm, results indicate that the coherence diameter for normal daytime turbulence is larger than 30 cm, even when propagation paths are several kilometers.

FRIED has also analyzed the deleterious effects of turbulence-induced, time dependent signal variations in heterodyne receivers [7.70]. This is called atmospheric modulation noise. His results are expressed as a product of two functions: the first is the variance in total optical power collected by the receiver, which decreases as the collector aperture increases; and the second is a wavefront distortion modulation factor, appropriate to the phase sensitive heterodyne receiver. The latter function departs from unity only when the aperture diameter is larger than the coherence diameter, defined in (7.75). For collector aperture diameters less than d_c, this analysis indicates that there is no excess modulation noise in addition to the normal scintillation noise which is common to "photon bucket", or phase insensitive receivers.

Beam wander effects, due to large blobs with a perturbed refractive index passing through the beam, are an important consideration if the radiation source is small, or if it has fine spatial structure. Experiments over path lengths up to five kilometers have indicated that beam wander is nearly independent of wavelength and proportional to the square root of the path length. Wander angles of the order of 10^{-5} radians are typical in a 5 km path length [7.71, 72]. If the heterodyne radiometer is viewing a source which is uniform over dimensions corresponding to a 10^{-4} radian field of view, beam wander should not be a problem. Slow scanning mirrors can be used to reduce the effects of beam wander, since its characteristic frequencies are near 1 Hz [7.71].

7.3 Heterodyne Techniques in Active Monitoring Systems

7.3.1 Sensitivity Improvements with Heterodyning

When a heterodyne radiometer is used to detect infrared laser radiation in active atmospheric monitoring systems, it is generally several orders of magnitude more sensitive than direct detection, both for cw and pulsed transmitter modes. The increased sensitivity of the detector makes possible the use of simple, diffusely-reflecting surfaces in long path, double-ended measurements. When heterodyne detection is employed in sensing thermal radiation, the low noise level of the receiver is partially offset by the phase-matching requirement, with its concomitant field-of-view restriction. This restriction is especially severe at shorter wavelengths. A laser transmitter can be designed such that its beam is entirely within the field-of-view of the heterodyne receiver at a pre-determined region in space; thus no loss of efficiency occurs. The heterodyne receivers in active systems do not necessarily lose their value if they do not perform with nearly quantum-limited sensitivity, whereas they lose their competitive edge as passive radiometers if their performance is not optimum.

cw Transmitters

If the laser transmitter were operating in a cw mode, the Dicke-type heterodyne radiometer would be an appropriate detector. Its noise-equivalent-power was given earlier by (7.46)

$$\mathrm{NEP} = (\pi/\sqrt{2})\,(hf/\eta)\,(B_{\mathrm{IF}}/\tau)^{1/2}\ .$$

The frequency stability of infrared gas lasers is such that the width of the beat signal between the transmitter and local oscillator is less than 1 MHz, even for very long pathlength systems. Frequency broadening effects of the scattering medium and atmospheric turbulence normally amount to less than 1 MHz also. Thus an IF bandwidth of 1 MHz is reasonable for an active system. According to (7.56), in this case the Dicke heterodyne detector noise level is

$$NEP = (9 \times 10^{-17})(1/\tau)^{1/2} [W] \qquad (7.76)$$

for 10 μm wavelength operation, with $\eta = 0.5$, $B = 10^6$ Hz, and τ the integration time. If the scattering medium is the atmospheric aerosol, heavy winds might produce frequency shifts of a few MHz, and a wider IF bandwidth would be necessary.

The noise-equivalent power of a direct photodetector in the 10 μm region is given by [7.26, 28]

$$NEP = (A/\tau)^{1/2}/D_\lambda^* \qquad (7.77)$$

where A is the detector area, τ is the integration time or inverse bandwidth. Cooled, 10 μm photodetectors have values for D_λ^* near the background noise limit, $D_\lambda^* \simeq 3 \times 10^{10}$ cm (Hz)$^{1/2}$ (W)$^{-1}$. Thus, if a small area $(A \simeq 10^{-3}$ cm) detector were used in order to reduce the noise level, the result would be

$$NEP \simeq (10^{-12})(1/\tau)^{1/2} [W]. \qquad (7.78)$$

The difference between heterodyne and direct detection in this case is about four orders of magnitude.

Pulsed Transmitters

If the laser transmitter were pulsed, a straight heterodyne receiver with no synchronous demodulation might be more appropriate, in which case the noise-equivalent-power is given by (7.45)

$$NEP = (hf/\eta)B.$$

It would probably be necessary to use a separate, cw local oscillator, since pulsed lasers generally have significant amplitude fluctuations for the duration of each pulse. Let us consider a lidar system with a transmitter having pulses of 10^{-8} s duration. If we assume that a bandwidth Δf equal to twice the pulse duration is necessary to resolve the pulse, the direct detection noise-equivalent-power at 10 μm wavelength is near 1.5×10^{-8} watts, according to (7.78). From (7.45), we calculate that a heterodyne receiver with a 500 MHz bandwidth (broad enough to allow for frequency drift of the transmitter) has a much lower noise level, near 4×10^{-11} watts. Although the heterodyne advantage decreases at shorter wavelengths, where D_λ^* values rise and quantum noise also rises, there remains a significant difference down to 3 μm wavelength. Analyses of lidar systems in atmospheric monitoring applications ([7.73, 74], and Chapts. 4 and 5) usually consider direct detection in the infrared. With suitable local oscillators, the use of heterodyne detection will increase the capabilities of lidar.

7.3.2. Atmosphere-Induced Degradation of Bistatic System Performance

Turbulence Effects

Beam wander is not as severe a problem in long path, bistatic systems when a large diffuse reflector is used in place of a small mirror or cube-corner reflector. The large refractive index inhomogeneities which pass through the beam typically cause angular deviations of a few times 10^{-5} radians over a kilometer path length [7.71, 72]. This can steer part of the beam off the small cooperative reflector, and it can cause movement of the image of the reflector in the receiver optics. These beam steering effects can be reduced by employing coherent optical adaptive techniques [7.75], but these represent significant added complexity. If the return is provided by means of diffuse reflection from a rough surface, beam wander does not present the same problems. Since the light transit time is much shorter than the characteristic beam steering periods, the large scale turbulence is "frozen" while the optical beam travels along its devious course to the scattering surface and back again. If the scattering surface is larger than beam wander distances, and if the transmitter and receiver apertures are coincident or very close, the receiver field of view will follow the wandering beam spot at the scattering surface. Another problem which must be considered is caused by the fact that the laser beam exhibits spatial and temporal coherence. The resulting speckle pattern from the diffuse surface must be taken into account. If beam wander distances are large compared with the instantaneous beam spot size, numerous lobes of the speckle diffraction pattern will pass over the receiver optics, causing sizeable signal fluctuations. The probability distribution of the received intensity would likely be a Rayleigh distribution, as in laser scattering from the roughened surface of a spinning disc [7.76, 77]. If the instantaneous beam spot size at the scattering surface is much larger than beam wander distances, the scattered radiation signals at different times will remain strongly correlated, and fluctuations will be substantially reduced.

In addition to the beam wander effects, scintillations due to small scale refractive index inhomogeneities are present. ISHIMARU has developed an expression for the log-amplitude variance at a point a distance L from a transmitted laser beam, in terms of the transmitter size and beam radius of curvature [7.78, 79]. For a bistatic system which transmits a collimated beam, with transmitter diameter nearly equal to $(\lambda L)^{1/2}$, his expression gives a variance which is slightly less than that predicted from an infinite plane wave [7.80]. These expressions can be used to estimate the size of intensity fluctuations which one might expect in a long path, bistatic system. For example, the expression in [7.78] predicts that for $\lambda = 10\,\mu m$, $L = 10^3\,m$, and transmitter and collecting apertures equal to $(\lambda L)^{1/2}$, the intensity fluctuations will be around 20% if $C_n^2 = 5 \times 10^{-14}\,m^{-2/3}$. This value for the refractive index structure constant is typical for daytime turbulence. These intensity fluctuations are reduced somewhat when either the transmitter and receiver apertures or the beam spot size at the scattering surface are much larger than the Fresnel radius $(\lambda L)^{1/2}$ [7.79, 81].

The temporal frequency spectrum of amplitude and phase fluctuations for both plane and spherical waves have been calculated, using the normal assumptions involved in the diffraction theory of turbulence [7.82]. Results of this theory can be used to indicate the fluctuation spectra in bistatic, differential absorption systems. According to the Taylor hypothesis [7.82], time variations of the refractive index are mainly due to large-scale mean motions of turbulent blobs, transported across the beam by the wind. The temporal power spectrum of log-amplitude and phase fluctuations is given in terms of the quantity f_t/f_w, where f_t is the fluctuation frequency and $f_w = v_\perp (2\pi\lambda L)^{-1/2}$, the factor $v_\perp$ being the component of wind velocity perpendicular to the propagation path. The spectra fall off at frequencies larger than f_w, being very much reduced at $\ln(f_t/f_w) = 1$. To calculate what might be expected in a typical bistatic system, consider a value for $v_\perp = 5\,m/s$. Then if $\lambda = 10\,\mu m$, $L = 10^3\,m$, we find $f_w = 20\,Hz$. The scintillations which have frequency components as high as 100 Hz are extremely small. Aerosols can cause scintillation at higher frequencies, but at $\lambda = 10\,\mu m$, this effect is also very small. See Subsect. 3.6.2 for further discussion.

From this discussion one must conclude that atmospheric turbulence causes noticeable signal fluctuations, and special care must be taken to reduce their effects on the measurement. Since the scintillation frequencies are predominantly in the range 1–50 Hz, long integration times will reduce their importance. Simultaneous propagation at two or more wavelengths, with appropriate signal processing, will help to minimize turbulence effects. The word "simultaneous" in this sense means in a time scale short compared with temporal changes in the refractive index pattern.

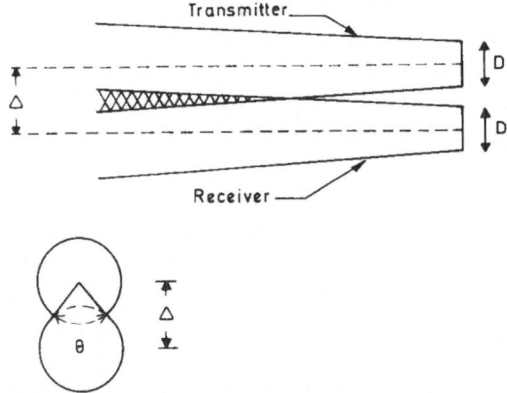

Fig. 7.6. Geometry for the calculation of aerosol backscatter intensity, see (7.79, 80). Transmitter and receiver diameters are D, and their initial separation is $\varDelta$

Aerosol Backscattering Effects

The atmospheric aerosol can interfere with a bistatic, differential absorption system by scattering the transmitted signal back into the receiver. If diffuse reflecting surfaces at long distances provide the return signal, a separation of the transmitting and receiving optics is required, unless the intervening atmosphere contains very little particulate matter. In order to calculate the power received from aerosol scattering, the geometry of the instrument's transmitter and receiver patterns must be analyzed. Figure 7.6 depicts a special case in which the transmitter and receiver aperture diameters D are equal, and the beam spread is diffraction limited. In this case, neglecting turbulence and near-field diffraction, the integrated scattered power which is collected by the receiver is

$$P_s = P_0(D^2\beta/4)\int_{R_0}^{R} \frac{\theta(r) - \sin[\theta(r)]}{r^2} \exp(-2\int_0^r \alpha dr')dr , \qquad (7.79)$$

where $R_0 = (\varDelta - D)(D/\lambda)$ is the range at which the far-field patterns begin to overlap, R is the distance to the scattering surface, P_0 is the transmitted power, β is the backscattering coefficient, and $\alpha = \alpha_a + \alpha_s$ is the total extinction coefficient. The angle θ, pictured in Fig. 7.6 is determined by the relation

$$\cos[\theta(r)/2] = (\varDelta/2)(D/2 + r\lambda D)^{-1} . \qquad (7.80)$$

If the beam patterns pictured are not parallel, then $\varDelta$ is a function of r. If the beams overlap exactly at range R, then $\varDelta(r) = \varDelta_0(1 - r/R)$, where $\varDelta_0$ is the initial separation. The integrand in (7.79) rises rapidly to a maximum at a distance for which substantial beam overlap begins to take place, and it falls off somewhat more slowly than $1/r^2$ at larger distances. If we assume, for example, $D = 10$ cm, $\Delta = 12$ cm, $\lambda = 10 \mu$m, the integrand peaks near a distance of 250 m. If $R = 1$ km, the value for the integral is about 4×10^{-3} m^{-1}, assuming the total extinction coefficient is very small.

If the aerosol backscattering coefficient β is known ahead of time, an expression such as (7.79) can be used to calculate the received signal from aerosol backscattering. Aerosol scattering coefficients have been measured at 10 μm wavelength under various visibility conditions, using a CO_2 laser system [7.83]. It was found that β ranged from 10^{-9} to 4×10^{-8} m^{-1} sr^{-1} as the visibility decreased from over 100 km to slightly less than 8 km. These results agree quite well with scattering calculations based on the Junge continental aerosol size distribution [7.48], described in detail in Chapter 4

$$N_p(a) \sim a^{-(\Lambda + 1)} \tag{7.81}$$

for $0.1 \leq a \leq 10.0$ μm, $\Lambda = 3.5$, assuming the particles' refractive index is predominantly due to water [7.84].

The ratio of received aerosol backscattered power to the received power from a remote scattering surface is

$$P_s(\text{aerosol})/P_s(\text{surface}) = (\beta \mathcal{I} R^2/4\varrho) \exp(2\int_0^R \alpha dr) \tag{7.82}$$

where ϱ is the reflectivity of the surface, and $\mathcal{I}$ the value of the integral in (7.79). Assume that the extinction coefficient is small enough so that the exponential is approximately unity. For a system with $\mathcal{I} = 4 \times 10^{-3}$ m^{-1}, $R = 10^3$ m, $\varrho = 0.01$, and for $\beta = 10^{-7}$ m^{-1} sr^{-1}, the ratio in (7.82) is about 0.01. This is near the desired upper limit for the ratio if one wishes to safely neglect the aerosol backscattering effect.

7.3.3 Pollutant Detection Experiments with a Bistatic System

We have performed a few ambient-air monitoring experiments in the Los Angeles, California area with a bistatic, long path differential absorption system, using both cube-corner and diffuse reflectors. Heterodyne detection was employed, with 10 cm transmitter and receiver aperture diameters. Two pollutants were monitored over a round trip path of nearly one kilometer: nitric oxide and ozone. The transmitter used in the nitric oxide monitoring experiments was a cw CO laser. The laser line chosen to interact with NO was the $7-6$ band, $P(15)$ line at 1927.299 cm^{-1} [7.58]. The transmitter used to monitor ozone was a cw CO_2 laser. The CO_2 laser $00°1$-II band, $P(14)$ line at $1052.194 \, cm^{-1}$ [7.85] interacts strongly with the ozone v_3 band. This line, and several other P-branch lines up to $P(24)$, were used in the experiments. Both lasers were stabilized with Invar cavity spacers and fitted with gratings for line-to-line tuning capability.

In order to appreciate the sensitivity of a heterodyne receiver in an experiment such as this, let us consider a typical example. The noise-equivalent-power of a typical Dicke radiometer is given by (7.46) and (7.76); for $\tau = 1$ s, NEP $\simeq 10^{-16}$ W. The signal return from a depolarizing, Lambertian reflector of reflectivity ϱ at a distance R from the receiver is

$$P_r = (\varrho P_0 A_r / 2\pi R^2) \exp(-2\int_0^R \alpha dr), \tag{7.83}$$

where P_0 is the transmitted power, A_r is the collecting aperture area, and α is the extinction coefficient. If $R = 500$ m, $A_r = 10^{-2} \, m^2$, and the extinction factor is nearly unity, then $P_r \simeq (6 \times 10^{-9}) \varrho P_0$. If the product ϱP_0 equals 10^{-5} watts, the receiver signal-to-noise ratio is 600.

Return laser signals from a stationary, diffusely reflecting surface at 500 m distance were detected using a heterodyne system with no frequency shift between the local oscillator and transmitted signal frequencies. Both an n-type HgCdTe photoconductor and a HgCdTe photodiode were used as mixers. This configuration is commonly called "homodyne" detection. The photocurrent power spectrum in this case can be represented by (7.41–43), where $\omega_S = \omega_{LO}$.

The detection technique used in these experiments differed somewhat from the standard technique used when $\omega_S \neq \omega_{LO}$. The transmitted signal was chopped at a frequency near 1 kHz, and the beat signal from the photomixer was passed through a bandpass amplifier tuned to the chopping frequency. Then the signal was detected with an ac voltmeter with variable output filter time constant. (Phase-synchronous detection was inconvenient because of phase fluctuations imposed on the transmitted signal by atmospheric effects.) In this case, the receiver NEP is given by (7.46), where B is the tuned amplifier bandwidth and τ is the ac voltmeter filter time constant. Since phase-synchronous detection is not used, the ac voltmeter will deflect in one direction due to the rectified noise, and the dc component of this deflection can be nulled. Gain variations degrade the sensitivity of a receiver such as this, and the true sensitivity is described by an expression similar to (7.14). Excess local oscillator noise near the chopping frequency degrades the sensitivity, since frequency components within the bandpass of the tuned amplifier add to the noise.

Nitric oxide was monitored during several days in the spring of 1973 and during several periods in 1974 [7.100]. Integration times near one

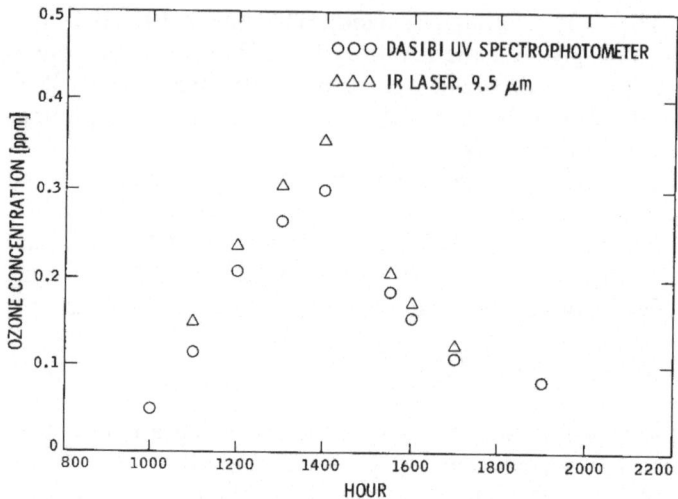

Fig. 7.7. Ozone measurements for 12 May 1975, using a 1-km round-trip path at the Jet Propulsion Laboratory, Pasadena, California. Also shown are simultaneous readings taken with a commercial point monitor for ozone (Dasibi ultraviolet spectrophotometer)

minute were usually sufficient to reduce turbulence fluctuations to the level at which a 2% differential absorption could be detected. Typical plots of the daily trend agreed with the classic pattern of NO buildup during early morning hours and subsequent decay via photochemical reactions [7.38]. Unfortunately, the accuracy of the early measurements was suspect because of the inability to accurately remove the effects of water vapor interference. The CO laser lines $7-6$ band, $P(14)$, $P(15)$, and $6-5$ band, $P(19)$ were used to monitor water vapor concentrations over the path, as well as NO concentrations. It was discovered that the water vapor absorption at these wavelengths was much larger than indicated from calculations based on spectral tables, a fact substantiated by independent measurements elsewhere [7.86]. The NO differential absorption, based on its absorption of the $P(15)$ line, amounted to 0.03 km^{-1} for an average concentration of 0.1 ppm. The typical water vapor differential absorption amounted to about 0.4 km^{-1} for average conditions (40% relative humidity, 298 K ambient temperature). During the course of these experiments, the water vapor absorption could not be determined with better than 10% accuracy. It was felt after this experience that: 1) accurate, independent measurements of water vapor absorption at various fixed laser frequencies were mandatory, and 2) the CO laser, $9-8$ band, $P(9)$ line was a better choice for nitric oxide monitoring, since the absorption coefficient at this wavelength

Table 7.2. Measured absorption coefficients for several gases at CO and CO_2 laser lines used for pollutant monitoring. The absorption coefficients listed for nitric oxide, ozone, and ethylene are in the units $(ppm\text{-}km)^{-1}$. The water vapor tabulation is given in units of km^{-1} for a partial pressure of 10 torr H_2O

Laser line	Wavelength [µm]	Absorption coefficient at 300 K			
		NO	O_3	C_2H_4	H_2O
CO:					
6–5, $P(19)$	5.164				1.86
6–5, $P(20)$	5.176	0.275			2.01
7–6, $P(13)$	5.166	0.175			1.45
7–6, $P(14)$	5.177				1.39
7–6, $P(15)$	5.187	0.250			2.16
9–8, $P(9)$	5.262	1.08			2.02
9–8, $P(14)$	5.317				2.58
CO_2:					
001-II, $P(14)$	9.504		1.25		0.11
001-II, $P(20)$	9.552		0.56		0.11
001-II, $P(24)$	9.586		0.08		0.09
001-I, $P(14)$	10.529			2.98	0.12
001-I, $P(16)$	10.549			0.46	0.12
001-I, $P(20)$	10.588			0.15	0.11

is four times that at the $7-6$ band, $P(15)$ line. The $9-8$, $P(9)$ line oscillates with adequate power at lower gas mix temperatures.

Measurements of ozone were also conducted using a CO_2 laser operating near 9.5 µm [7.100]. A typical ozone buildup during the middle of the day was observed on several occasions, as exemplified in Fig. 7.7. During the course of these measurements, independent ambient ozone readings were being taken with a calibrated ultraviolet photometer point monitor near the optical path. The agreement between the two sets of measurements was very good in a qualitative sense; however, the laser ozone readings were usually slightly higher. There was no reason to expect absolute agreement, but the discrepancy was puzzling. Careful measurements of differential absorption due to water vapor were conducted in order to eliminate this as a source of error.

Table 7.2 is a compilation of absorption coefficients due to atmospheric species at several gas laser lines which have been used to monitor pollutants. Nitric oxide, ozone, and ethylene appear in this table. Notice that water vapor absorbs at all of the listed sensing wavelengths, so that one must be aware of water vapor effects when using the infrared bistatic, differential absorption technique for pollutant monitoring. Although not listed in Table 7.2, carbon dioxide is a weak absorber at the CO_2 laser wavelengths and must also be considered. The water vapor absorption coefficients were determined using a resonant spectrophone absorption cell made of Pyrex. Use of this technique to measure water vapor absorption at CO_2 laser wavelengths has recently been reported [7.103]. Absorption coefficients of the other species listed were measured using standard cells in the differential transmission mode. Where applicable, these results agree to within a few percent with measurements made by others [7.36, 86].

7.3.4 The Airborne Laser Absorption Spectrometer

An airborne, downward-pointing laser transmitter and heterodyne receiver can be used to monitor pollutant concentrations in the vertical path between the instrument and the ground [7.87, 88]. A salient characteristic of this type of instrument, which measures differential absorption at preselected wavelengths, is the high sensitivity to low-altitude pollutants. For example, if ozone were trapped beneath an inversion layer at 1 km altitude, then an average ozone concentration of 10 ppb would produce a 1% signal change if the CO_2 laser $P(14)$ line at 1052.194 cm^{-1} were transmitted. Passive radiometric sensors are often ineffective in this application because the low-altitude pollutants are at nearly the same temperature as the earth's surface. With the use of a heterodyne radiometer, modest transmitter power is adequate.

Since the airborne laser absorption spectrometer would normally be moving over varied terrain, the received signal will fluctuate. Simultaneous transmission of at least two wavelengths is necessary—by "simultaneous", we mean on a time scale which is small compared with fluctuation times. The power spectrum of the fluctuations will depend on the velocity of the instrument and the size of the beam spot on the ground. The rolloff frequency for this and for turbulence-induced fluctuations would normally be below 100 Hz. A receiver front end with high dynamic range capability, followed by an AGC circuit, would be appropriate.

Pollutants in the lower atmosphere which absorb at infrared frequencies have absorption lines described by the Lorentz shape (7.51, 52). At higher altitudes it is necessary to use the Voigt profile (7.53). The altitude at which Doppler broadening becomes comparable to pressure broadening depends on the transition frequency and molecular weight. For the ozone 9.5-μm band, this altitude is 30 to 35 km. For the nitric oxide fundamental at 5.2 μm, it is near 20 km.

If the airborne laser absorption spectrometer uses a transmitter which spectrally overlaps an absorption line, and if the laser can be tuned over a range equivalent to the linewidth, then altitude profiles can be obtained. Alternatively, profiles can be obtained if several laser transmitter frequencies overlap different absorption lines at various frequency displacements from their line centers. When a frequency v_i is transmitted downward from an airborne platform, the return signal intensity at frequency v_i can be expressed as

$$I_r(v_i) = I_0 \exp[-2\tau(v_i)],\tag{7.84}$$

where I_0 is the transmitted intensity multiplied by surface albedo, optical collection efficiency, etc., and $\tau(v_i)$ is the one-way optical thickness due to a particular atmospheric constituent

$$\tau(v_i) = (1/g)\int_{p_s}^{\bar{p}} k_i[p, T(p)]q(p)dp,\tag{7.85}$$

where g is the gravitational force on a unit mass, p_s and $\bar{p}$ are the atmospheric pressure at the surface and at the instrument, respectively, and $q(p)$ is the mixing ratio of the constituent. The units for k_i are $cm^{-1}/gm\ cm^{-3}$.

The inverse solution to (7.85) is desired. The absorption coefficient at v_i is due to one or more absorption lines whose center frequencies must be known. For each line, the dependence of S and α on both p and T must also be known. The dependence of linewidth on T, for temperature ranges encountered in the region below 30 km altitude, is slight, and for most cases a line should be chosen whose intensity depends only slightly on temperature, i.e., the lower level energy should not be much larger than kT.

Assuming the necessary information about line strengths and widths is known beforehand, the accuracy and altitude resolution of inverse solutions to (7.85) depends on the shapes of the weighting functions, $k_i[p, T(p)]$. Information about the mixing ratio at a certain altitude should be obtained from a weighting function which peaks near that altitude. The width of the weighting functions determine the altitude resolution, as in the passive radiometric measurements discussed in Section 7.2.

In Fig. 7.8 are shown seven normalized weighting functions which were chosen for the reconstruction of a few typical ozone profiles. These weighting functions were calculated based on the AFCRL line parameter compilation [7.89]. The seven sounding frequencies which were chosen are as follows: 1) $1043.188\ cm^{-1}$, 2) $1043.186\ cm^{-1}$, 3) $1043.184\ cm^{-1}$, 4) $1043.180\ cm^{-1}$, 5) $1046.875\ cm^{-1}$, 6) $1046.870\ cm^{-1}$, 7) $1045.039\ cm^{-1}$. The first four of these sounding frequencies are near the $P(24)$ line of the $C^{12}O_2^{16}$ laser, oscillating in the 9.4 μm band. This line frequency is $1043.163\ cm^{-1}$. The next two sounding frequencies are near the $P(20)$ line at $1046.854\ cm^{-1}$, and the last one is near the $P(22)$ line at

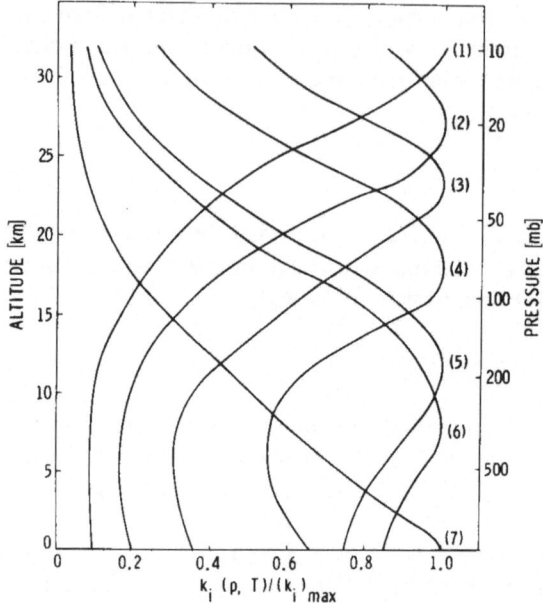

Fig. 7.8. Normalized weighting functions for ozone profile determination. See text for the sounding frequencies which produce these functions. (Note: 1 mb pressure $\simeq 0.750$ Torr)

1045.022 cm^{-1}. (These laser frequencies can be found in [7.85].) The maximum frequency displacement from laser line center to sounding frequency occurs for channel 1), and it is about 750 MHz. Compact, sealed-off waveguide CO_2 lasers have been tuned 600 MHz from line center [7.90], and further developments should increase this capability. If sounding frequencies 3) through 7) are used to obtain ozone profiles in the 0–25 km altitude region, a maximum tuning capability of 600 MHz would be required.

Figure 7.9 illustrates how ozone profiles can be reconstructed using the above sounding frequencies and an iterative technique to solve (7.85) for $q(p)$. This iterative technique was originally developed by CHAHINE to solve the radiative transfer equation for temperature profile information [7.91, 64]. It produces unique solutions which converge after a number of iterations, the rate of convergence depending somewhat on the initial guess for the profile. Good agreement between the solution and the actual profile depends, of course, on the weighting function overlap and instrument noise. In Fig. 7.9 are depicted fits to two different ozone profiles. A single initial guess is used for both profile reconstructions. One profile corresponds to what might be expected over

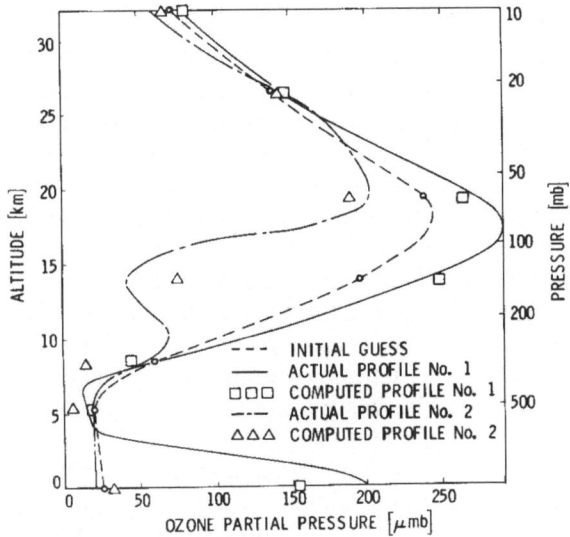

Fig. 7.9. Ozone profile reconstructions, using an initial guess which inlcudes the stratospheric band

an urban area, with high concentrations near the ground. The profile reconstruction is quite good at all altitudes. The fit to the other profile is also good near the ground and at the upper altitudes, but there is not enough altitude resolution to follow the inflection in the 8 to 16 km region. The weighting functions used for this altitude region are quite broad and overlap extensively.

It is apparent that the differential absorption technique promises to be useful in obtaining altitude profiles of pollutants, with high sensitivity to ground level concentrations. A number of factors which would influence instrument performance have not been discussed here: e.g., effects of instrument noise on profile reconstruction, sensitivity of the technique to temperature profile inaccuracies, and laser frequency stability requirements. Some of these factors were discussed in [7.88].

7.4 Future Improvements in Spectral Flexibility

7.4.1 Infrared Diode Lasers

Injection diode lasers, made of binary and ternary lead chalcogenide mixtures, have been used in several atmospheric monitoring experiments (see Chapter 6). They are attractive as local oscillators in heterodyne detection because of the ability to chemically tailor their spectral output, the variety of ways to "fine tune" them over smaller frequency intervals, and their compactness and low power requirements. Recent developments have produced 10-μm diode lasers which can operate cw at liquid nitrogen temperature, with single mode output power near 1 mW [7.92]. Advanced commercial units with this capability may soon be available.

A PbSe diode laser has been used as a local oscillator in recent heterodyne radiometer experiments [7.93]. Using a 200 MHz total IF bandwidth and an 8-channel filter bank, an absorption line profile of N_2O was observed near 8.5 μm using a laboratory blackbody radiation source. Thermal emission from the moon was also observed. Difficulties were encountered due to the low laser power ($\simeq 100$ μW per mode) and various instabilities. The heterodyne sensitivity was not limited by the local oscillator quantum noise as a result.

To obtain optimum performance using a diode laser local oscillator, it must emit in a single spatial and longitudinal mode. It is not sufficient to use a dispersive instrument as a filter for a multi-mode output laser. In Subsection 7.1.4, we discussed the effects of local oscillator amplitude fluctuations on heterodyne performance and stated that a single-mode laser exhibits very little excess noise. Competition effects in multimode lasers produce amplitude fluctuations, such that even though the total power output of the laser may be stable, the individual mode outputs are quite noisy [7.20, 94].

7.4.2 High Pressure Gas Lasers

Dramatic advances have occurred during the past few years in the development of higher pressure gas lasers. Compact, cw CO_2 waveguide lasers have recently been line tuned with a grating and fine tuned over 1200 MHz intervals [7.90]. These lasers operate single-mode and exhibit the output power and stability properties which are required of local oscillators. They can be used in both active and passive systems for studying complete absorption line shapes in such applications as remote altitude profiling of pollutants.

Pulsed operation in larger, high pressure CO_2 and CO systems has been discussed in a recent review paper [7.95]. These lasers might be used as transmitters in heterodyne lidar systems which employ cw gas or diode local oscillators. Several features of this technique, as applied to atmospheric monitoring, were discussed in another recent paper [7.105]. Tremendous output powers can be achieved with these lasers. Pulse energies in the range of several joules per liter of active gas volume are common. These pulses generally contain a large fraction of their energy in bursts of a few hundred nanoseconds duration; thus lidar systems with less than 100 m range resolution can be achieved. Pulsed operation of photodissociation-excited CO_2 lasers, operating at a few atmospheres total pressure, appears attractive for atmospheric monitoring systems because of its relative simplicity.

7.4.3 Parametric Oscillators

Parametric oscillators provide tunable, pulsed output over wide spectral ranges in the infrared [7.96–7]. The most advanced and reliable representative of this class uses lithium niobate as the nonlinear element and oscillates in the near infrared out to about 4 µm. Work is in progress to extend reliable, efficient operation to the 8–12 µm region [7.97–8]. In this longer wavelength region, which is attractive for atmospheric monitoring, further developments should result in tunable operation with narrow spectral width ($\simeq 0.05$ cm^{-1}). When this is accomplished, heterodyne detection should provide sensitivity improvements in many applications.

7.4.4 Photodiode Varactor Mixers

Recent experiments using HgCdTe photomixers to which microwave signals are applied, indicate impressive sensitivity in detecting laser beat frequencies up to 60 GHz [7.99]. With the use of CO_2 lasers employing a combination of isotopes, the entire 8.9 to 11.5 µm region can be covered with no more than 60 GHz separation between lines. (See [7.57] for a listing of frequencies from several pure isotope CO_2 lasers.) Since the CO_2 laser lines can serve as benchmark frequencies, photodiode varacter mixers will be useful in heterodyne spectroscopy experiments which employ tunable lasers in this spectral region.

7.5 Conclusion

It is evident that heterodyne detection techniques can be employed in a variety of ways for atmospheric monitoring applications. Although progress has seemed painfully slow at times, results of laboratory experiments are encouraging. The time has come to move a few specialized instruments employing heterodyne detection out of the laboratory, and to test their performance in both ground-based and airborne experiments. The next few years will witness the use of such systems which employ gas laser local oscillators. Continued laboratory work with lasers which have more spectral flexibility should lead to their use in practical systems a few years from now.

It is important to realize that a great deal of high-resolution laboratory spectroscopy is necessary before laser systems can be used for remote atmospheric sensing of gaseous pollutants. Laser systems offer potential as highly selective sensors, but experience has shown that accurate spectral data on both the constituent of interest and all interfering species must be available in order to interpret measurement results correctly. Necessary spectral data must be amassed during the next few years, concurrent with laser component developments. When this information is in hand, opportunities will be greatly expanded.

Acknowledgments. The author is pleased to acknowledge the helpfulness of Prof. S.T.ENG and other members of the Department of Electrical Measurements at Chalmers University of Technology, where the author resided during the writing of this chapter. Numerous contributions from M. S. SHUMATE at the Jet Propulsion Laboratory are also appreciated. A large part of the work described in this chapter was carried out at the Jet Propulsion Laboratory, California Institute of Technology, under Contract NAS 7-100, sponsored by the National Aeronautics and Space Administration.

References

7.1   E.D.HINKLEY, P.L.KELLEY: Science **171**, 635 (1971)
7.2   R.T.MENZIES: Appl. Opt. **10**, 1532 (1971); also U.S. Patent 3, 761, 715, California Institute of Technology (1971)
7.3   E.D.HINKLEY: Opto-Electron. **4**, 69 (1972)
7.4   R.T.MENZIES: Opto-Electron. **4**, 179 (1972)
7.5   L.D.LANDAU, E.M.LIFSHITZ: *Statistical Physics* (Pergamon Press, London 1959)
7.6   E.SCHRÖDINGER: *Statistical Thermodynamics* (Cambridge University Press, New York 1969)
7.7   S.S.PENNER: *Quantitative Molecular Spectroscopy and Gas Emissivities* (Addison-Wesley, Reading, Mass 1959)
7.8   J.D.KRAUS: *Radio Astronomy* (McGraw-Hill, New York 1966), with the chapter on receivers written by M. E. TIURI
7.9   B.M.OLIVER: Proc. IEEE **53**, 436 (1965)
7.10   R.H.DICKE: Rev. Sci. Instr. **16**, 268 (1946)
7.11   M.E.TIURI: IEEE Trans. AP-**12**, 930 (1964)
7.12   L.I.SCHIFF: *Quantum Mechanics,* 3rd ed. (McGraw-Hill, New York 1968)
7.13   A.E.SIEGMAN: Proc. IEEE **54**, 1350 (1966)
7.14   J.L.LAWSON, G.E.UHLENBECK: *Threshold Signals* (McGraw-Hill, New York 1950)
7.15   H.Z.CUMMINS, H.L.SWINNEY: In *Progress in Optics*, vol. 8, ed. by E.WOLF (North-Holland, Amsterdam 1970) p. 134
7.16   F.T.ARECCHI, V.DEGIORGIO: in *Laser Handbook*, vol. 1, ed by F.T.ARECCHI and E.O.SCHULZ-DUBOIS (North-Holland, Amsterdam 1972) p. 191
7.17   L.MANDEL, E.WOLF: Rev. Mod. Phys. **37**, 231 (1965)
7.18   R.J.GLAUBER: In *Laser Handbook,* vol. 1, ed. by F.T.ARECCHI and E.O.SCHULZ-DUBOIS (North-Holland, Amsterdam 1972) p. 1
7.19   C.FREED, H.A.HAUS: Phys. Rev. **141**, 287 (1966)
7.20   J.A.ARMSTRONG, A.W.SMITH: In *Progress in Optics*, vol. 6, ed. by E. WOLF (North-Holland, Amsterdam 1967) p. 213

7.21   C. Freed: IEEE J. Quant. Electron. QE-4, 404 (1968)
7.22   K. M. Van Vliet: Appl. Opt. 6, 1145 (1967)
7.23   R. A. Smith, F. E. Jones, R. P. Chasmar: *The Detection and Measurement of Infrared Radiation*, 2nd ed. (Oxford University Press, London 1968)
7.24   P. W. Kruse, R. D. McGlauchlin, R. B. McQuistan: *Elements of Infrared Technology* (J. Wiley and Sons, New York 1962)
7.25   R. J. Keyes, T. M. Quist: In *Semiconductors and Semimetals*, vol. 5, ed. by R. K. Willardson and A. C. Beer (Academic Press, New York 1970) p. 321
7.26   M. Ross: *Laser Receivers* (J. Wiley and Sons, New York 1966)
7.27   F. H. Murcray, D. G. Murcray, W. J. Williams: Appl. Opt. 3, 1373 (1964)
7.28   P. Bratt, W. Engeler, H. Levinstein, A. MacRae, J. Pehek: Infrared Phys. 1, 27 (1971)
7.29   B. J. Peyton, A. J. DiNardo, G. M. Kanischak, F. R. Arams, R. A. Lange, E. W. Sard: IEEE J. Quant. Electron. QE-8, 252 (1972)
7.30   G. Fiorito, G. Gasparrini, F. Svelto: Appl. Phys. Lett. 23, 448 (1973)
7.31   J. Marine, C. Motte: Appl. Phys. Lett. 23, 450 (1973)
7.32   M. Abramowitz, I. A. Stegun, (editors): *Handbook of Mathematical Functions* (U.S. Govt. Printing Office, Washington, D. C. 1964)
7.33   C. B. Ludwig, M. Griggs, W. Malkmus, E. R. Bartle: Appl. Opt. 13, 1494 (1974)
7.34   P. L. Hanst: In *Advances in Environmental Science and Technology*, vol. 2, ed. by J. N. Pitts and R. L. Metcalf (J. Wiley and Sons, New York 1971)
7.35   R. T. Menzies, N. George, M. L. Bhaumik: IEEE J. Quant. Electron. QE-6, 800 (1970)
7.36   R. R. Patty, G. M. Russwurm, W. A. McClenney, D. R. Morgan: Appl. Opt. 13, 2850 (1974)
7.37   P. A. Leighton: *Photochemistry of Air Pollution* (Academic Press, New York 1961)
7.38   J. N. Pitts: In *Advances in Environmental Sciences*, Vol. 1, ed. by J. N. Pitts and R. L. Metcalf (J. Wiley and Sons, New York 1969)
7.39   E. Robinson, R. C. Robbins: *Sources, Abundances, and Fate of Gaseous Atmospheric Pollutants*, SRI Report PR-6755 (1968); Supplement, June, 1969
7.40   H. Johnston: Science 173, 517 (1971)
7.41   R. A. Toth, C. B. Farmer, R. A. Schindler, O. F. Raper, P. W. Schaper: Nature 244, 7 (1973)
7.42   M. Loewenstein, J. P. Paddock, I. G. Poppoff, H. F. Savage: Nature 249, 817 (1974)
7.43   C. K. N. Patel: IEEE Quantum Electronics Conference, San Francisco, Calif June 1974
7.44   J. C. McConnell, M. B. McElroy, S. C. Wofsky: Nature 233, 187 (1971)
7.45   H. Levy II: J. Geophys. Res. 78, 5325 (1973)
7.46   E. R. Stephens: In *Advances in Environmental Sciences*, vol. 1, ed. by J. N. Pitts and R. L. Metcalf (J. Wiley and Sons, New York 1969)
7.47   S. K. Friedlander, California Institute of Technology (private communication)
7.48   C. E. Junge: *Air Chemistry and Radioactivity* (Academic Press, New York 1963)
7.49   D. G. Murcray, A. Goldman, A. Csoeke-Poeckh, F. H. Murcray, W. J. Williams, R. N. Stocker: J. Geophys. Res. 78, 7033 (1973)
7.50   A. J. Broderick (editor): Proceedings of the Second Conference on the Climatic Impact Assessment Program, U.S. Department of Transportation, Washington, D. C. (1972)
7.51   W. Chameides, J. G. Walker: J. Geophys. Res. 78, 8751 (1973)
7.52   C. Verie, M. Sirieix: IEEE J. Quant. Electron. QE-8, 180 (1972)
7.53   F. R. Arams, E. W. Sard, B. J. Peyton, F. P. Pace: IEEE J. Quant. Electron. QE-3, 484 (1967)
7.54   J. Gay, A. Journet: Appl. Phys. Lett. 22, 448 (1973)

7.55 R. T. Menzies: Appl. Phys. Lett. **22**, 592 (1973)
7.56 R. T. Menzies, M. S. Shumate: Science **184**, 570 (1974)
7.57 C. Freed, A. H. M. Ross, R. G. O'Donnell: J. Mol. Spectrosc. **49**, 439 (1974)
7.58 A. W. Mantz, J. K. G. Watson, K. Narahari Rao, D. L. Albritton, A. L. Schmelt-kopf, R. N. Zare: J. Mol. Spectrosc. **39**, 180 (1971)
7.59 G. A. Vanasse, H. Sakai: In *Progress in Optics*, vol. 6, ed. by E. Wolf (North-Holland, Amsterdam 1967) p. 261
7.60 R. A. Hanel, B. Schlachman, D. Rogers, D. Vanous: Appl. Opt. **10**, 1376 (1971)
7.61 J. H. McElroy: Appl. Opt. **11**, 1619 (1972)
7.62 S. R. King, D. T. Hodges, T. S. Hartwick, D. H. Barker: Appl. Opt. **12**, 1106 (1973)
7.63 J. T. Houghton, F. W. Taylor: Rept. Progr. Phys. **36**, 827 (1973)
7.64 M. T. Chahine: J. Atmos. Sci. **29**, 741 (1972)
7.65 C. B. Ludwig, E. R. Bartle, M. Griggs: *Study of Air Pollutant Detection by Remote Sensors*, NASA CR-1380, N79-31961 (1969)
7.66 B. J. Peyton (Airborne Instruments Laboratory): private communication; see also Proc. Int. Telemetry Conf., Vol. X, p. 403 (October 1974)
7.67 D. G. Murcray, T. G. Kyle, F. H. Murcray, W. J. Williams: J. Opt. Soc. Am. **59**, 1131 (1969)
7.68 D. L. Fried: Proc. IEEE **55**, 57 (1967)
7.69 V. I. Tatarski: *Wave Propagation in a Turbulent Medium* (McGraw-Hill, New York 1961)
7.70 D. L. Fried: IEEE J. Quant. Electron. QE-**3**, 213 (1967)
7.71 J. A. Dowling, P. M. Livingston: J. Opt. Soc. Am. **63**, 846 (1973)
7.72 T. J. Gilmartin, J. Z. Holtz: Appl. Opt. **13**, 1906 (1974)
7.73 H. Kildal, R. L. Byer: Proc. IEEE **59**, 1644 (1971)
7.74 R. L. Byer, M. Garbuny: Appl. Opt. **12**, 1496 (1973)
7.75 W. B. Bridges, S. Hansen, L. Horwitz, S. P. Lazzara, T. R. O'Meara, J. E. Pearson, T. J. Walsh: J. Opt. Soc. Am. **64**, 541 (1974)
7.76 M. C. Teich: In *Semiconductors and Semimetals*, vol. 5, ed. by R. K. Willardson and A. C. Beer (Academic Press, New York 1970) p. 361
7.77 G. Gould, S. F. Jacobs, J. T. LaTourette, M. Newstein, P. Rabinowitz: Appl. Opt. **3**, 648 (1964)
7.78 A. Ishimaru: Proc. IEEE **57**, 407 (1969)
7.79 J. R. Kerr, J. R. Dunphy: J. Opt. Soc. Am. **63**, 1 (1973)
7.80 R. S. Lawrence, J. W. Strohbehn: Proc. IEEE **58**, (1970)
7.81 D. L. Fried: J. Opt. Soc. Am. **57**, 169 (1967)
7.82 S. F. Clifford: J. Opt. Soc. Am. **61**, 1285 (1971)
7.83 R. A. Brandewie, W. C. Davis: Appl. Opt. **11**, 1526 (1972)
7.84 D. B. Rensch, R. K. Long: Appl. Opt. **9**, 1563 (1970)
7.85 K. M. Baird, H. D. Riccius, K. J. Siemsen: Opt. Commun. **6**, 91 (1972)
7.86 R. K. Long, F. S. Mills, G. L. Trusty: The Ohio State University (Rome Air Development Center TR 73–126, March, 1973)
7.87 R. K. Seals, Jr., C. H. Bair: 2nd Joint Conf. on Sensing of Environmental Pollutants, Washington, D.C., (1973)
7.88 R. T. Menzies, M. T. Chahine: Appl. Opt. **13**, 2840 (1974)
7.89 J. S. Garing, R. A. McClatchey: Appl. Opt. **12**, 2545 (1973)
7.90 R. L. Abrams: Appl. Phys. Lett. **25**, 304 (1974)
7.91 M. T. Chahine: J. Atmos. Sci. **27**, 960 (1970)
7.92 S. H. Groves, K. W. Nill, A. J. Strauss: Appl. Phys. Lett. **25**, 331 (1974)
7.93 M. Mumma, T. Kostiuk, S. Cohen, D. Buhl, P. C. von Thuna: Report X-691-74-237, Goddard Space Flight Center, Greenbelt, Maryland (1974); see also, Nature **253**, 514 (1975)

7.94 D.E.McCumber: Phys. Rev. **141**, 306 (1966)

7.95 O.R.Wood: Proc. IEEE **62**, 355 (1974)

7.96 S.E.Harris: Proc. IEEE **57**, 2096 (1969)

7.97 R.G.Smith: In *Laser Handbook,* vol. 1, ed. by F.T.Arecchi and E.O.Schulz-Dubois (North-Holland, Amsterdam 1972) p. 837

7.98 R.L.Byer, M.M.Choy, R.L.Herbst, D.S.Chemla, R.S.Feigelson: Appl. Phys. Lett. **24**, 65 (1974)

7.99 D.L.Spears, C.Freed: Appl. Phys. Lett. **23**, 445 (1973)

7.100 R.T.Menzies, M.S.Shumate: IEEE/OSA Conf. on Laser Engineering and Applications, Washington, D.C. 1975

7.101 D.W.Peterson, M.A.Johnson, A.L.Betz: Nature **250**, 128 (1974)

7.102 B.J.Peyton, A.J.DiNardo, S.C.Cohen, J.H.McElroy, R.J.Coates: IEEE J. Quant. Electron. QE-11, 569 (1975)

7.103 M.S.Shumate, L.G.Rosengren, R.T.Menzies, J.S.Margolis: IEEE/OSA Conf. on Laser Engineering and Applications, Washington, D.C. (1975) paper 9.3

7.104 M.J.Molina, F.S.Rowland: Nature **249**, 810 (1974)

7.105 T.Kobayasi, H.Inaba: Opt. Quant. Electron. **7**, 319 (1975)

Additional References with Titles

Remote Detection of Gases, Particles, Temperature, Wind Velocity

K.G.BARTLETT, C.Y.SHE: Remote measurement of wind speed using a dual beam backscatter Doppler velocimeter. Appl. Opt. **15**, 1980 (1976)

T.A.CLARK, D.J.W.KENDALL: Far infrared emission spectrum of the stratosphere from balloon altitudes. Nature **260**, 31 (1976)

A.COHEN, J.A.COONEY, K.N.GELLER: Atmospheric temperature profiles from lidar measurements of rotational Raman and elastic scattering. Appl. Opt. (to be published)

B.D.GREEN, J.I.STEINFELD: Absorption coefficients for fourteen gases at CO_2 laser frequencies. Appl. Optics **15**, 1688 (1976)

B.D.GREEN, J.I.STEINFELD: Laser Absorption Spectroscopy: A Method for Monitoring Complex Trace Gas Mixtures. Environmental Science and Technology, November 1976

P.L.HANST: Optical measurement of atmospheric pollutants: Accomplishments and problems. Opt. Quant. Electron. **8**, 87 (1976)

J.E.HARRIES, D.G.MOSS, N.R.W.SWANN, G.F.NEILL, P.GILDWARG: Simultaneous measurements of H_2O, NO_2, and HNO_3 in the daytime stratosphere from 15 to 35 km. Nature **259**, 300 (1976)

R.K.KAKAR, J.W.WATERS, W.J.WILSON: Venus: Microwave detection of carbon monoxide. Science **191**, 379 (1976)

M.KERKER, D.D.COOKE: Remote sensing of particle size and refractive index by varying the wavelength. Appl. Opt. **15**, 2105 (1976)

R.T.MENZIES (chairman, working group): Global and regional monitoring from airborne and satellite platforms. Opt. Quant. Electron. **8**, 185 (1976)

R.T.MENZIES, M.S.SHUMATE: Remote measurements of ambient air pollutants with a bistatic laser system. Appl. Opt. **15**, 2080 (1976)

M.MUMMA, T.KOSTIUK, S.COHEN, D.BUHL, P.C.VON THUNA: Heterodyne spectroscopy of astronomical and laboratory sources at 8.5 µm using diode laser local Oscillators. Space Sci. Rev. **17**, 661 (1975)

E.R.MURRAY, R.L.BYER: Remote Measurement of Air Pollutants. 3rd Annual Progress Report for the National Science Foundation, February 1976

E.R.MURRAY, R.D.HAKE,JR., J.E.VAN DER LAAN, J.G.HAWLEY: Atmospheric water vapor measurements with an infrared (10-µm) differential-absorption lidar system. Appl. Phys. Lett. **28**, 542 (1976)

E.R.MURRAY, J.E.VAN der LAAN, J.G.HAWLEY: Remote measurement of HCl, CH_4, and N_2O using a single-ended chemical-laser lidar system. Appl. Opt. (to be published)

S.R.PAL, A.I.CARSWELL: Multiple scattering in atmospheric clouds: Lidar observations. Appl. Opt. **15**, 1990 (1976)

J.SHEWCHUN, B.K.GARSIDE, E.A.BALLIK, C.C.Y.KWAN, M.M.ELSHERBINY, G.HOGEN-KAMP, A.KAZANDJIAN: Pollution monitoring systems based on resonance absorption measurements of ozone with a 'tunable' CO_2 laser: Some criteria. Appl. Opt. **15**, 340 (1976)

H.TANNENBAUM (chairman, working group): Long-path monitoring of atmospheric pollutant gases. Opt. Quant. Electron. **8**, 194 (1976)

A.P.WAGGONER, A.J.VANDERPOL, R.J.CHARLSON, S.LARSEN, L.GRANAT, C.TRAGARDH: Sulphate—Light scattering ratio as an index of the role of sulphur in tropospheric optics. Nature **261**, 120 (1976)

J.W.WATERS, W.J.WILSON, F.I.SHIMABUKURO: Microwave measurement of mesospheric carbon monoxide. Science **191**, 1174 (1976)

Laser Spectroscopy and Atmospheric Propagation (Including Optoacoustic Detection)

T.G.ADIX, V.N.AREFYEV, V.I.DIANOV-KLOKOV: The effect of molecular absorption on the propagation of CO_2 laser radiation in the Earth's atmosphere, (Review). Soviet J. Quant. Electron. **2**, 885 (1975)

E.N.ANTONOV, V.G.KOLOSHNIKOV, V.P.MIRONENKO: Using a tunable cw dye laser for obtaining absorption spectrum of free air within the laser cavity. Sov. J. Quant. Electron. **2**, 171 (1975)

V.N.AREFYEV, V.I.DIANOV-KLOKOV, N.I.SISOV: Labortory measurements of attenuation of CO_2 laser radiation by pure water vapor. Opt. Spectrosc. **39**, 982 (1975)

P.C.CLASPY, YOH-HAN PAO, SIULIT KWONG, EUGENE NODOV: Laser optoacoustic detection of explosive vapors. Appl. Opt. **15**, 1506 (1976)

S.F.CLIFFORD: Physical properties of the atmosphere in relation to laser probing. Opt. Quant. Electron. **8**, 105 (1976)

J.HÄGER, W.HINZ, H.WALTHER: High-resolution spectroscopy of ethylene by means of a spin-flip-Raman laser. Appl. Phys. **9**, 35 (1976)

R.T.MENZIES, M.S.SHUMATE: Acousto-optic measurements of water vapor absorption at selected CO laser wavelengths in the 5-μm region. Appl. Opt. **15**, 2025 (1976)

C.K.N.PATEL: Spectroscopic measurements of the stratosphere using tunable infrared lasers. Opt. Quant. Electron. **8**, 145 (1976)

H.PREIER, W.RIEDEL: NO spectroscopy by pulsed $PbS_{1-x}Se_x$ diode lasers. J. Appl. Phys. **45**, 3955 (1974)

R.E.RICHTON: NO line parameters measured by CO laser transmittance. Appl. Opt. **15**, 1686 (1976)

M.S.SHUMATE, R.T.MENZIES, J.S.MARGOLIS, L.G.ROSENGREN: Water vapor absorption of CO_2 laser radiation. Appl. Opt. (November 1976)

D.J.SPENCER, G.C.DENAULT, H.H.TAKIMOTO: Atmospheric gas absorption at DF laser wavelengths. Appl. Opt. **13**, 2855 (1974)

R.A.TOTH (chairman, working group): Infrared spectral properties of atmospheric molecules. Opt. Quant. Electron. **8**, 191 (1976)

L.S.VASILENKO, A.A.KOVALEV, A.S.PROVOROV, V.P.CHEBOTAEV: The P2O line collision broadening of $00°1–10°0$ CO_2 transition, measured with a tunable CO_2 laser. Soviet J. Quant. Electron. **2**, 2528 (1975)

K.O.WHITE, W.R.WATKINS, S.A.SCHLEUSENER: Holmium 2.06 mm laser spectral characteristics and absorption by CO_2 gas. Appl. Opt. **14**, 16 (1975)

Advances in Lasers and Laser Particle Sizing Instrumentation

P.BURLAMACCHI, R.PRATESI: GHz tuning of a planar dye laser with single dispersive element. Appl. Phys. Lett. **28**, 124 (1976)

F.R.FAXVOG: Detection of airborne particles using optical extinction measurements. Appl. Opt. **13**, 1913 (1974)

F. R. Faxvog: New laser particle sizing instrument. SAE Automotive Engineering Congress & Exposition, Detroit, Michigan, U.S.A. Feb. 8—March 4, 1977

J. C. Hill, G. P. Montgomery, Jr.: Diode lasers for gas analysis: Some characteristics. Appl. Opt. **15**, 748 (1976)

G. W. Kattawar, D. A. Hood: Electromagnetic scattering from a spherical polydispersion of coated spheres. Appl. Opt. **15**, 1996 (1976)

W. Lo, G. P. Montgomery, Jr., D. E. Swets: Ingot-nucleated $Pb_{1-x}Sn_xTe$ diode lasers. J. Appl. Phys. **47**, 267 (1976)

A. Mooradian: Recent advances in tunable lasers. Sov. J. Quant. Electron. **6**, 420 (1976)

V. T. Nguyen, E. G. Burkhardt: cw tunable laser-sideband generation from 5.5 to 6.5 µm by light scattering from spin motion in a spin-flip Raman laser. Appl. Phys. Lett. **28**, 187 (1976)

C. K. N. Patel, T. Y. Chang, V. T. Nguyen: Spin-flip Raman laser at wavelengths up to 16.8 µm. Appl. Phys. Lett. **28**, 603 (1976)

H. Preier, M. Bletcher, W. Reidel, H. Maier: Double heterojunction $PbS-PbS_{1-x}Se_x-PbS$ laser diodes with cw operation up to 96 K. Appl. Phys. Lett. **28**, 669 (1976)

J. Reid, K. Siemens: New CO_2 laser bands in the 9–11 µm wavelength region. Appl. Phys. Lett. **29**, 250 (1976)

J. N. Walpole, A. R. Calawa, S. R. Chinn, S. H. Groves, T. C. Harman: Distributed feedback $Pb_{1-x}Sn_xTe$ double-heterostructure lasers. Appl. Phys. Lett. **29**, 307 (1976)

J. N. Walpole, A. R. Calawa, T. C. Harman, S. H. Groves: "Double-heterostructure PbSnTe lasers grown by molecular-beam epitaxy with cw operation up to 114K". Appl. Phys. Lett. **28**, 552 (1976)

Atmospheric Chemical Processes and Modeling

P. L. Hanst (chairman, working group): Stratospheric chemistry and measurement techniques. Opt. Quant. Electron. **8**, 187 (1976)

H. S. Johnston: Photochemistry in the stratosphere. International Conf. on Tunable Lasers and Applications, June 7–11, 1976, Leon, Nordfjord, Norway. In *Springer Series in Optical Sciences*, ed. by D. L. MacAdam, Vol. 3 (Springer, Berlin, Heidelberg, New York 1976) p. 259

J. N. Pitts, Jr., B. J. Finlayson-Pitts: Tropospheric photochemical and photophysical processes. International Conf. on Tunable Lasers and Applications, June 7–11, 1976, Leon, Nordfjord, Norway. In *Springer Series in Optical Sciences*, ed. by D. L. MacAdam, Vol. 3 (Springer, Berlin, Heidelberg, New York 1976) p. 236

Resonance Fluorescence—Characteristics and Discussion of Recent Measurements

M. Jyumonji, T. Kobayasi, H. Inaba: Nonlinear resonance effects in atomic gases— Spontaneous and stimulated electronic Raman scattering of Na atoms excited resonantly at D lines by a tunable dye laser. Sov. J. Quant. Electron. **6**, 430 (1976)

G. Megie: Paper presented at the IAGA Symposium on Optical Sensing and Probing of the Atmosphere, Grenoble, France, August 1975; also, postdeadline paper presented at the 7th International Laser Radar Conf., XI-19, Menlo Park, California, USA, November 1975
In the above technical presentation, Megie reported the first simultaneous measurements of sodium and potassium density in the upper atmosphere. The laser radar system consisted of two dye lasers, tuned to the D_2 line of the sodium atom and to the resonance line of the potassium atom, and an 82-cm-diameter Coudé telescope together with photon-counting apparatus. One dye laser, pumped by two flashlamps,

delivered 0.8–1 J output power with ~ 8 pm spectral width and 0.5–1 Hz repetition rate, while the other laser, pumped by a Q-switched ruby laser, emitted ~1 J output power with ~7 pm width and 0.1 Hz repetition rate. The same height of around 91 km for the density maximum for the two atomic constituents was observed over 3 nights during the end of July 1975, whereas another profile obtained at the beginning of July indicated the sodium maximum to be 3 km higher. He also reported that the abundance ratio Na/K ~ 58^{+20}_{-15} is consistent only with a sea water source. Furthermore, a sporadic increase in both sodium and potassium densities was observed during the night of July 29, 1975, which appeared to be attributable to the Perseids permanent shower. A temperature distribution corresponding to the thermal equilibrium of atmospheric neutral species was also derived from the Doppler width of the scattered line at the sodium D_2 transition in the range of 85–100 km.

D. L. ROUSSEAU, G. D. PATTERSON, P. F. WILLIAMS: Resonance Raman scattering and collision-induced redistribution scattering in I_2. Phys. Rev. Lett. **34**, 1306 (1975)

Y. R. SHEN: Distinction between resonance Raman scattering and hot luminescence. Phys. Rev. **B9**, 622 (1974)

J. R. SOLIN, H. MERKELO: Resonant scattering or absorption followed by emission. Phys. Rev. **B12**, 624 (1975)

A. SZÖKE, E. COURTENS: Time-resolved resonance fluorescence and resonance Raman scattering. Phys. Rev. Lett. **34**, 1053 (1975)

Author Index

Standard numbers indicate pages in text where an author's name is mentioned; numbers in *italics* indicate the pages containing References

Subject Index

Applied Physics

A monthly journal

Board of Editors	**S. Amelinckx,** Mol. · **V. P. Chebotayev,** Novosibirsk
	R. Gomer, Chicago, III. · **H. Ibach,** Jülich
	V. S. Letokhov, Moskau · **H. K. V. Lotsch,** Heidelberg
	H. J. Queisser, Stuttgart · **F. P. Schäfer,** Göttingen
	A. Seeger, Stuttgart · **K. Shimoda,** Tokyo
	T. Tamir, Brooklyn, N.Y. · **W. T. Welford,** London
	H. P. J. Wijn, Eindhoven

Coverage

application-oriented experimental and theoretical physics:

Solid-State Physics	*Quantum Electronics*
Surface Physics	*Laser Spectroscopy*
Chemisorption	*Photophysical Chemistry*
Microwave Acoustics	*Optical Physics*
Electrophysics	*Integrated Optics*

Special Features

rapid publication (3–4 months)
no page charge for **concise** reports
prepublication of titles and abstracts
microfiche edition available as well

Languages

Mostly English

Articles

original reports, and short communications
review and/or tutorial papers

Manuscripts

to Springer-Verlag (Attn. H. Lotsch), P.O. Box 105 280
D-69 Heidelberg 1, F.R. Germany

Place North-American orders with:
Springer-Verlag New York Inc., 175 Fifth Avenue, New York. N.Y. 10010, USA

Springer-Verlag
Berlin Heidelberg New York

Laser Spectroscopy

Proceedings of the 2nd International Conference, Mégève, France, June 23–27, 1975
Edited by *S. Haroche, J. C. Pebay-Peyroula, T. W. Hänsch, S. E. Harris*
230 figures, 30 tables. X, 468 pages (5 pages in French). 1975
(Lecture Notes in Physics, Vol. 43)

Springer Series in Optical Sciences
Editor: D. L. MacAdam

Vol. 1   W. Koechner
Solid State Laser Engineering
XI, 620 pages. 1976

Vol. 2   R. Beck, W. Englisch, K. Gürs
Table of Laser Lines in Gases and Vapors
IV, 130 pages. 1976

Vol. 3   **Tunable Lasers and Applications**
Proc. of the Loen Conf., Norway (1976). Editors: A. Mooradian, T. Jaeger,
P. Stokseth, VIII, 404 pages. 1976

Vol. 4   V. S. Letokhov, V. P. Chebotayev
Nonlinear Laser Spectroscopy

Topics in Current Physics, Vol. 1

Beam-Foil Spectroscopy   S. Bashkin (editor)

S. Bashkin: Introduction. — *S. Bashkin:* Instrumentation. — *I. Martinson:* Wavelengths Measurements and Level Analysis. — *L. Curtis:* Lifetime Measurements. — *I. Sellin:* Autoionizing Levels. — *H. Marrus:* Studies of H-like and He-like Ions of High Z. — *W. Whaling, L. Heroux:* Applications to Astrophysics. — *Sinanoglu:* Fundamental Calculation of Level Lifetimes. — *W. Wiese:* Systematic Effects in Z-Dependence of Oscillator Strengths. — *J. Macek, D. J. Burns:* Coherence, Alignment, and Orientation Phenomena

Springer-Verlag Berlin Heidelberg New York